Integrating Research on the Graphical Representation of Functions

STUDIES IN MATHEMATICAL THINKING AND LEARNING

A Series of volumes edited by
Alan Schoenfeld

Carpenter/Fennema/Romberg • Rational Numbers: An Integration of Research

Romberg/Fennema/Carpenter • Integrating Research on the Graphical Representation of Functions

Schoenfeld • Mathematical Thinking and Problem Solving

Integrating Research on the Graphical Representation of Functions

Edited by

Thomas A. Romberg
Elizabeth Fennema
Thomas P. Carpenter
University of Wisconsin–Madison

LAWRENCE ERLBAUM ASSOCIATES, PUBLISHERS
1993   Hillsdale, New Jersey   Hove and London

The preparation of this book was supported by the Office for Educational Research and Improvement, United States Department of Education (Grant Number R117G10002) and the Wisconsin Center for Education Research, School of Education, University of Wisconsin-Madison. The opinions expressed in this publication do not necessarily reflect the views of the Office of Educational Research and Improvement or the Wisconsin Center for Education Research.

Copyright © 1993 by Lawrence Erlbaum Associates, Inc.
All rights reserved. No part of this book may be reproduced in any form, by photostat, microform, retrieval system, or any other means, without the prior written permission of the publisher.

Lawrence Erlbaum Associates, Inc., Publishers
365 Broadway
Hillsdale, New Jersey 07642

Library of Congress Cataloging-in-Publication Data

Romberg, Thomas A.
    Integrating research on the graphical representation of functions / Thomas A. Romberg, Elizabeth Fennema, Thomas P. Carpenter.
      p. cm.
    Includes bibliographical references and index.
    ISBN 0-8058-1134-6
    1. Functions. 2. Algebra—Graphic methods. I. Fennema, Elizabeth. II. Carpenter, Thomas P. III. Title.
QA331.3.R66 1993
515—dc20   92-21138
                                                                       CIP

Books published by Lawrence Erlbaum Associates are printed on acid-free paper, and their bindings are chosen for strength and durability.

Printed in the United States of America
10 9 8 7 6 5 4 3 2 1

Contents

Preface   ix

1   Toward a Common Research Perspective   1
    Thomas A. Romberg, Thomas P. Carpenter, and Elizabeth Fennema

I   CONTENT

2   Graphing in the K-12 Curriculum: The Impact of the Graphing Calculator   11
    Frank Demana, Harold L. Schoen, and Bert Waits

3   Seizing the Opportunity to Make Algebra Mathematically and Pedagogically Interesting   41
    Michal Yerushalmy and Judah L. Schwartz

4   Aspects of Understanding: On Multiple Perspectives and Representations of Linear Relations and Connections Among Them   69
    Judit Moschkovich, Alan H. Schoenfeld, and Abraham Arcavi

II STUDENT THINKING

5 Functions and Graphs—Perspectives on Student Thinking 101
Sharon Dugdale

III TEACHING THINKING

6 Teachers' Thinking About Functions: Historical and Research Perspectives 131
Thomas J. Cooney and Melvin R. Wilson

IV TEACHER KNOWLEDGE

7 Integrating Research on Teachers' Knowledge of Functions and Their Graphs 159
F. Alexander Norman

V CLASSROOM INSTRUCTION

8 Functions, Graphing, and Technology: Integrating Research on Learning and Instruction 189
Carolyn Kieran

VI CURRICULAR IMPLICATIONS

9 Curricular Implications of Graphical Representations of Functions 239
Randolph A. Philipp, William O. Martin, and Glen W. Richgels

VII REACTIONS

10 The Urgent Need for Proleptic Research in the
Representation of Quantitative Relationships **279**
James Kaput

11 Some Common Themes and Uncommon Directions **313**
Steven R. Williams

Author Index **339**

Subject Index **345**

Preface

This volume is concerned with the integration of research on teaching, learning, curriculum, and assessment with respect to an increasingly important mathematical domain, the *graphical representation of functions*. The effort at integration reflects an ambitious attempt, from a common perspective, to focus the research of a number of scholars on critical problems in the study of this domain. Research in this area is sparse, but with the increasing use of computers and graphing calculators, a coherent body of knowledge about how the connections are developed among tables, graphs, and the algebraic expressions related to functions is desperately needed.

The primary audience for this book consists of researchers in mathematics learning and teaching. We anticipate that university professors and their graduate students will use the volume as a basis for a rich set of studies during the coming decade. However, because of the growing importance of the topics covered, we expect many teacher educators, textbook authors, and teachers of mathematics will find the chapters enlightening.

The strategy that was followed in planning this book is one that has been previously used with considerable success. Over ten years ago, the Wisconsin Center for Education Research sponsored a conference that brought together researchers from around the world who had studied children's addition and subtraction concepts and skills. An outcome of that conference was the book, *Addition and Subtraction: A Cognitive Perspective* (Carpenter, Moser, & Romberg, 1982). However, more importantly, the conference served to initiate an ongoing research effort that, during the ensuing decade, clarified our understanding of how addition and subtraction concepts develop in children and demonstrated that this knowledge has major implications for classroom instruction. This book is the outcome of a similar conference. The chapters have been prepared by a distinguished group of researchers concerned about the subject, the impact of technology on the domain, and the integrations of research on teaching, learning, curriculum, and assessment. We can only hope that, in the next decade, work in this area will be productive in a manner similar to what has been accomplished in the area of addition and subtraction.

This is a companion volume to a book titled *Rational Numbers: An Integration of Research* (Carpenter, Fennema, & Romberg, 1993). It too is the product of a conference of respected scholars involved in the study of a specific domain. However, it provides another perspective on the integration of research in that content area. Research on the teaching and learning of rational number concepts is about where addition and subtraction research was ten years ago. There is an emerging consensus regarding some of the most critical problems in studying teaching and learning in this domain, and researchers are beginning to share a common perspective in the study of them. There are, however, important differences between the current state of research in the two areas. There is a more extensive body of research on children's conceptions of rational number than that which exists on the graphical representation of functions. The semantic analysis of rational numbers and previous research on children's rational number concepts play central roles in the integration of programs of research on rational number, and there is a substantial research base in this area to build on. The current interest in the graphical representation of functions, however, is concerned with modes of representation that were not readily available a few years ago, which means we lack an extensive research base. Instead, the work on the graphical representation of functions is driven by advances in technology that now make it possible to readily construct graphical representations of functions that previously were cumbersome to create. Thus, these companion volumes provide somewhat different perspectives on integrating research on learning, teaching, curriculum, and assessment. Nevertheless, the two volumes are not simply about the study of teaching and learning in each domain. They are about the construction of integration programs of research that are concerned with content analysis, learning, teaching, curriculum, and assessment.

We are indebted to a number of people and organizations responsible for making the publication of this volume possible. The writing of individual chapters and the editing and preparation of the volume was supported by the Office of Educational Research and Improvement of the U. S. Department of Education through the support of the National Center for Research in Mathematical Sciences Education. The Wisconsin Center for Education Research provided the ancillary services that are so necessary for this type of project. We thank Andrew Porter, the director of the Wisconsin Center, and Jerry Grossman, business manager of the Center, for their continued support. We also thank Joan Pedro for attending to many of the administrative details involved in its publication. We would especially like to thank Jean von Allmen for the tremendous amount of work she did—retyping manuscripts, drawing figures, and generally doing whatever was necessary to keep the project on track. We also owe a special thanks to Margaret Powell for the careful editing that contributed immeasurably to the clarity and quality of writing in the book.

—Thomas A. Romberg
—Elizabeth Fennema
—Thomas P. Carpenter

1 Toward a Common Research Perspective

Thomas A. Romberg
Thomas P. Carpenter
Elizabeth Fennema
University of Wisconsin-Madison

In the last few years there has been significant progress in integrating programs of research in school mathematics. This has occurred through the cooperation of scholars working on content analysis, student learning, teaching, curriculum, and assessment in particular mathematical domains. Furthermore, the products of such integration have demonstrated considerable potential for influencing curricular reform. In this volume a different approach to scholarly integration has been followed. The impact of technology on the way mathematical functions are represented provides a common focus for the research discussed here.

This volume was conceived as an initial attempt to create a scholarly community with a common approach to the study of functions and their graphs. There is general consensus that functions are among the most powerful and useful notions in all mathematics. Nevertheless, the learning and teaching of functions are understudied in comparison to other areas of mathematical instruction. Leinhardt, Zaslavsky, and Stein (1990), based on their review of research, found that "actual studies of teaching [functions] at either the elementary or secondary level are quite rare and, in general, unconnected to the knowledge that a student develops" (p. 54).

The common approach to teaching functions in American schools follows the historical evolution of the ideas in the domain. Although the formal discussion of functions is a 19th century development, its roots go back centuries. Mesopotamians in the 17th century BC constructed tables of related values for two variables (e.g., taxation and flow of goods) and in some cases were able to express, in a primitive manner, that relationship (Säljö, 1991). Their efforts are similar to "guess my rule" exercises common in elementary texts. The concept of variable, with the problem of how to represent the relationship between two variables, underlies this domain.

The ancient Greeks by the 3rd century BC used ratios and proportions to express a wide variety of relationships, such as the corresponding sides of similar triangles, the radius of a circle and its circumference, and weights and lengths on a balance beam. They were able to develop considerable mathematics based on this way of considering relationships. Given that all ratios [y:x] can be considered as simple linear functions of the form $y = mx$ (where m is the constant of proportionality) ratios, proportions, and proportional reasoning are still considered as necessary background skills to the study of functions even if these connections are rarely pointed out in contemporary texts.

In spite of the mathematical insights these ancients were able to develop they were hampered by lack of a suitable notational system. This was remedied in the 14th century AD with the importation of the Hindu-Arabic numerals and algebraic notation from the Mideast at the end of the era of the Crusades. The use of a letter (e.g., X) to represent a variable and of an equation to relate two variables now made it possible to express in a compact form a vast number of relationships. Since that time the core of work on relations and functions has grown as a consequence of the introduction of this powerful notational system. A second effective way of representing the relationship between two variables was invented by René Descartes in the 17th century: By picturing related sets of values on a coordinate system, it becomes possible to visualize the relationship between the variables via a graph.

In summary, to express the relationship between two variables humans have invented tables of related values, algebraic expressions, and graphs on a coordinate system. For many years students have been taught both how to construct such representations and how to interpret them. However, it is basically the algebraic representations and the subsequent methods of manipulating those representations that have been emphasized in most mathematics courses in traditional systems—from beginning algebra to calculus and beyond.

Today, the impact of technology on the way mathematical functions can be represented and manipulated is forcing scholars to reconsider the way functions are used and taught. The algebraic instructional emphasis is being challenged in most of the chapters in this book. Technology makes it possible to deal with functions in new ways and to explore new ideas in curriculum and classroom practice. The abstractness of the algebraic expressions and the variety of transformations of such expressions have proved to be difficult for many students to fathom. In the past, graphical representations of special functions were often employed to motivate the use of an algebraic expression. However, graphs often were difficult and cumbersome for teachers and students to create or manipulate. Today, with the advent of the computer and graphing calculator, these representations not only are easy to create, but they too can be transformed in a variety of ways. The belief is that emphasizing graphical representations will make functions easier to learn and use for most students. The chapters in this volume represent an attempt by the authors to substantiate the basis of this belief by examining (1) the shift to graphical representations of functions using new technological tools and (2) the impact of this shift on content, learning, teaching, and assessment in mathematics classrooms.

The development of a common research perspective with respect to the teaching and learning of any mathematical domain is relatively recent (Romberg & Carpenter, 1986). It has become apparent that a more unified program of research is needed if we are to acquire an understanding of teaching and learning in schools that will inform curriculum development and assessment (Carpenter & Peterson, 1988; Fennema, Carpenter, & Lamon, 1991). Attempting to integrate a number of different research perspectives is a complex task, and ways need to be found both for organizing a group of scholars and for constraining the task so that complexity will be reduced without sacrificing integration.

One strategy for organizing a group has been to hold a conference that brings to the same venue researchers with a common interest. This book is a product of such a conference—one at which scholars interested in the graphical representation of functions had a chance to share their work. Precedence for such a gathering came from the Wingspread Conference on addition and subtraction held in 1979 (Carpenter, Moser, & Romberg, 1982). At that conference, researchers interested in the learning of early number concepts and operations found that there was general agreement as to the basic questions of interest regarding what had been accomplished, and what had yet to be done. The participants were able to come to terms with a common vocabulary and a common agenda for future work. Thus, they were able to integrate their research endeavors into a common framework. As Sowder (1989) pointed out, it is difficult to find research in the learning of early number concepts that has not been positively affected by the Wingspread Conference.

The common method of structuring the work of such a research group has involved studying specific mathematical domains. The domain knowledge strategy (Romberg, 1987) is based on the philosophic premise that the power of mathematics lies in the fact that a small number of symbols and symbolic statements can be used to represent a vast array of different problem situations. The properties of a domain include a set of problem situations that make a set of related concepts meaningful (problems that give rise to the need to relate two variables), a set of invariants that constitute the concept (e.g., the concept of a variable), and a set of symbolic situations used to represent the concept, its properties, and the situations it refers to (tables, equations, and graphs). In the past scholars have started with a semantic analysis of the problem situations. This has been done for such domains as addition and subtraction of whole numbers, multiplication and division of whole numbers, rational numbers, algebra, and geometry. This strategy has helped researchers coalesce their work because:

1. They were able to agree on the mathematical and semantic structure underlying a particular domain.
2. They could relate student and teacher cognitions directly to this underlying structure.
3. They could use this knowledge to enhance learning in classrooms.

In this volume a different point of departure has been chosen—the focus is the impact of technology on the way functions are represented. This choice was made due to the fact that the number of situations in which even simple linear functions can apply is phenomenally large. Thus, nothing akin to the semantic analysis of problem types seemed realistic for functions. Instead, we chose to confine work in this domain to an examination of the impact of graphing technology on the domain. Whether this will prove to be a viable structure for organizing a program of research is open to question. The perspectives presented in this volume illustrate the potential for adopting this starting point. However, its full potential will not be realized until problems related to three aspects of the teaching and learning of functions are addressed and resolved. These are agreement about what constitutes the domain of mathematical functions, the impact of graphing technology on the domain, and research on the teaching and learning of functions.

THE DOMAIN: MATHEMATICAL FUNCTIONS

The evolution of functions in mathematics clearly reflects the fact that mathematical objects are created in response to social problem situations (Romberg, 1992). Furthermore, their development involves invention and abstraction, and even today mathematicians' views about the domain are changing (Kleiner, 1989). Several chapters in this volume (those by Moschkovich, Schoenfeld, & Arcavi [Chapter 4], Cooney & Wilson [Chapter 6], and Kieran [Chapter 8]) describe aspects of this evolution both conceptually and pedagogically. The initial situations that gave rise to functions were problems involving related variables. For example, the distance an object moves in relationship to its speed, a child's weight in relationship to his or her age, and the amount of postage required to mail a letter in relationship to its weight are typical situations. How best to consider functions has not been agreed upon. However, most authors agree that a functional relationship involves some sort of rule that assigns to each element of a given set a unique element of some other, not necessarily distinct, set. Freudenthal (1983) claimed that there are two essential features of the function domain that have evolved with the structure of the sets and the rules relating them — arbitrariness and univalence.

The arbitrary nature of functions refers to both the relationship between the two sets on which the function is defined and the sets themselves. The first means that functions do not have to exhibit some regularity — that is, functions do not have to be described by any specific algebraic expression or by a graph of a particular shape. For example, the relationship between time and temperature is an illustration of this kind of function. The arbitrary nature of the two sets means that functions do not have to be defined on any specific sets of objects; in particular, the sets do not have to be sets of numbers. A rotation of the plane is an example of this type of function, because it is defined on sets of points.

Whereas the arbitrary nature of functions is implicit in the common definition of a function, the univalence requirement, that for each element of the first set the rule assigns a unique element of the second set, is explicit. Freudenthal (1983) attributed this requirement to the desire of mathematicians to keep things manageable. Keeping track of meanings of multivalued symbols (such as $\sqrt{}$), and taking care that they have the same meaning in the same context, became too messy in advanced analysis. Thus, the univalent restriction on functions became an accepted part of all definitions of functions.

Pedagogically, the notion that the relationship does not have to have any regularity causes both teachers and students problems. It is clear from the chapters in this volume by Yerushalmy and Schwartz (Chapter 3), Moschkovich, Schoenfeld, and Arcavi (Chapter 4), Cooney and Wilson (Chapter 6), and Kieran (Chapter 8) that the way one considers the "rule or recipe" relating elements from one set to elements of another is critical. Although indeed logically a relationship may be arbitrary, in almost all practical situations there is an implicit dependency relationship. For example, the area of a square can be expressed as the length of a side squared ($A = s^2$). Such a functional relationship between the length of a side of a square and its area is not arbitrary. The current pedagogic view as strongly endorsed by Thorpe (1989) is to emphasize such practical relationships in school mathematics.

In order to develop an integrated research program for school mathematics in this domain, agreement needs to be reached on how best to characterize the domain. The semantic analyses used as a starting point for many other mathematical domains would be unwieldy for functions, but a conceptual and pedagogic framework to tie scholarly work together is still needed.

GRAPHING TECHNOLOGY

There are three primary ways of representing a functional relationship — by presenting ordered pairs of values in a table, by creating algebraic expressions to represent the relationship, and by constructing graphs to picture the relationship. For purely arbitrary functions, tables of related values (e.g., time and temperature) are commonly used. For most mathematical applications, such sets of ordered pairs are of interest if one suspects the existence of a pattern or regularity. The identification of such a pattern is then described. At this point the pattern itself may become the object of study. As Moschkovich, Schoenfeld, and Arcavi (Chapter 4) point out, this is a critical step in the mathematical modeling process, but it shifts one's attention from the relationship to the model.

The identification of patterns, while based on the observed ordered pairs, involves generalization to unobserved data. The primary means of representing such patterns in the past has been to use algebraic equations. In fact, it is the creation of such equations that many view as the most important pedagogic feature of functions. Unfortunately, as several authors in this volume (e.g., Kieran, Chapter 8; Philipp, Martin, & Richgels, Chapter 9) point out, too often the study of functions is limited to the study of a small number of given algebraic functions. In fact, students are not expected to create an expression for a set of data; they are only expected to learn the properties of certain recipes. Even and Ball (1989) showed that many mathematics teachers and students think that functions must be expressible as algebraic formulas and, indeed, must express a regularity.

The third method of representing functional relationships, via graphs, is the theme of this book. Graphs were initially created from the ordered pairs in a table to facilitate discovery of some regularity. Also, they were sometimes created to picture common algebraic functions (e.g., $y = 3x - 4$ as a straight line; or $y = sin\ x$ as a wave with a particular shape). This is still usually done pedagogically for motivational purposes.

However, until recently the creation of a graph for most functions was difficult. Many points needed to be plotted and sketched. The ready availability of graphing technology has changed that. This fact informs all of the chapters in this volume. If one has an algebraic representation and enters it on a graphing utility, its graph is immediately created. Similarly, it is now easy to graph data sets. The easy creation of such graphs should lead to the introduction of such important topics as curve fitting, and why some classes of functions (e.g., polynomial, exponential, logarithmic) are important families of functions in mathematical modeling. Such technologies promise to provide a more natural pedagogical approach to the teaching and learning of the functional domain because the introduction of an algebraic formula can be made secondary to the visual picture of a functional relationship.

The problems facing scholars involved in studying the uses of graphing technologies are compounded by the rapid development of new technologies. As Kaput (Chapter 10) and Williams (Chapter 11) point out, all current research is limited by the available technology. This does not mean that investigations of the teaching and learning of functions using available technology is not important. The chapters by Demana, Schoen, and Waits (Chapter 2) and by Yerushalmy and Schwartz (Chapter 3) demonstrate this fact. However, if an integrated approach to investigating issues in this domain is to proceed, we also must consider the power and capability of new technologies.

IMPLICATIONS FOR RESEARCH ON THE LEARNING AND TEACHING OF FUNCTIONS

The need to create a conceptual and pedagogic framework for the functional domain and the need to consider the power and capability of graphing technologies are at the core of an integrated program of research in the domain. If these needs can be met, then research can proceed on understanding students' and teachers' thinking about functions, and on identifying important goals of instruction. Moschkovich, Schoenfeld, and Arcavi (Chapter 4) present a preliminary framework for examining this domain. It includes consideration of the method of representation (tabular, symbolic, or graphical), and whether the process of relating variables is emphasized or whether a function is treated as an object. If one adds to that whether students are expected to construct a representation or interpret a given representation, it could be considered to be a reasonable initial research framework for this domain.

During the past decade there has been a major shift in the paradigm adopted by researchers studying teaching—from a process-product paradigm (Brophy & Good, 1986) to paradigms that emphasize students' and teachers' thinking (Clark & Peterson, 1986). Most studies on teaching now take into account student learning. However, they have not been organized around content themes. Only recently have researchers begun to recognize that the content being taught needs to be taken into account in studying classroom teaching. For the functional domain, a systematically derived framework would provide the basis for this potential line of research.

Although current research on student and teacher thinking is sparse, Moschkovich, Schoenfeld, and Arcavi (Chapter 4), Dugdale (Chapter 5), and Kieran (Chapter 8) present perspectives about student thinking on functions and graphs, and Cooney and Wilson (Chapter 6) and Norman (Chapter 7) do the same for teachers.

In a similar manner, graphing technologies force educators to rethink curricular and instructional procedures. Philipp, Martin, and Richgels (Chapter 9) and Norman (Chapter 7) examine some of the issues involved. For example, there is a fundamental question about how a rich set of informal, instructional experiences with graphical images of functions should be related to the formal knowledge of symbols, concepts, and procedures that are taught in school. The issue is whether knowledge in a mathematical domain, like functions, should be developed locally in contexts that students readily understand (e.g., common dependency relations), or whether ways need to be found for introducing more complete conceptions of the domain via graphical techniques so that the symbols and procedures can be introduced as formalizations of those broad conceptual frames.

At the end of their review Leinhardt, Zaslavsky, and Stein (1990) raised the question: "What do we need to further our communal understanding of this area?" (p. 54). This volume is in part an attempt to answer that question. If we can successfully develop an integrated program of research on teaching, learning, curriculum, and assessment that is based on a common framework for functions that considers the power of graphical technologies, then there is a genuine potential for this research to have a significant impact on the current reform movement in school mathematics.

REFERENCES

Brophy, J. E., & Good, T. L. (1986). Teacher behavior and student achievement. In M. C. Wittrock (Ed.), *Handbook of research on teaching* (3rd ed., pp. 328-375). New York: Macmillan.

Carpenter, T. P., Moser, J., & Romberg, T. A. (1982). *Addition and subtraction: A cognitive perspective*. Hillsdale, NJ: Lawrence Erlbaum Associates.

Carpenter, T. P., & Peterson, P. I. (1988). Learning mathematics from instruction [Special issue]. *Educational Psychologist, 23*(2).

Clark, C. M., & Peterson, P. L. (1986). Teachers' thought processes. In M. C. Wittrock (Ed.), *Handbook of research on teaching* (3rd ed., pp. 255-297). New York: Macmillan.

Even, R., & Ball, D. L. (1989, March). *How do prospective secondary mathematics teachers understand the univalence of functions?* Paper presented at the Annual Meeting of the American Educational Research Association, San Francisco, CA.

Fennema, E., Carpenter, T. P., & Lamon, S. (1991). *Integrating research on teaching and learning mathematics*. Albany, NY: SUNY Press.

Freudenthal, H. (1983). *Didactical phenomenology of mathematical structures*. Dordrecht, The Netherlands: D. Reidel.

Kleiner, L. (1989). Evolution of the function concept: A brief survey. *College Mathematics Journal, 20*(4), 282–300.

Leinhardt, G., Zaslavsky, O., & Stein, M. K. (1990). Functions, graphs, and graphing: Tasks, learning, and teaching. *Review of Educational Research, 60*(1), 1–64.

Romberg, T. A. (1987). *The domain knowledge strategy for mathematical assessment.* Project Paper #1. Madison, WI: National Center for Education Research.

Romberg, T. A. (1992). Problematic features of the school mathematics curriculum. In P. W. Jackson (Ed.), *Handbook of research on curriculum* (pp. 749–788). New York: Macmillan.

Romberg, T. A., & Carpenter, T. P. (1986). Research on teaching and learning mathematics: Two disciplines of scientific inquiry. In M. C. Wittrock (Ed.), *Handbook of research on teaching* (3rd ed., pp. 297–314). New York: Macmillan.

Säljö, R. (1991). Learning and mediation: Fitting reality into a table. *Learning and Instruction, 1*(3), 261–272.

Sowder, J. T. (1989). *Setting a research agenda.* Reston, VA: National Council of Teachers of Mathematics.

Thorpe, J. A. (1989). Algebra: What should we teach and how should we teach it? In S. Wagner & C. Kieran (Eds.), *Research issues in the learning and teaching of algebra.* Reston, VA: National Council of Teachers of Mathematics.

CONTENT

2 Graphing in the K–12 Curriculum: The Impact of the Graphing Calculator

Frank Demana
Ohio State University

Harold L. Schoen
University of Iowa

Bert Waits
Ohio State University

The graphical representation of functions is an important topic for several reasons. Historically, it is a key development in mathematics that has paved the way for other important discoveries including calculus. In the mathematical education of school children, it has long represented a key and often difficult leap in understanding. Today, computer graphing software and graphing calculators have added a dimension to this topic that has attracted the interest of many researchers and curriculum developers in mathematics education. The rapidly growing body of research in this area is the subject of three recent reviews from different perspectives (Herscovics, 1989; Kaput, 1989; Leinhardt, Zaslavsky, & Stein, 1990). Not another research review, this chapter focuses rather on the graphical content of school mathematics. It begins with an analysis of the content on graphing in typical mathematics textbooks prior to algebra—that is, Grades 1-8—and argues that many common student difficulties with graphing can be understood in light of the inadequacy of this content. Then a discussion of techniques that can be used to introduce graphing in the primary grades is presented. Finally, the ways in which graphing utilities have begun to change the role of graphing in the school curriculum and in mathematics itself and the adjustments in content in prealgebra and algebra required by this change are analyzed.

GRAPHICAL CONTENT IN TEXTBOOKS PRIOR TO ALGEBRA

Graphs are clearly very important in mathematics and in other school subjects such as science and social studies, and graphical displays appear in newspapers and magazines often enough to make it desirable that informed citizens have a basic understanding of their meaning. This section analyzes the content that involves graphs in a typical mathematics textbook series Grades 1-8, content that presumably is meant to provide a basic understanding of graphical representations and to lay the foundation for understanding graphing in Algebra 1. This analysis is also based on the assumption, well supported by research on teaching (e.g., Porter, Floden, Freeman, Schmidt, & Schwille, 1988), that the content of textbooks guides teachers' topic selection to the extent that the teacher covers some, but usually not all, of the textbook topics and rarely covers topics not in the textbook. All graphing activities (with the exception of those involving number lines only), textual presentation, examples, and exercises in each book in the Houghton Mifflin (1987) series were included in this analysis.

A perusal of several major publishers' mathematics textbooks suggested that they contained graphical treatment similar to that of Houghton Mifflin. To check this informal perception of the similarity of these series, the graphical content of books of another publisher, Scott, Foresman, was carefully compared at the third- and sixth-grade levels to the corresponding Houghton Mifflin books. There was little difference in graphical content, and the trends and conclusions were the same.

Objective textbook adoption data are not readily available, but it is safe to say that a very large percentage of American school children are using mathematics textbooks with graphical content that is essentially the same as that described herein (Leinhardt et al., 1990). (The graphical content of some innovative textbook series, such as the series developed by the University of Chicago School Mathematics Project, does not fit the description that follows, but is only beginning to gain widespread use.)

The categorization of graphing tasks or activities by Leinhardt and her colleagues (1990) provides a useful language for this discussion. This categorization, given in Table 2.1, was developed for graphing tasks used by researchers, but it also captures well the types of instructional activities in textbooks. Activities are classified by type of action (interpretation or

TABLE 2.1
Action-Task Classification for Graphing and Functions

| *Actions* | *Tasks* | | | |
	Prediction	*Classification*	*Translation*	*Scaling*
Interpretation	—	T2	T3	T5
Construction	T1	—	T4	T6

Note. The design of this table is based on constructs of Leinhardt, Zaslavsky, & Stein (1990). Used with permission.

construction) crossed with kind of task (prediction, classification, translation, and scaling).

Interpretation refers to the action by which a student makes sense or gains meaning from a given graph or a portion of the graph. Interpretation can be global, referring to properties of the entire graph or major portions of the graph, or local, referring to properties of a point of the graph. A student may be asked to decide issues of pattern (e.g., For what x-intervals is y increasing?), continuation (e.g., interpolation or extrapolation of a graph), rate (e.g., How does area change when length increases by 4 cm?), or to determine when specific events or conditions are met (e.g., At what point is the picture of the cat located? Which is the shortest child? How much more in earnings did company A have than company B in 1989?).

Construction refers to activities in which students build a graph from an equation or from data points, or, in the other direction, build an algebraic relation that represents a given graph. Construction is quite different than interpretation. Whereas interpretation requires reaction to a given graph, construction requires the student to generate new graphs or equations that are not given. In some easier, local construction activities, students may only have to plot points on a given set of axes from a given table of ordered pairs. In a fuller, more global sense of construction, students may be required to construct their own graph, including appropriate placement and scaling of axes, directly from the algebraic equation, or to construct an algebraic relation from a given graph.

Prediction tasks are those that require students to conjecture from a graph where other points (not explicitly given or plotted) of the graph should be located or how other parts of the graph should look. Prediction tasks necessarily involve construction of a missing part of the graph, rather than interpreting a given graph.

Classification refers to activities that involve deciding whether a given graph is that of a function or a special kind of function (e.g., one-to-one or continuous), or identifying a function or a special kind of function among given graphs. Classification activities mainly involve an interpretation that connects the formal definition of function and of special kinds of functions to their graphs.

Translation activities are those that require students to (1) recognize the same function in different representations (e.g., data from a real situation, graph, equation, or table of ordered pairs), (2) identify for a specific transformation of a function in one representation its corresponding transformation in another representation, (3) construct a graph from an equation or set of ordered pairs, or (4) construct an algebraic representation from a graph. The typical elementary graph interpretation questions (e.g., Which child is tallest? How much more money did Alice earn than Ernie?) are classified in the translation category as instances of (2).

Scaling activities require students to attend to the axes, their scales, and the units that mark them. A simple scaling activity involves deciding whether the number of units each interval represents is one or two and whether both axes have the same scale. The issue of scale becomes fundamental when students use graphing utilities (Demana & Waits, 1990a; Goldenberg, 1988). Scaling activities can be either construction (e.g., constructing appropriate scales and "graphing windows" to graph a given function) or interpretation (e.g., identifying which of several given graphs with different scales represent the same function).

The graphing content analysis for K-8 is divided into three main sections: Grades 1-6, which include some graphing content but precede formal attempts to prepare students for graphing functions; Grades 7 and 8, the prealgebra years, which lay the foundation for the study of algebra including the graphing of functions; and finally, a summary of the findings as they relate to recent research on misconceptions and difficulties.

Grades 1 Through 6

As Table 2.2 indicates, mathematics textbooks for Grades 1-6 do not emphasize graphs. In each of these grades, the number of textbook pages with graphing content represents 1-2% of the textbook pages. In Grades 1-4, all the graphing content is in special sections labeled "Enrichment" or

TABLE 2.2
Numbers of Activities (Homework Exercises) in Grades 1-6

Activity	Grade 1	Grade 2	Grade 3	Grade 4	Grade 5	Grade 6	Total
Bar graphs	3(15)	1(2)	1(6)	3(18)	4(15)	1(4)	13(60)
						1(7)	*1(7)*
Pictograph	1(4)	1(8)	3(15)	1(7)			6(39)
					1(4)		*1(4)*
Rectangular coordinates		1(9)	2(18)	2(5)	1(5)	1(4)	7(41)
				1(22)	*2(39)*	*2(48)*	*5(109)*
Line graphs			1(10)	1(9)	1(10)		3(29)
						1(9)	*1(9)*
Circle graphs				1(6)			1(6)
					1(7)	*1(8)*	*2(15)*
Total enriched	4(24)	3(19)	7(49)	8(45)	6(30)	2(8)	30(175)
Total mainstream				*1(22)*	*4(50)*	*5(72)*	*10(144)*

Note. Values in italics are number of exercises.

"Problem Solving." Much of it remains in special sections in Grades 5 and 6 as well, and so it is questionable whether even the sparse graphical content that is in the textbooks is discussed in class.

The instructional sequence for bar graphs begins in Grade 1 with pictures (e.g., various pets) on the horizontal axis and numbers on the vertical. The vertical column for each pet is marked in units from 1 to 10 and shaded to indicate the number of that pet. Students are expected to use the bar graph to determine how many of each pet is represented. In later grades, pictures of pets are replaced by words, and the units indicate more than one of the objects under consideration and, later still, decimal parts of objects. Finally in Grades 5 and 6 double bar graphs are introduced in which one bar represents, for example, "sales with a coupon" and the other "sales without a coupon."

A similar sequence is followed for pictographs, although they receive less attention than bar graphs. Line graphs with clearly marked axes appear in special sections in Grades 3, 4, and 5. In Grades 5 and 6, pictographs, line graphs, and circle graphs are included as mainline content. All bar graph, line graph, pictograph, and circle graph activities involve interpretation and translation from a given graph to a given "real" context. Students are never expected to construct any of these graphs from data they collect or even from given data. Concerning the focus of the homework exercises, only 7 of the 319 are global in nature; all others can be answered by reference to one or a few points on the graph.

Rectangular coordinates appear as enrichment in Grades 2 and 3, but only in the context of an $N \times N$ grid with pictures (e.g., pets or food items) at some of the vertices. Students are asked to tell the location of each picture in terms of a number "across" and a number "up." In an enrichment activity in Grade 3, students also plot given ordered points on a given set of axes and then join the dots to see an interesting picture. A map with a coordinate overlay is the setting for a Grade 5 enrichment activity, and students are asked to give the whole-number coordinates of various points on the map. Plotting ordered pairs of numbers, including negative rationals, on a given set of axes is the main graphing activity in Grades 5 and 6. In Grade 6, slides and flips are explored by respectively changing signs and adding numbers to the coordinates of the vertices of given triangles.

In exercises for students throughout these grades, the emphasis is on plotting given pairs on given coordinate systems or naming the coordinates of given points. Students are not expected to generate pairs from given information or to construct and scale their own coordinate system in order to graph certain data. Through Grade 6, no mention is made of a connection between a numerical relationship and a graph, and no situations involving continuous curves are encountered. Furthermore, in 27 of the 40 graphing settings, one of the variables is nominal (e.g., names of pets or days of the week), and in only 2 settings (both involving rectangular coordinates in Grade 6) are the two variables continuous.

Grades 7 and 8

Graphing receives a good deal more attention in Grades 7 and 8 than it does in the earlier grades, as Table 2.3 indicates. Only about 3% of the textbook pages in these grades contain graphing content, but this is twice as much as in any of the earlier grades. Bar graphs, pictographs, line graphs, and circle graphs continue to appear regularly, and in Grade 8 histograms and scattergrams are discussed. As in the earlier grades, however, students are rarely expected to construct any of these graphs. Rather, the graphs are given and the students are asked interpretive questions, such as, "Find the height of a student who weighs 96 pounds." One of the variables that is graphed in these ways is either nominal or discrete (21 of the 34 lessons or features involving graphs).

Most of the graphing content in Grades 7 and 8 focuses on the rectangular coordinate system (e.g., 68.5% of all graphing exercises deal with the rectangular coordinate system, exclusive of the types of graphs just mentioned). Presumably this content is intended to prepare students for algebra in Grade 9. Both variables that are graphed are continuous in these

TABLE 2.3
Numbers of Activities (Homework Exercises) in Grades 7 and 8

Activity	Grade 7	Grade 8	Total
Bar graphs	2(10)		2(10)
	2(8)	*3(13)*	*5(21)*
Pictographs	2(7)		2(7)
Rectangular coordinates	1(5)	1(7)	2(12)
	5(105)	*6(144)*	*11(249)*
Line graphs	1(9)		1(9)
	2(20)	*3(18)*	*5(38)*
Circle graphs			
	2(12)	*1(4)*	*3(16)*
Scattergrams			
		1(9)	*1(9)*
Normal curve		1(2)	1(2)
Latitude and Longitude			
		1(8)	*1(8)*
Total enrichment	6(31)	2(11)	8(42)
Total mainstream	*11(145)*	*15(196)*	*26(341)*

Note. Values in italics are number of exercises.

lessons. While the graphing content described in the previous paragraph is set in a "real-world" context, this material is not. Overall, about 53% of the graphing exercises in these two grades are presented in a context. Following a good deal of practice in plotting ordered pairs, all repeated from earlier grades, the first reference to the use of a graph to represent an equation in two variables is made. This seventh-grade lesson is also the first exposure of students to the idea of a line being continuous and infinite—ideas that the textbook hardly mentions. "When an equation has two variables, you can use a chart to show some of its solutions." An equation, $x + y = 5$, is then given with a table of eight ordered pairs that satisfy it. The lesson goes on to say, "Notice that the points lie on a straight line. You draw the line connecting them to show that the list of solutions is endless." No mention is made that the line is continuous or that a point is on the line if and only if its coordinates satisfy the equation. The exercises call for students to graph equations (all linear), given the equation and a table of several ordered pairs satisfying the equation. The student does not construct the table of ordered pairs.

In later exercises, students are expected to graph linear equations and inequalities and systems of linear equations, given only the equations or inequalities, but continuity of the line is never explicitly taught. No reference is made at either grade level to nonlinear equations or inequalities. Exercises that required students to construct a graph were classified as global, that is, requiring students to attend to the entire graph. In Grade 7 and Grade 8, about 14% and 37% of the exercises were global, respectively, considerably more than in the earlier grades where almost all exercises had a local focus.

Summary of Prealgebra Graphing Content

The following generalizations summarize the graphical content in typical mathematics textbooks in Grades 1-8.

1. In Grades 1 through 6, students have almost no experience constructing a graph of any kind. Most American junior high students have not been taught or been expected to learn how to construct a graph. Both Leinhardt and her colleagues (1990) and Herscovics (1989) cited several research studies that have found that junior high students do indeed have difficulty constructing a graph, including proper scaling and placing of axes. As Herscovics (1989, p. 68) pointed out, "It is easy to underestimate the level of sophistication involved in the construction of the Cartesian plane schema. Since most students manage to provide the coordinates of points in the plane and, conversely, locate points when given coordinates, we conclude that they generally understand the structure involved." Yet research indicates that this conclusion is not warranted.

Nor are American students the only ones who find it difficult to construct a graph. Kerslake (1981) found that British secondary school students aged 13, 14, and 15 years had trouble choosing an appropriate scale to plot points. After two years of algebra, 38.7% of the 15-year-olds could not provide suitable scales to plot pairs like (20, 15), (−14, 3), and (5, −12) on a sheet of graph paper. Over 34% of these same students could not correctly position their coordinate axes, often not locating the origin at (0, 0).

2. In Grades 1 through 6, variables that are graphed are nominal or sometimes discrete, with variable values placed at equidistant points on the axis, and this continues to be true, to a lesser extent, in Grades 7 and 8. Textbook activities involving graphs have been almost exclusively limited to the interpretation of graphs displaying nominal or discrete data in which the axes are always marked by equally spaced points, each representing a value of the variable. Prior to Grade 7, students would have seen few if any counterexamples to this. That junior high students do indeed have this misconception is supported by research. Herscovics (1989) noted that many junior high students represent data from real settings on a line simply in terms of the relative order of the quantities, without accounting for their measures. Furthermore, many students fail to set the distance between the points to represent a measure of the relative difference in quantity between consecutive numbers. For instance, 16.3 cm, 19.6 cm, and 28.8 cm would be placed in the proper order, but equally spaced on a line.

3. In Grades 1 through 6 graphs that students encounter have a limited number of points to be interpreted, with the intervals between those points being meaningless, and this continues to be true, to a lesser extent, in Grades 7 and 8. This is precisely the case with bar graphs and line graphs, which comprise a major part of the students' experience with graphs. Through Grade 6, the only content involving rectangular coordinates has been plotting ordered pairs and, infrequently, joining the resulting points to form a figure. However, even in these latter activities the segments only serve to form the picture, and the "points" on the segment other than those plotted are not salient. This understanding about graphs is probably also reinforced by the fact that nearly all the graphical exercises in the textbooks have a local, rather than global, focus. Leinhardt and her colleagues (1990) called this difficulty the continuous versus discrete misconception.

In a study that described this misconception, Kerslake (1981) presented students with a set of ordered pairs and instructed them to plot the points and connect them with a straight line. She then asked the students questions about the points on the line that were not marked. About one-fifth of the 15-year-olds indicated that there were infinitely many points on the line. Many students thought that there were no other points than those they had plotted, other students thought the midpoint was the only other point, and still others counted the number of points where the line crossed the grid.

(All line graphs in the elementary textbooks examined in preparation for the present article showed only the points at which the line graph crossed the vertical grid lines.)

4. In Grades 1 through 6, graphs are almost always presented as existing entities to be interpreted point by point in connection with some "real" context, and no suggestion is made that graphs may be connected to numerical relationships. This generalization is clearly supported by the vast majority of the activities in the mathematics textbooks and is probably reinforced by the social studies and science curricula, which, beginning at about Grade 5, include graphs of "real data" as well. On the whole, it appears that the curriculum in Grades 1-6 provides essentially no preparation, beyond plotting ordered pairs to students, for graphing functions. In particular, typical early graphing experiences are, at best, of no help in elucidating the connection between a numerical relation and its graph, a connection that continues to be difficult for students through high school and into college (Schoenfeld, in press).

5. Other than when graphing linear equations and inequalities in Grades 7 and 8, students are not expected to make global or qualitative interpretations of graphs. Virtually none of the graphing activities through Grade 6 involve a global interpretation of a graph. Furthermore, almost as rare are activities that engage students in constructing a graph, an exercise that could naturally lead them to confront global and qualitative issues. Not until Grade 7 are students expected to construct the graph of an equation (and then it's always a linear equation). It seems understandable that students would have great difficulty with the global interpretation tasks presented by researchers. An often cited finding of this type is that many students interpret travel graphs as the path of the actual journey, rather than as a time-versus-distance or velocity graph (Janvier, 1981; Kerslake, 1981). This is a particularly important concern today, because graphing utilities automatically require a global view of the graph of a function.

6. Through Grade 8, students have no experience with graphs of nonlinear functions. Through sixth grade, the textbooks contain no activities that suggest a connection between a graph and a numerical relationship. In Grades 7 and 8, students are expected to graph linear equations in two variables, but there is no suggestion that the graphs of some equations are nonlinear. Not surprisingly, researchers found that students do tend to have a strong tendency to identify functions with linear functions (e.g., Karplus, 1979; Lovell, 1971; Markovits, Eylon, & Bruckheimer, 1983). Markovits et al. (1983) also report that many students reject constant functions as not being legitimate. In our analysis, constant functions first appeared in two exercises in Grade 8, with no accompanying discussion, and that was the extent of their treatment prior to algebra.

In a review of recent research, Herscovics (1989) described some of the obstacles many students encounter as they attempt to understand functions, graphs, and the complex connection between a function and its graph. He used the term *cognitive obstacles*, attributing the difficulties to the need for the student to construct new cognitive structures to accommodate these higher order concepts. Leinhardt et al. (1990) distinguished between misconceptions (i.e., incorrect features of student knowledge that are repeatable and explicit) and difficulties. Our content analysis illustrates that many of the misconceptions students have about graphs are legitimate, understandable generalizations of the graphing content in typical mathematics textbooks in Grades 1–8. This finding does not conflict with Herscovics' cognitive obstacles theory or the analysis of Leinhardt and her colleagues. On the contrary, these misconceptions have developed with the support of eight years of the textbook curriculum, and there is good reason to believe that students would indeed have great difficulty overcoming them.

Teachers introducing computer graphing in a high school algebra course must be made aware of the students' misconceptions about graphs. There will have to be a period of adjustment while students investigate and explore the effect of scaling, global relationships, and the connections among fundamental numerical, algebraic, and graphical representations. In the future, as the *Curriculum and Evaluation Standards for School Mathematics* (National Council of Teachers of Mathematics [NCTM], 1989) are implemented, these misconceptions will be addressed in Grades K–8 by introducing appropriate graphing activities. Examples of suitable activities are illustrated in the next section of this chapter.

BRIDGING THE GRAPHING GAP

Graphs of both functions and relations can now be obtained quickly with the use of technology. This calls for graphing activities to start very early in the curriculum and to be a major part of the curriculum. This section indicates how paper, pencil, and calculator graphing can be started in the primary grades and includes a discussion of how ordinary hand-held algebraic logic calculators can be used to start graphing and build a base for the later use of computer-based graphing utilities such as those available in powerful graphing calculators like the TI-81. The approach described here also shows how the curriculum can be infused with problem solving and how function can become a major theme of the curriculum.

Establishing Functions Numerically and Graphically

Graphing should be started early in the elementary mathematics curriculum—well before the study of negative numbers—and should be used to solve problems graphically. The early start can be achieved by restricting graphs to the first quadrant. Significance and meaning can be established about graphing by using graphs to represent problem situations.

Prior to making graphs, tables can be constructed and used to explore and investigate problem situations numerically. The tables can be used to help students develop understanding about arithmetic processes, to foreshadow the study of algebra, and to solve problems numerically. Later the tables will be used to make graphs. This approach is illustrated in the following example.

Example 1. Mr. Thompson's third grade class is having a candy sale to raise money for a field trip. The class receives $0.25 profit for each candy bar sold.
(a) Complete the following table:

Number of candy bars sold	Profit ($)
50	
100	
150	
200	

(b) How many candy bars must be sold to make a profit of $45.00?

Teachers can control the numbers used in the first column, and the per bar profit can be selected to fit the arithmetic being studied. For example, students could be selling boxes of candy yielding $1 profit per box so that the profit column and the first column can be restricted to whole numbers.

There are many benefits to using problems as a vehicle to develop understanding about mathematical concepts and to using a numerical approach to analyze problems (e.g., Demana & Leitzel, 1988). Using problems to introduce mathematical ideas helps students learn to value mathematics—an important goal of the NCTM *Standards*. This approach helps integrate problem solving into the study of mathematics and establishes problem solving as a major focus of mathematics. Exploring problems numerically builds understanding about problem situations that can be exploited when these problems are returned to in algebra. With the aid of calculators, students can do numerous computations in a given problem situation quickly. Repeated calculator-based computation in a given problem situation provides the necessary experience that allows students to establish understanding about arithmetic processes. For example, if more lines are added to the table in Example 1 about selling candy, and other problems similar to this one are investigated, students will master the important relationship:

Total profit = (profit per item sold) × (number of items sold)

In the current curriculum, practice is limited because of total dependence on paper-and-pencil-computation, and the practice occurs out of context so that such understandings are hard to establish. Because of this limited experience, students all too often must ask teachers whether to add, subtract, multiply, or divide in the rare applications that do occur. Traditional paper-and-pencil drill and practice should be replaced by calculator-based drill and practice. The computational load will be reduced while computational experience, estimation skills, and understanding are enhanced.

Additionally, the frequent use of tables helps establish function as a major theme of mathematics, a theme that is missing from the typical early curriculum as was shown earlier in this chapter. Each table constructed gives students a concrete example of a function presented numerically. These tables can be used to construct graphs to represent functions geometrically and can be used to write expressions that represent functions algebraically.

Problem situations can be chosen to expose students arithmetically to the many expressions they need to deal with formally in algebra (e.g., the expressions for perimeter of a rectangle, volume, area of a rectangle, sale price, population, amount of money in a savings account, and so forth). Repeated exposure to these expressions arithmetically helps students understand the syntax of the algebraic expressions that will occur later in the study of algebra (e.g., see Comstock & Demana, 1987). This rich arithmetic activity builds a base of student understanding from which teachers and students can draw in the later more formal studies in algebra.

Problem Solving Numerically

Students can use the completed lines in Example 1 to get started on the solution to (b) of the example. The completed lines of the table can be used to estimate that the answer to (b) is between 150 and 200 bars of candy and a little closer to 200 bars. Then a guess-and-check approach can be used to determine that 180 bars is the answer to (b). Besides developing understanding about arithmetic processes, this rich arithmetic activity gives meaning to and understanding about finding solutions to such problems. Students can be guided to see that solving this problem numerically amounts to finding a line of the table with an entry in the second column equal to 45. This activity can be exploited later when students return to such problems and solve them algebraically.

Tables and Variables

Many students entering algebra think that a variable stands for only one or two numbers to be determined. Unfortunately, most of the exercises concerning variables in prealgebra and beginning algebra courses reinforce this belief. Students need deeper understanding about variables to be successful in mathematics. This deeper understanding about variables can be developed early by adding a line containing a variable to tables such as the one in Example 1. Suppose the following line is added to that table.

Number of candy bars sold	Profit ($)
N	

The use of lines with variables in tables helps students see that a variable can have an infinite number of replacements. In this case, N can stand for any whole number—the possible number of candy bars sold. Moreover, using variables in tables helps students write simple algebraic representations such as the equation $P = 0.25N$ for the candy bar problem situation. Completing the second column in this new line ($0.25N$) helps students develop an understanding of syntax in algebra.

Various problem situations can be constructed to give students experience with a wide range of algebraic expressions; for example, the expression $2W + 2(W + 4)$ would come up in a problem about perimeters of rectangles with length 4 more than the width. The expression $W(W + 4)$ would come up in a problem about area of such rectangles. Exposing students to these algebraic expressions in numerical tables helps them understand the meaning of the expressions and improves their ability to manipulate them in algebra courses.

Tables and Graphs

Tables similar to the one in Example 1 can be used to introduce students to graphing in concrete problem situations. Initially, graphing should be restricted to the first quadrant because real problem situations usually only involve nonnegative numbers. For Example 1, "Number of Candy Bars Sold" can be used as the label for the horizontal axis and "Profit ($)" as the label for the vertical axis (Fig. 2.1). Then each line of the table corresponds to a point in the first quadrant of the coordinate plane.

Coordinates of points in the plane are more meaningful in concrete

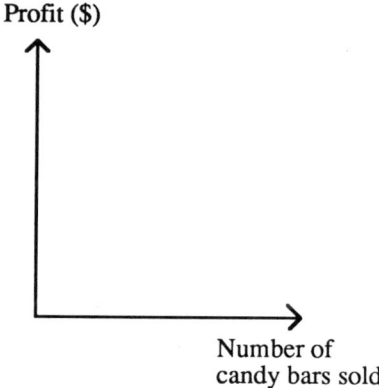

FIG. 2.1. The candy bar problem.

situations. In this case, the first coordinate gives the number of candy bars sold and the second coordinate the corresponding profit. Constructing a table of values and the corresponding points for a function such as $y = 2x + 3$ is not nearly as meaningful as constructing a table of values and the corresponding points for a concrete problem situation such as the candy bar example.

Kaput (1989) pointed out how important tables are in the development of the function concept. In fact, he described how computer graphics can make it possible to show dynamically the link of the tabular, symbolic, and graphical representations of a function.

Scale

Determining a scale for each axis in Fig. 2.1 is an important and difficult activity for students and one that the present elementary school curriculum ignores. Scale is fundamental when constructing and interpreting graphs with technology (Demana & Waits, 1990a; Leinhardt et al., 1990). Concrete problems can help students understand issues involved in scale. An interesting start-up activity for students is to give them an ordinary piece of graph paper and let them choose their own scale and construct a graph. Some students will choose a scale that causes the graph to go off the paper, others will choose a scale that crowds their graph near the origin, and some will interchange the axes. These beginner choices provide rich material for classroom discussion. The fact that many choices for scale are reasonable builds flexibility into student thinking.

Different-sized pieces of graph paper can be used to help students build arithmetic understanding. For example, suppose the table for the candy bar example is expanded to show the profit for the sale of 10, 20, ..., 200 candy bars. Twenty points corresponding to the 20 lines of the table will need to be plotted. Suppose, further, that a piece of graph paper with 10 scale marks on the horizontal axis other than the origin is given to students to graph the points in the expanded table for the candy bar problem (Fig. 2.2). By experimenting, students will see that a good choice for the distance between consecutive scale marks on the horizontal axis is 20. This is not to say that choosing the distance between consecutive scale marks to be 40 is wrong. In fact, given the right problem situation, other choices for the distance between consecutive scale marks could be more natural.

Notice in Fig. 2.2 that the number of vertical scale marks is also 10. By studying the entries in the second column of the expanded table, students will see that 5 is a good choice for the distance between consecutive vertical scale marks because 50 would be the largest number in the second column. The 20 points corresponding to the expanded table are graphed in Fig. 2.3. We have observed middle-school students become very adept at choosing scale so that the entire piece of graph paper is used. Later, when we want students to extend their graph and make projections, students find, with appropriate teacher help, that other scale choices are preferred.

Rich (1990) found that students who used a graphing calculator for the entire year of precalculus were far better able to deal with issues of scale on a graph than comparison students in traditional instruction. In tests and interviews at the end of the year, the students were able to choose appropriate scales for graphing a variety of functions and to explain the appearance of a graph in terms of the scale of the calculator's viewing window. For example, 95% of students who had used a graphing calculator for the year were able to choose correctly from among four given parabolas the pair that was identical except for scale. In interviews, most of these students correctly gave the difference of scale as the reason for their choice.

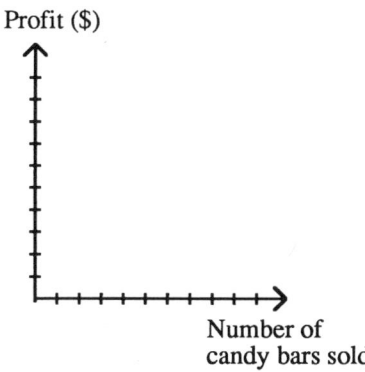

FIG. 2.2. Scaling in the candy bar problem.

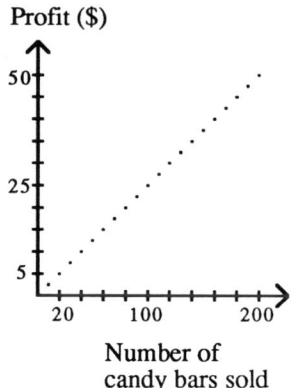

FIG. 2.3. Plotting points in the candy bar problem.

Only 75% of the comparison students responded correctly in the same situation. The other 25% had no idea how to begin or were persuaded by the similarity of the shape of two parabolas that were, in fact, different.

THE IMPACT OF THE GRAPHING CALCULATOR ON THE CURRICULUM

The Role of Paper, Pencil, and Calculator Point Plotting

There needs to be a good deal of research about how much point plotting by hand and with the aid of a calculator is necessary before students are turned loose with modern computer-based graphing tools such as a graphing calculator. Considerable point plotting is certainly necessary in Grades K-8, but if used as much as it presently is, it may well interfere with students' understanding of the continuous nature of most graphs. As evidence of this, consider the fact that early experiences with graphing are now heavily dominated by point plotting, as illustrated earlier in this article. This experience leaves many students with the misconception that curves are not continuous—that is, they believe the graph only contains the plotted points (Herscovics, 1989; Leinhardt et al., 1990). Although we do not know how much point plotting is needed, we do know that the steady diet of point plotting in today's early curriculum fails to have the desired effect on students' understanding of graphs.

Certainly students need to see how the graph in Fig. 2.3 consisting of 20 points can be extended to a complete graph of the algebraic representation

$P = 0.25N$ (a continuous function) of the candy bar problem situation. It should be noted that a complete graph of the problem situation is not a continuous function.

Complete Graphs

A complete graph of a problem situation is a graph that indicates all of the points and only the points corresponding to the problem situation. The complete graph of the candy bar problem is the set of all points of the form $(N, 0.25N)$ where N is a positive integer. Eventually students need to determine complete graphs by themselves. Teacher-guided classroom discussion can help students determine complete graphs. In the candy bar problem, for example, teachers can probe with questions like, "Can the number of candy bars sold be different from the ones represented in Fig. 2.3? Can the number of candy bars sold be 15, or 22?" This type of questioning will help students see that the complete graph of the problem situation consists of the discrete set of points $(N, 0.25N)$ where N is a positive integer. This helps students establish a relationship between numerical tables and graphs.

Students in early grades will not be able to describe the complete graph using the above description. As a first experience, students can plot numerous points between those plotted in Fig. 2.3 and discover that the complete graph appears to be a straight line. This straight line is a *geometric representation* of the algebraic representation $P = 0.25N$ for the candy bar situation. It contains a complete graph of the problem situation (Fig. 2.4). Of course, not every point on the straight line makes sense in the candy bar problem. The relationship between a complete graph of an algebraic

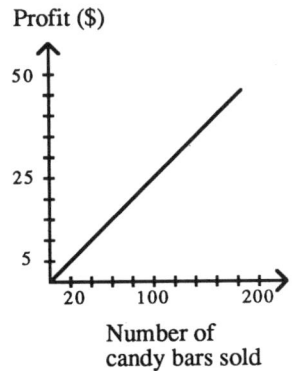

FIG. 2.4. A geometric representation of the algebraic representation $P = 0.25N$ of the candy bar problem.

representation and a complete graph of the associated problem situation is subtle but very important. However, knowing that the points representing the problem situation lie on this straight line is very useful.

Students acquire deeper understanding about graphs of problem situations and associated algebraic and geometric representations as they progress through the curriculum. Eventually, we would expect students to know that the points in Fig. 2.3 lie on the straight line $P = 0.25N$, to know that a point lies on the graph if and only if its coordinates satisfy the equation, and to have complete understanding about the class of linear functions. In the primary grades, we would be content with students observing that the points appear to lie on a straight line, or even to have students fit their graphed data with a "best fitting" straight line.

A major goal of the Grades 9-12 mathematics curriculum is for students to achieve in-depth understanding about important classes of functions. Prior to the study of calculus, this understanding would need to come from exploring numerous graphs quickly with the aid of technology. Teachers can guide this exploration so that students will actually conjecture statements that are true. In this way, students feel ownership in the mathematics. Teachers need to provide a few pitfalls to be sure that students use care when making conjectures (Demana & Waits, 1988a; Dion, 1990). However, this is not to say that students should be inundated with pathological examples.

A *complete graph* of a function $y = f(x)$ is a graph that shows all its important behavior (Demana & Waits, 1990b). What constitutes the important behavior of a function depends on where the student is in the mathematics curriculum. Important behavior of a function includes its y intercept, zeros, relative extrema, and end behavior. In algebra and precalculus, teachers will have to tell students that a cubic polynomial function has zero or two relative extrema. Then students can find the coordinates of such extrema using graphical methods. In calculus, we would expect students to know *why* there are always exactly zero or two relative extrema for a cubic polynomial. Finding points of inflection would also probably be reserved for calculus.

The *end behavior* of $y = f(x)$ is its behavior for x large in absolute value. For example, the end behavior of $y = ax^2 + bx + c$ ($a \neq 0$) is $f(x) \to \infty$ as $|x| \to \infty$ if $a > 0$, and $f(x) \to -\infty$ as $|x| \to \infty$ if $a < 0$. Notice that the end behavior of $y = ax^2 + bx + c$ and that of $y = ax^2$ are the same because the values of $ax^2 + bx + c$ are dominated by the values of ax^2. In other words, $y = ax^2$ gives a model of the end behavior of $y = ax^2 + bx + c$. We say that $y = ax^2$ is an *end behavior model* of $y = ax^2 + bx + c$. More precisely, g ($\neq 0$) is an end behavior model of f if and only if $f/g \to 1$ as $|x| \to \infty$. The end behavior model concept is also very visual: For example, the graphs of $y = -2x^2$ and $y = -2x^2 + 5x - 7$ in the viewing window

[−1,000, 1,000] by [−1,000,000, 500,000]—that is, in the rectangular region of the place given by $-1{,}000 \leq x \leq 1{,}000$ and $-1{,}000{,}000 \leq y \leq 500{,}000$—are visually identical. This is convincing evidence that the behavior of $y = -2x^2 + 5x - 7$ is essentially the behavior of the simpler function $y = -2x^2$ for large values of $|x|$.

We would expect that prior to the study of calculus students will, through the power of visualization, have in-depth understanding about polynomial functions, radical functions, rational functions, exponential and logarithmic functions, trigonometric functions, conics (relations), and polar and parametric equations. The class of rational functions, for example, provides important foreshadowing activity for the later study of calculus, as illustrated in Example 2.

Example 2. Determine a complete graph of

$$f(x) = \frac{x^4 + x^3 - 6x^2 + 6}{x^2 + x - 6}.$$

Fig. 2.5 gives the graph of f in the viewing rectangle $[-7, 7]$ by $[-10, 30]$. Notice that this graph illustrates the end behavior of f because it suggests that $f(x) \to \infty$ as $|x| \to \infty$. Some beginning students will need more detail near $x = -3$ and $x = 2$ to be sure that f has vertical asymptotes at $x = -3$ and $x = 2$. The graph of f in Fig. 2.6 gives strong evidence that f has only two real zeros and three local extrema in $-3 < x < 2$. Notice that Fig. 2.5 shows f has two other relative extrema outside this interval for a total of five extrema. Prior to the study of calculus, teachers will need to guide students to discover that the graphs in Figs. 2.5 and 2.6, together with more detail around $x = -3$ and $x = 2$, if needed, constitute a complete graph of f. This example illustrates that more than one graph is often necessary to show all of the important behavior of a function.

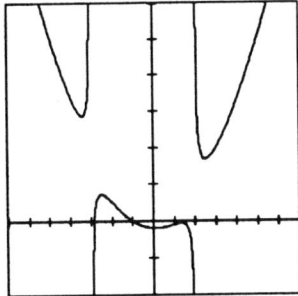

FIG. 2.5. The graph of $f(x) = \dfrac{x^4 + x^3 - 6x^2 + 6}{x^2 + x - 6}$ in $[-7, 7]$ by $[-10, 30]$.

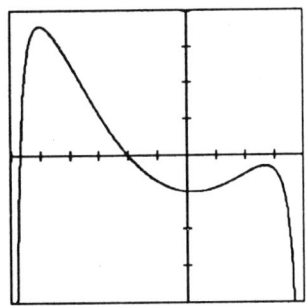

FIG. 2.6. The graph of $f(x) = \dfrac{x^4 + x^3 - 6x^2 + 6}{x^2 + x - 6}$ in $[-3, 2]$ by $[-4, 4]$.

Students can use algebraic manipulation to help understand the graph of f. Using long division and factoring, f can be rewritten as follows:

$$f(x) = \frac{x^4 + x^3 - 6x^2 + 6}{x^2 + x - 6} = x^2 + \frac{6}{(x-2)(x+3)}$$

In this form, students become convinced that the graph of f looks like the graph of $y = x^2$ away from $x = -3$ and $x = 2$ and "blows up" near $x = -3$ and $x = 2$, as suggested by Fig. 2.5. This form also suggests that $y = x^2$ is an end behavior model of f, because it is easy to see that $f/x^2 \to 1$ as $|x| \to \infty$. An end behavior model of f can also be discovered graphically using *zoom out*, that is, viewing the graph of f in large viewing rectangles. The graphs of f and $y = x^2$ will appear to be nearly identical in the $[-100, 100]$ by $[-10,000, 10,000]$ viewing rectangle.

In calculus, students can compute the derivative of f to see that it is also a rational function with the polynomial of degree 5 as the numerator. Because a polynomial of degree 5 has at most five real zeros, the calculus student can now conclude that, based on the graph of f in Fig. 2.5, f has exactly five relative extrema. In calculus, conjectures about the behavior of functions made in precalculus can be proven. This is a modern role of calculus.

Results from Recent Research

Rich (1990) found that students who are taught precalculus using a graphing calculator better understand the connections between an algebraic representation and its graph and that they view graphs more globally, in that they understand the importance of a function's domain, the intervals where it increases and decreases, its asymptotic behavior, and its end behavior. Browning (1988) found that high-school precalculus students who used graphing calculators for one year exhibited a significantly increased ability to deal with graphing at the more advanced Van Hiele levels of analysis and ordering. Farrell (1989) also observed that precalculus students who were taught the use of graphing calculators demonstrated greater facility with higher order thinking skills than traditional students. Further, Dunham (1990) observed that in college algebra classes requiring graphing calculators, gender-related differences in performance on graphing items were eliminated, whereas pretest performance on graphing items indicated that females performed at a lower level than males.

Solving Problems Graphically

Once students obtain a complete graph, the graph can be used to solve problems: for example, using the graph in Fig. 2.4, we can answer the question in (b) of Example 1. To determine the number of candy bars that must be sold to realize a profit of $45, we draw a horizontal line from 45 on the vertical axis to the line and then drop a perpendicular line to the horizontal axis (Fig. 2.7). We can read from the horizontal axis to find that selling 180 candy bars produces a profit of $45.

Students may well read a number other than 180 from their graphs as the number of candy bars to be sold to produce a profit of $45. However, students can use their number as an initial estimate and then use a guess-and-check approach to determine an exact answer, or an answer with prescribed accuracy. In fact, exact answers will be rare with the realistic problems that are possible with the use of technology in mathematics.

The graphical problem-solving process illustrated in Fig. 2.7 is a significant foreshadowing of the important concept of inverse relations. This process is very natural, even for young children, when described in the context of a concrete problem situation.

After students know how to draw graphs using paper, pencils, and calculators, graphing utilities such as graphing calculators and computer graphing software should be introduced. First-year algebra is a natural place to introduce graphing calculators.

Because of the speed and power of graphing utilities, we can solve (b) of Example 1 another way. The solution is the intersection of the lines $y = 0.25x$ and $y = 45$. Of course, students would need to have progressed far enough in the mathematics curriculum to be able to write this algebraic formulation of the problem. Now the students can draw a graph of the two functions $y = 0.25x$ and $y = 45$ and use *zoom-in* (a process described later in this chapter) to determine the x coordinate of the point of intersection.

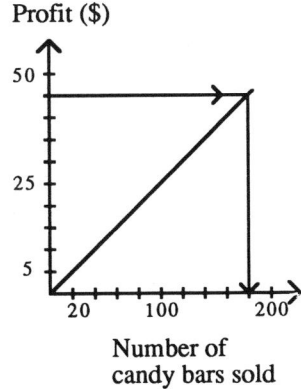

FIG. 2.7. Solving the candy bar problem graphically.

The x coordinate 180 of the point of intersection of $y = 0.25x$ and $y = 45$ gives the number of candy bars that must be sold to produce a profit of $45.

Solving Equations and Inequalities: The Modern Role of Computer-Based Graphing

In the past, students would use a careful numerical and analytical investigation to produce a correct graph. The work required was time-consuming, and these graphs were rarely used. With modern graphing calculators, students can obtain accurate graphs quickly. Through the use of graphing technology it becomes very natural to add visualization to the standard algebraic techniques used in the curriculum. Suppose, for example, we want to solve the equation $2x^2 - 5x - 3 = 0$. Students can use the quadratic formula or factor as $(2x + 1)(x - 3) = 0$ to see that the solutions are $-\frac{1}{2}$ or 3. Without technology it is not efficient to solve this equation graphically because of the time required for most students to draw the graph of $f(x) = 2x^2 - 5x - 3$ by hand. With technology it is easy to obtain a complete graph of f (Fig. 2.8).

Because graphs are easily obtained, it is reasonable to emphasize that finding all the real solutions of $f(x) = 0$ is the same as finding all the x intercepts of the graph of f. Using Fig. 2.8, students can conjecture that the x intercepts of f are $-\frac{1}{2}$ and 3. Comparing these with the solutions to $f(x) = 0$ helps establish the connection between solutions of $f(x) = 0$ and x intercepts of the graph of $y = f(x)$.

Limiting ourselves to algebraic techniques seriously restricts the types of equations students can solve. In the conventional curriculum, students solve linear equations, quadratic equations, easily factored contrived higher order polynomial equations, and other contrived equations whose form is special. However, using computer graphing, students can easily solve very complex equations, even equations that do not admit an algebraic solution. In the following example, we solve an equation that does not typically occur

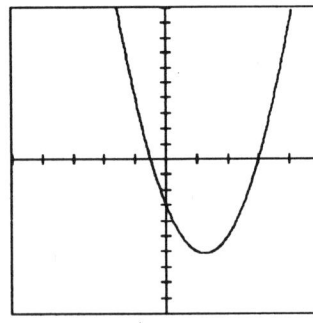

FIG. 2.8. The graph of $f(x) = 2x^2 - 5x - 3$ in $[-5, 5]$ by $[-10, 10]$.

2. IMPACT OF THE GRAPHING CALCULATOR, K-12

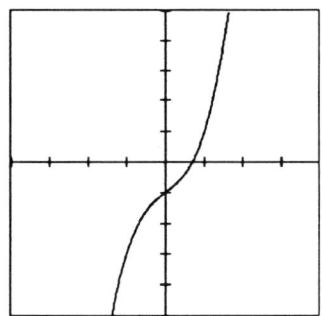

FIG. 2.9. The graph of $f(x) = x^3 + x - 1$ in $[-4, 4]$ by $[-5, 5]$.

in the school algebra curriculum, but should be included and is routine with graphing utilities.

Example 3. Solve $x^3 + x = 1$.

It can be shown that the graph of $f(x) = x^3 + x - 1$ in Figure 2.9 is complete. Thus, the equation has one real solution. Because the scale marks are one unit apart in Fig. 2.9, the one real solution lies between 0 and 1 and is about 0.7. We can create a decreasing nested sequence of viewing rectangles that converge to the x intercept and that allows us to determine the x intercept with accuracy within the limits of the precision of the machine in use.

The scale marks on the horizontal axis in Fig. 2.10 are 0.1 apart. This allows us to read the solution to the equation as 0.69 with error of at most 0.1, the distance between the horizontal scale marks. Similarly, we can read the solution as 0.683 with error of at most 0.01 from Fig. 2.11, and as 0.6824 with error of at most 0.001 from Fig. 2.12. The process illustrated by Figs. 2.10–2.12 is called *zoom-in*. Modern graphing calculators, such as the TI-81, have automatic zoom-in as a feature. However, the careful selection of viewing rectangles suggested by Figs. 2.10–2.12 is a good beginning

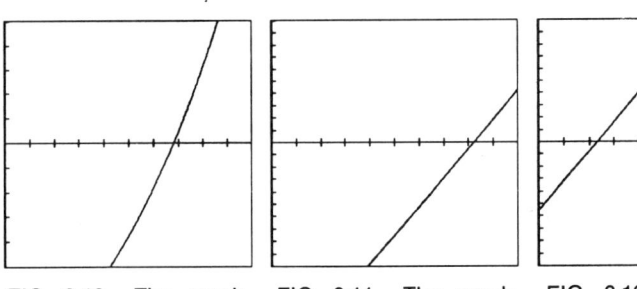
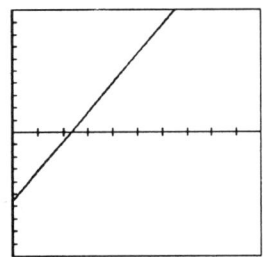

FIG. 2.10. The graph of $f(x) = x^3 + x - 1$ in $[0, 1]$ by $[-0.5, 0.5]$.

FIG. 2.11. The graph of $f(x) = x^3 + x - 1$ in $[0.6, 0.7]$ by $[-0.1, 0.1]$.

FIG. 2.12. The graph of $f(x) = x^3 + x - 1$ in $[0.68, 0.69]$ by $[-0.01, 0.01]$.

activity for students using graphing utilities. This process can be continued until we determine the solution of the equation with desired accuracy within the limits of machine precision.

The situation for solving inequalities is very similar to solving equations. The algebraic techniques are limited, causing a corresponding restriction on the types of inequalities that can be solved. Again, solving inequalities graphically allows a wider variety of complicated inequalities to be solved: For example, to solve $x^3 + x < 1$ (or $x^3 + x - 1 < 0$), we need to determine the values of x for which the graph of $f(x) = x^3 + x - 1$ lies below the x axis. Figure 2.12 permits us to read the solution to the inequality as $(-\infty, 0.6824)$ with an error of at most 0.001.

Technology allows algebra students to solve equations such as $30 = x(11 - 2x)(8.5 - 2x)$, $x^3 - 2x = 2 \cos x$, $0.1x = \sin x$, and $x \tan x - \sin x = \cos x$. And some of these equations are the algebraic representations of interesting, real problem situations. These are equations that we have not been able to ask students in school mathematics to solve without graphing calculators. Solutions are now possible because of the graphical techniques available with today's technology.

Rich (1990) found that precalculus students in traditional instruction made almost no use of graphs outside units on graphing. This is consistent with Kaput's (1989) description of graphs as "display notation" when no technology is available. Graphing technology allows the graph to become the vehicle for a powerful, visual approach to problem solving. In Kaput's words, technology allows graphs to become "action notation." Rich (1990) found that precalculus students who had used a graphing calculator for the year made frequent use of functions and their graphs in algebraic problem-solving situations. They were also more confident than a comparison group of students about their ability to solve verbally stated applications.

Pitfalls in Computer-Based Graphing

Students and teachers must be careful and must analyze graphs to determine whether they are correct: for example, Fig. 2.13 gives the graph of $f(x) = \cos(31x)$ in $[-10, 10]$ by $[-2, 2]$. Notice that there appears to be between 3 and 4 periods of the graph of f shown. However, the period of f is $2\pi/31$, which is approximately 0.2. Thus, there should be about $20/0.2 = 100$ periods in this viewing rectangle. This graph is incorrect because at least three distinct points are needed to display each period of the graph of f and there are not enough pixels on the screen to show the at least 300 distinct points necessary to illustrate the 100 periods of f. We can make the situation even worse if we try to graph $y = \cos ax$, with a much larger than 31 in the same viewing rectangle. We are simply asking the technology to do something that it is not capable of. The main point, though, is that the

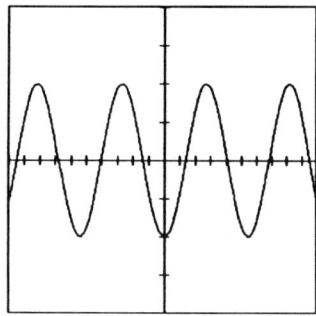

FIG. 2.13. An incorrect graph of f(x) = *cos(31x)* in [−10, 10] by [−2, 2].

modern approach to graphing should involve obtaining graphs quickly with technology and then checking to see that the graph is correct. However, it is not necessary for students to check each and every graph. Teacher guidance is important with this approach.

The Modern Role of Algebraic Manipulation

Much of the present K-12 curriculum consists of arithmetic drill and algebraic manipulative practice. The NCTM *Standards* call for this type of activity to be substantially reduced and replaced by activities designed to foster higher order thinking skills. This is not to say that arithmetic and algebraic manipulation should be completely removed from the curriculum. In light of technology, we now need to ask ourselves why we do these things and we must try to give students reasons (Demana & Waits, 1988b). Students will view these skills as important if they grow out of applications or out of situations where other mathematical understanding is obtained. In other words, these rote activities should not be the focus of the majority of lessons. Rather, they need to be used in more interesting activities.

Earlier in this chapter, for example, we illustrated how the long division of polynomials could be used to obtain and understand a complete graph of a rational function. Students are willing to do long division of polynomials when it is used to illustrate something interesting or when it can be used to answer one of their questions. In fact, one question raised by students while studying end behavior of rational functions is, "How can you tell what the end behavior is without actually using graphing zoom-out?"

Factoring is another rote activity that receives considerable attention in the current curriculum. We must ask ourselves why we spend so much time on this topic. How is it to be used later? Is it still important enough to warrant this much attention today?

With the use of graphing technology, we should introduce students to the Rational Zeros Theorem early in algebra and use this theorem to organize factoring activities. This requires that we establish the important connections among linear factors, zeros, and x intercepts of related graphs. The list of possible rational zeros of a polynomial with rational-number coefficients can be quite long. Using synthetic division or synthetic substitution to test each of the possibilities is time-consuming. However, the graph of a polynomial function f will usually allow us to eliminate many of the possibilities for its rational roots. If r appears to be a zero of f—that is, an x intercept of the graph of f—then we can check using synthetic division. If r is, in fact, a zero, then the synthetic division process also gives the coefficients of the polynomial $q(x)$ so that $f(x) = (x - r) q(x)$.

Additional manipulation can be built into the curriculum by using graphing utilities to obtain graphs of relations such as $Ax^2 + Bxy + Cy^2 + Dx + Ey + F = 0$ (conics). Such relations are quadratic in y and, except for a few special cases, can be solved for y in terms of x using the quadratic formula. The two resulting expressions for y can be entered into a graphing utility and their combined graphs give a graph of the conic. Notice that this technique can be applied when $B \neq 0$ to easily graph conics that have both translated and rotated axes. Although the algebraic manipulation can be complicated, students are more willing to do it because they see a reason for the manipulation.

The Role of Realistic Problems

One of the NCTM Standards speaks to the issue of assisting students in learning to value mathematics. Mathematical ideas introduced in the context of realistic problem situations are appreciated by students: For example, problems involving growth and decay are interesting to students in all grades. Primary students can use calculators and a numerical approach to investigate and solve problems about topics such as population and compound interest. These problem situations can be returned to many times in the curriculum as more sophisticated methods and techniques to solve these problems are introduced. Returning to important problem situations allows opportunities for review and reinforcement, as well as providing a familiar setting to learn new techniques.

Teachers must be prepared to spend extended time on rich problem situations. This time is well spent because important review can be built in, and concrete reasons can be provided for many of the techniques used in mathematics and science. Real life situations make excellent mathematical problems and help students see that mathematics is important for everyone.

Questions For Further Research

Evidence like the textbook analysis reported in this chapter suggests that the present elementary school curriculum does not lay an adequate foundation for the understanding of functions and their graphs. There is an urgent need to study ways to enrich this early experience, and then to develop and test curriculum materials designed to accomplish such an enrichment. It seems feasible that students' common misconceptions concerning graphs that researchers are now discovering are artifacts of the present inadequate curriculum. An alternative, enriched curriculum would allow researchers to determine whether this is, in fact, the case.

The need to develop and test curriculum materials also carries over to higher grade levels. For example, it is not clear at what point and in what ways a graphing calculator should first be used as a teaching tool. It is certainly a very powerful tool in the hands of someone who understands functions, but in what ways can and does it affect initial learning of graphical representations? There are many more specific questions that follow from this general issue. For example, how much point plotting is necessary? How important is it to go from numerical, to geometric, and then to algebraic? That is, can one go from numerical, to algebraic, to geometric just as successfully? When should students start to use graphing utilities? How early in the curriculum can graphing be started? What pitfalls must we be aware of in the use of technology? How do we help students learn to use technology appropriately?

Researchers interested in studying teachers and classrooms should consider the following directions suggested by preliminary findings. For example, the use of graphing calculators in precalculus classes seems to enable and even entice teachers to use more group problem solving and less lecturing in their classes. It seems to change the way in which they plan for many of their classes and even the way in which they view mathematics and their role as a teacher. Much research is needed to help understand the extent to which this phenomenon is generally true and the reasons for it.

Summary

Graphing receives very little attention in the K–8 curriculum. Graphing utilities have and will continue to have a profound effect on the upper curriculum. Given this expanded use of graphing in the upper curriculum, it is imperative that graphing get started in the early grades, as stated at the beginning of this chapter. Students will need practice, beginning in the primary grades, with making graphs and reading from graphs. Then students need to learn how to use graphing as a problem-solving tool. The curriculum needs to be infused with realistic problems to help all students learn to value mathematics. The use of graphing calculators will help make the vision of the new NCTM *Standards* a reality.

REFERENCES

Browning, C. (1988). *Characterizing levels of understanding of functions and their graphs.* Unpublished doctoral dissertation, Ohio State University, Columbus.
Comstock, M., & Demana, F. (1987). The calculator is a problem solving concept developer. *Arithmetic Teacher, 34*(6), 48-51.
Demana, F., & Leitzel, J. (1988). Establishing fundamental concepts through numerical problem solving. In A. F. Crawford & A. P. Shulte (Eds.), *The ideas of algebra, K-12 1988 yearbook* (pp. 61-68). Reston, VA: National Council of Teachers of Mathematics.
Demana, F., & Waits, B. K. (1988a). Pitfalls in graphical computation, or why a single graph isn't enough. *College Mathematics Journal, 19*(2), 177-183.
Demana, F., & Waits, B. K. (1988b). Manipulative algebra—The culprit or the scapegoat? *Mathematics Teacher, 81*(5), 332-334.
Demana, F., & Waits, B. K. (1990a). Enhancing mathematics teaching and learning through technology. In T. J. Cooney & C. R. Hirsch (Eds.), *Teaching and learning mathematics in the 1990's, K-12 1990 yearbook* (pp. 212-222). Reston, VA: National Council of Teachers of Mathematics.
Demana, F., & Waits, B. K. (1990b). *Precalculus mathematics, a graphing approach.* Reading, MA: Addison-Wesley.
Dion, G. (1990). The graphics calculator: A tool for critical thinking. *Mathematics Teacher, 83*(7), 564-571.
Dunham, P. (1990). *Mathematical confidence and performance in technology-enhanced precalculus: Gender-related differences.* Unpublished doctoral dissertation, Ohio State University, Columbus.
Farrell, A. (1989). *Teaching and learning behaviors in technology-oriented precalculus classrooms.* Unpublished doctoral dissertation, Ohio State University, Columbus.
Goldenberg, E. P. (1988). Mathematics, metaphors, and human factors: Mathematical, technical, and pedagogical challenges in the educational use of graphical representation of functions. *Journal of Mathematical Behavior, 7,* 135-173.
Herscovics, N. (1989). Cognitive obstacles encountered in the learning of algebra. In S. Wagner & C. Kieran (Eds.), *Research issues in the learning of algebra* (pp. 60-86). Reston, VA: National Council of Teachers of Mathematics; Hillsdale, NJ: Lawrence Erlbaum Associates.
Janvier, C. (1981). Difficulties related to the concept of variable presented graphically. In C. Comiti & G. Vergnaud (Eds.), *Proceedings of the Fifth International Conference for the Psychology of Mathematics Education* (pp. 189-192). Grenoble, France: Laboratoire I.M.A.G.
Kaput, J. J. (1989). Linking representations in the symbol systems of algebra. In S. Wagner & C. Kieran (Eds.), *Research issues in the learning of algebra* (pp. 167-194). Reston, VA: National Council of Teachers of Mathematics; Hillsdale, NJ: Lawrence Erlbaum Associates.
Karplus, R. (1979). Continuous functions: Students' viewpoints. *European Journal of Science Education, 1*(4), 397-413.
Kerslake, D. (1981). Graphs. In K. M. Hart (Ed.), *Children's understanding of mathematics* (pp. 11-16, 120-136). London: John Murray.
Leinhardt, G., Zaslavsky, O., & Stein, M. K. (1990). Functions, graphs, and graphing: Tasks, learning, and teaching. *Review of Educational Research, 60*(1), 1-64.
Lovell, K. (1971). Some aspects of growth of the concept of a function. In M.F. Rosskopf, L. P. Steffe, & S. Taback (Eds.), *Piagetian cognitive development research and mathematical*

education (pp. 12-33). Washington, DC: National Council of Teachers of Mathematics.

Markovits, Z., Eylon, B., & Bruckheimer, M. (1983). Functions: Linearity unconstrained. In R. Hershkowitz (Ed.), *Proceedings of the Seventh International Conference of the International Group for the Psychology of Mathematics Education* (pp. 271-277). Rehovot, Israel: Weizmann Institute of Science.

National Council of Teachers of Mathematics. (1989). *Curriculum and evaluation standards for school mathematics.* Reston, VA: Author.

Porter, A., Floden, R., Freeman, D., Schmidt, W., & Schwille, J. (1988). Content determinants in elementary school mathematics. In D. A. Grouws & T. J. Cooney (Eds.), *Effective mathematics instruction* (pp. 96-113). Reston, VA: National Council of Teachers of Mathematics.

Rich, B. S. (1990). *The effect of the use of graphing calculators on the learning of functions concepts in precalculus mathematics.* Unpublished doctoral dissertation, University of Iowa, Iowa City.

Schoenfeld, A. H. (in press). Mathematics, technology, and higher order thinking. In R. S. Nickerson & P. P. Zodhiates (Eds.), *Technology in education: Looking toward 2020.* Hillsdale, NJ: Lawrence Erlbaum Associates.

3 Seizing the Opportunity to Make Algebra Mathematically and Pedagogically Interesting

Michal Yerushalmy
University of Haifa

Judah L. Schwartz
*Massachusetts Institute of Technology &
Harvard Graduate School of Education*

After a brief review of some of the sources of difficulty that students and teachers have with the learning and teaching of algebra, this chapter presents the outline of a new approach to the subject that is based on the centrality of the function and is deeply rooted in the use of multiple linked representations. Instantiating these representations in microcomputer software provides an opportunity for direct manipulation of both symbols and graphs. Early response to these materials has been encouraging.

The concept of function is a central one in mathematics. Moreover, it is a concept that grows in importance as one progresses in the depth and breadth of one's understanding of the subject. Functions are in principle an important part of the current secondary algebra curriculum. In practice, however, explicit attention to functions often comes quite late in the instructional sequence of algebra. Our position on the question of what the content of algebra should be and what the role of functions within that content should be is a good deal broader and more radical than we can fully address in this chapter. Suffice it to say that we believe that the function is the fundamental object of algebra and that it ought to be present in a variety of representations in algebra teaching and learning from the outset. Many entailments about the structure and sequence of an algebra course flow from this axiom. In the following paragraphs we sketch some of these entailments so that the reader may have a sense of what motivates many of the things we subsequently discuss.

The main lines of our approach to algebra can be summarized in the following statements:

1. A strong distinction should be made very early between number as a mathematical object and number recipe (or function) as a mathematical object. We take the position that algebra is fundamentally about functions.
2. Although there are many ways of representing functions, commonly functions are represented symbolically. Students learn to manipulate functions by manipulating the symbols that represent them. But functions can also be represented graphically, and students can also learn to manipulate functions by manipulating the graphs that represent them, using such graphical operations as translation, dilation, and reflection.

We argue that allowing students to use a rich set of operations, some of which operate on functions symbolically and some of which operate on functions graphically, builds a deeper and richer understanding of the mathematics. Other mathematical constructs, such as identities, equations, inequalities, and relations, can be seen and understood to be comparisons of functions. This chapter starts with some of the sources of difficulties that students have in understanding functions. We think that an artificial and illogical sequencing of concepts, dull examples, and noncreative learning could be amended and replaced by a pedagogy of inquiry learning augmented and empowered by appropriately crafted computer environments.

Tall (1989) discussed a variety of "cognitive obstacles" in the learning of algebra. We adopt his categories here in order to discuss the sources of the difficulties students encounter in learning about functions and their graphs.

Obstacles Rooted in Sequencing

Despite the fact that there are major conceptual differences between arithmetic and algebra, research on the learning of algebra points to a major difficulty caused by the presentation of algebra as "generalized arithmetic."

As Kieran (1990) suggested, "Arithmetic is primarily procedural. Strings of numbers and operations are not dealt with as mathematical objects but as processes for arriving at answers. In algebra, however, written symbolic representations are often considered as objects in their own right and do not necessarily represent specific procedures to solve concrete problems" (p. 99). One didactic approach to settling this conflict about the degree of meaning involves minimizing the amount of symbolic manipulation and at the same time introducing algebraic situations in less abstract environments. Recent works present evidence that the availability of symbolic manipulators, which free the student from spending time on performing algorithms, enables students to pay more attention to algebraic concepts presented within the context of concrete situations.

Heid (1988) reported on the "Algebra with Computers" curriculum in which students focused on applications. Heid's students were engaged with the analysis of a variety of realistic problem situations, and once they learned the concepts of variable and function, they used a function grapher and a symbol manipulator program. As a result, these students outperformed their peers from conventional algebra classes on each of the problem-solving goals of the new curriculum.

Nevertheless, algebra is the language of symbols. Therefore, we ought to look for representations that allow students to assign meaning to the symbols at the same time as they learn to operate with the symbols (Goldin, 1987; Janvier, 1987).

Although there is certainly individual variation among people and their abilities to use representations meaningfully and powerfully, it is usually assumed that most people benefit from visual representations. It is odd, therefore, that high-school mathematics makes use of visual representations only in very limited domains. In geometry it is now clear, although this has not always been true, that problems should be represented by diagrams and that the various actions that are part of the process of problem solving should be represented in the diagram.

In algebra, the role of graphs in the understanding of functions is well established. However, the nature of the algebra curriculum is such that the problems we offer students are for the most part limited to those problems that can be readily solved within the framework of symbolic representations alone. As a result, visual representation is not perceived as necessary by most students when engaged in mathematics problem solving.

Within the traditional process of learning algebra in the secondary schools, the learning of graphs of functions usually occurs after a long period of numerical and symbolic manipulations and is normally introduced as a final stage of the subject. We think it is quite likely that certain difficulties observed in the understanding of functions in various representations (numerical, visual, and symbolic representations) might be grounded in this form of learning. Dreyfus and Eisenberg (1987) revealed that high-school students who were able to solve standard problems in both symbolic and graphical representations still only vaguely understood the relationship between the two representations. Dufour-Janvier, Bednarz, and Belanger (1987) reported on high-school students who were reluctant to combine graphical representation within the algorithm they had for solving equations. The authors conclude that "Even if the children have studied the mathematical representations and can produce and use them on demand, they do not have the attitude of turning to these as tools to help them solve problems" (p. 113). More evidence about the tendency of students to reduce the mathematics related to functions into a collection of algorithms while not considering visual images can be found in Dreyfus and Vinner (1989). Thus, it would seem that this sequential learning of the symbolic and graphical representations does not promote a tendency on the part of the learner to move between them. Each of the representations presents a separate symbol system for the learner with no mutual and constructive interaction.

Concentrating on Trivial Cases

Often when we teach a complicated concept, particularly when more than one representation is involved, we tend to simplify the concept and the representations used to express it by choosing examples from a restricted and special domain. Vinner (1981) observed this process in the teaching of geometry, and it is plausibly established as the cause of several prototypic misconceptions. In algebra, we demonstrate the behavior of functions, using a limited number of very simple cases, and thus limit and restrict students' understanding. For example, students do not accept the idea that two different rules might be needed to define a function [such as in $f(x) = 0, x < = a$ and $f(x) = 1, x > a$], or that there might be more than one definition of the same function (Dreyfus & Vinner, 1989). They misunderstand slopes in the context of differentiation (Orton, 1983), probably because slope is encountered by algebra students only in the context of straight lines, and this in turn has a direct impact on the understanding of related concepts in calculus (Tall, 1987).

We often oversimplify and therefore misdefine connections between representations. For example, we define functions with equal slope as parallel graphs, although only linear functions behave as Euclidian parallels. Learners face great difficulty when they try to understand the connections between the intersection of two curves and the solutions of equations (Herscovics, 1989). We usually graph a function only as the final product of a process of simplifying expressions and present equations in the form $f(x) = 0$. As a result, students come to think of the zeros of $f(x)$ as solutions of an equation and have difficulty seeing this as a particular case of a more general phenomenon—that is, that the solution set of the equation $f(x) = g(x)$ are the abscissas of the intersections of the functions f and g. Over and over we try to simplify instruction by presenting symbolically simple expressions and graphically prototypical pictures. Our habitual use of the same scale for domain and range (Kerslake, 1981) often promotes a serious misconception—that a given expression, when plotted, has a particular form.

Inactive Learning

Dubinsky (1988) argued that according to Piaget understanding is best fostered when a learner develops the concept of function in a multistage fashion. First, the user performs an action on an object, usually the symbolic expression. Second, the object of the action is turned into a process by virtue of the evaluation of the function. Finally, the process is encapsulated into a new object, which again is a subject for action, and so forth. The ability to move nimbly and to generalize readily between and among these various stages (in any of several representations) is the essence of understanding. There are a number of studies that demonstrate that most students do not reach the stage of perceiving the function or its graph as an entity. In the learning and teaching of many topics of algebra such as solving equations and simplifying expressions, we simply do not go beyond the first stage of action on an object.

We conjecture that providing an environment in which functions can be manipulated as entities or objects and in which the actions of evaluating and graphing are automated should help students reach the final Piagetian stage of understanding of function—the encapsulation of function as process into function as entity.

We feel that supporting evidence for this conjecture can be found in the work of Goldenberg (1987), who suggested that the ability of the computer to deal easily with families of functions may introduce a new concept of function: a graphical-object-valued function (p. 27). Arcavi, Tirosh, and Nachmias (1990) reported on the work of mathematics teachers with a parallel axes representation (PAR) microcomputer environment. The teachers were able to enlarge their repertoire of concept images of functions, and, as a result, they analyzed functions by relating to a function as an object: a single point that is the focus point in PAR. Dugdale (1989) studied two groups of trigonometry students who worked with the *Green Globs* grapher. In this study, those students who experimented and actively participated in the development of mathematical ideas and who learned to visually conjecture before formalizing procedures were able to develop a qualitative perspective of functions as objects as well as the operations on them. Dubinsky (1988) described situations in which students who program with ISETL are able to overcome serious cognitive difficulties related to the function as an object. Our own experience with the *Geometric Supposer* (Schwartz & Yerushalmy, 1980, 1985, 1988, 1990; Schwartz, Yerushalmy, & the Education Development Center, 1990) suggests that students whose learning is reinforced by an environment that allows them to easily operate on a geometric shape tend to create a sequence of diagrams and generalize it into a new and more general geometric entity. (See also Vinner & Hershkowitz, 1983.)

Having outlined some of the difficulties students have in understanding functions, we now describe the possibilities offered by suitably crafted microcomputer environments that may help to overcome some of the major obstacles to understanding that are normally encountered with traditional teaching approaches and content sequences.

VISUAL ALGEBRA: AN APPROACH USING SOFTWARE AND INQUIRY

How might technology affect these teaching and learning difficulties we have discussed?

The use of graph-plotting software frees users from the need to collect numerical information and the physical action of plotting. This, in turn, may permit them to shift the focus of their attention away from the detail of the process of making the graph and allow them to benefit from the contemplation and manipulation of the graph itself. Lesh (1987) studied environments that included symbolic manipulators and graph plotters and compared the behavior of two groups of ninth graders: One group worked with the symbolic manipulator (SAM) and received instruction on the graphical representation of algebraic equations at each step, while the other group also worked with a computer but performed the operations associated with each step toward a solution themselves and received graphical feedback only when the correct solution had been found. On the basis of the interviews and tests at the end of the experiment, Lesh concluded that dynamic visual representations became "conceptual amplifiers" for learners, who not only outperformed their peers but also developed ideas about important algebraic ideas such as transformations and invariance.

However, Dreyfus and Eisenberg (1987) showed that even with the availability of plotting software, students might not benefit extensively from the multiple representations of functions that such environments offer. The graph remained an adjunct additional representation rather than a central and focal entity. One possible reason could be that the participants in the studies were influenced by the limitations and difficulties of prior algebra learning that we have already discussed. Another explanation for this result could be that in the environments tested, an asymmetry existed between the symbolic and graphic representations—that is, although the symbolic representation of functions could be manipulated symbolically, the visual representation could not be manipulated graphically (visually). Microcomputer technology now allows us to fashion new software tools that enable a user to manipulate the graphical representation itself (for example, "stretch" or "squeeze" graphs) and to view the impacts on the numerical and symbolic representations that result from this manipulation. Other tools allow us to manipulate numerical representations and view results on the graphs and in the symbolic expressions. The objective in using such tools is the manipulation of the function (in any of its representations). Further, by providing the user with an ability to combine functions in such environments, using the arithmetic operations and composition, we hope to make it possible for the learner to move in the direction of viewing the function as an entity without losing sight of its interpretation as a process.

The essential features of the kinds of software environments we are discussing are that:

1. Varied representations are linked.
2. Each representation can be manipulated in ways that reflect the structure of that representation.

The major conceptual foci of the traditional algebra curriculum can be investigated in depth in settings that stress the symbolic representation of functions augmented by their graphical representations. At the same time, newly developed software environments such as the Visualizing Algebra series and the Function Supposer series allow students and teachers who think and analyze visually, rather than symbolically, to approach the subject in ways never before possible.

The presence of simultaneous multiple representations of functions allows for the development of new kinds of conceptual understanding that go well beyond those understandings normally associated with simple manipulative skills (Yerushalmy, 1991). For example, rather than merely thinking of functions as collections of ordered pairs of numbers, students can come to think of them as processes, and then, finally, learn to think of these processes as entities. The procedure of plotting a large number of points according to a specific rule and the resulting plotted points becomes an entity, one that can be expressed in a variety of representations. These functions, each of which can now be thought of as an entity in itself, can be grouped into families in many ways. Appropriately designed software environments make it possible to treat these families of functions themselves as entities, thereby deepening still further the understanding that users develop about algebra.

Each graph plotted represents, of necessity, a specific function. The graph, therefore, is a representation of only a single member of one or more families of functions. On the other hand, the symbolic representation of functions often lends itself more readily to generalizing to families of functions. Mathematical phenomena that are observed in visual settings and whose truth is suspected can then be profitably explored within symbolic representations. It is in this way that visual experiments in the Visualizing Algebra and Function Supposer environments allow for a more accessible pedagogic path from concreteness to abstraction and from particularity to generality.

In the past, operations combining and composing functions were almost universally taught and learned in environments that made use of symbolic representations only. Similarly, roots of functions, factoring, composition of functions, equations, and inequalities have traditionally been thought of as topics to be explored in a world of symbols. It is possible, however, to develop new meanings for and understandings of these issues when they are examined in environments of the sort we are describing here.

In the past few years we have developed a range of software environments that enable us to better understand (a) the nature of the difficulties that students have in learning the subject and (b) how we might ameliorate some of these difficulties. We should add at once that we were interested in difficulties with the conceptual structure of the subject of algebra and not particularly in the difficulties that are artifactually derived from the way the curriculum happened to be organized and taught. This attitude permitted us the luxury of rethinking what the subject of algebra was really about and what the teaching and learning of algebra in the schools might become.

We next describe these environments in some detail for two reasons. First, we know them better than we know the work of others, and thus, we know the reasons for their particular form and approach. If we were to attempt a review of extant microcomputer software, we would have to infer the intentions of the designers. Second, we believe that this collection of software environments, taken as a whole, constitutes a coherent rethinking of what algebra teaching and learning might be.

The Visualizing Algebra Series. This software series, published by Sunburst Communications, provides users with a graded sequence of environments, culminating in the Function Analyzer (Schwartz, Yerushalmy, & Education Development Center, 1988), that introduce the notion of multiple, linked manipulable representations of functions. In this environment, the user works primarily with a single function that can be transformed either graphically or symbolically.

The Function Supposer: Symbols & Graphs. This is a three-part environment in which the emphasis is on the symbolic manipulation of functions. The GRAPHER, the first section of this environment, is a flexible graphing package. The second section, the TRANSFORMER, has been used to study students' understanding of symbolic manipulation in algebra. In this environment, users receive graphical feedback about the significance and appropriateness of their symbolic manipulations of a function. The third section, the COMPARATOR, has been used to study students' conceptual understanding of equations, inequalities, and identities. In this environment, the comparison operators $<$, $>$, and $=$ are used to compare two functions. Users can explore the legitimacy of various graphical and symbolic transformations of the function comparisons—that is, the equations and inequalities.

The Function Supposer: Explorations in Algebra. This is a broad-ranging software environment that provides the user with a rich set of tools for making and exploring conjectures about functions. It combines many of the features of the other environments in a format that encourages users to move in the direction of thinking of functions as entities and in concatenating these entities into increasingly rich structures. Specifically, users may transform single functions using the unary graphical operations, or operate on two functions using binary symbolic operations of addition, subtraction, multiplication, division, and composition.

Visualizing Algebra: The Function Analyzer. This software (Schwartz et al., 1988) provides an environment for plotting and manipulating functions in which students have simultaneous access to, and ability to modify, the various representations of the function they are working on: symbolic, graphical, and numerical. The Analyzer was designed to promote the intensive use of visual mathematical thinking. The following major options of the tool support the ability to benefit from multiple representations, while understanding the special advantages and drawbacks of each representation:

1. Plotting: The *Analyzer* includes a plotter of algebraic functions. The exposure of the learner to many graphs is assumed to catalyze the ability of conceiving the graph as an entity by itself, without the necessity of seeing the graph as a collection of plotted numerical data. The simultaneous availability of tables of ordered pairs makes it possible for the user to move readily between the function as entity, as reflected in the graph, and, function as process, as reflected in the table of values.

2. Scaling: The shape of a graph is determined to a large extent by the choice of scale in the coordinate system. Using the scaling options of the *Analyzer,* learners are exposed to various pictures of the same function simultaneously. This permits exploration of the invariants of the graph of the function, such as the slope of the function in units of the graph, the number of roots of the function, and the nature of the asymptotic behavior, as contrasted with the artifactual properties of the graph, such as the angle of the slope at any point.

3. Modifying expressions: This option allows users to explore the graphical role of each of the numerical parameters that appears in the symbolic representation of the function. As the user generates a family of functions by incrementing or decrementing a parameter, each newly created function is graphed on the same screen. This offers users an opportunity to investigate the properties of the family of functions that has been created by this parametric variation. The main power of this feature lies in the presentation of the linkage between particular numerical parameters that are localized in the symbolic representation of the function and the global effect of these parameters on the behavior of the graph of the function.

4. Modifying graphs: This feature allows users to perform various kinds of geometric transformations on a graph in order to create a family of related functions. Because of the linking of representations, the symbolic and numerical transformations of the functions are automatically displayed as the geometric transformations are carried out. For example, one has the option of translating the graph in four directions, stretching or compressing the graph vertically or horizontally, or reflecting it in each of the coordinate axes. It will be noted that all of these transformations are unary transformations of functions that preserve the nature of the function.

5. Values and points: In addition to plotting graphs of functions, the *Analyzer* allows the user to place ordered pairs on the coordinate plane. This facility makes it possible for a user to analyze a function by creating tables of values of the user-inputted ordered pairs x,y and ordered pairs $x,f(x)$ that lie on the plotted function. The user can thus explore the locations of critical points such as roots and intersections, as well as the value of the function at extrema and points of inflection. Numerical comparison of two functions and an exploration of the relations between them can also be readily carried out.

The *Function Analyzer* was designed to be used as a tool of inquiry and instruction in a variety of learning settings such as whole-group exploration, among pairs of students in the computer laboratory, or as a personal tool to aid in the learning of algebra.

The first section of the *Function Supposer, Symbols & Graphs,* the GRAPHER, is a very flexible graphing package. As such, it differs little from other good graphing packages and is included in the software environment primarily for the sake of completeness.

The second section of the *Function Supposer, Symbols & Graphs,* the TRANSFORMER, is mainly an algebraic notepad that allows the input of any expression whose syntax is acceptable in algebra (numbers, variables, and operations). The program allows the user to carry out a series of operations that transforms one expression into another, usually equivalent, expression, allowing the user to investigate the result of each transformation along the way.

The TRANSFORMER can work in two modes: a "free" mode in which the user can enter any transformed expression, and a "target" mode in which the expressions are part of a library that a teacher, researcher, or the student enters and stores in the program at any time. Such a file may include expressions, or pairs of expressions: a given, and a target to reach. The target expression is written as representing a family of expressions [for example, the given function might be $x^2 - 2x - 35$ and the target might be $A(x - r_1)(x - r_2)$].

The program provides an editor to facilitate making changes in the expressions, and it displays the results of each transformation in a graphical display window. At each stage, the graphical display shows the original expression, the current transformed expression, and the difference between (or ratio of) these two expressions. Because any transformation of an expression that does not change the function—for example, factoring, expanding parentheses—does not affect the graph of the expression, the graph-of-difference function provides qualitative and quantitative information about any departure from correct symbolic manipulation. All the steps in the series of transformations of expressions and graphs are available to the user at any time. Even though the TRANSFORMER includes tools to act on the graphs (such as scaling and zooming operations), the goal of this environment and the studies carried out with it is to use the graphs as feedback on symbolic manipulation and not as an environment on which to act.

The third and final section of the *Function Supposer, Symbols & Graphs,* is the COMPARATOR. The central idea of this section of the environment is that an equation in one unknown, for instance, can be thought of as a comparison of two functions of one variable. The equation $f(x) = g(x)$ compares the two functions f and g and implicitly poses the question, what value(s) of x have the property that $f(x)$ and $g(x)$, when evaluated, yield the same number? Similarly, inequalities and identities can be thought of as comparisons of functions.

When the user enters a comparison of functions, the software presents the two functions, f and g, plotted on the same graph. In the case of an equation, the solution set is determined by the intersections of the two curves, which occur at those values of the independent variable for which the two functions yield the same value. A related interpretation of intersection clearly obtains in the case of inequalities and identities.

Because of the presence of the graphical display, a user is provided with the basis for judging what operations may legally be performed on comparisons of functions. A necessary and sufficient condition for an operation to be legal on a comparison of functions is that it leave the solution set invariant. Thus, it is immediately apparent that there are difficulties with adding a nonzero quantity to one side of a comparison and not to the other. It is also clear that any comparison is a member of an equivalence class of comparisons all of which have the same solution set. Even subtler matters become clear in this environment: For example, although adding x (or any other polynomial) to both sides of a comparison does not change the solution set, multiplying an equation by x (or any other polynomial that has roots) does. Similarly, dividing an inequality by a nonnegative number leaves the solutions set invariant, but dividing by -1 clearly changes the solution set of the comparison and thus is a move that must be understood in some detail.

Finally, we give only the briefest description of the *Function Supposer: Explorations in Algebra.* All of the facilities for manipulating functions that are present in the *Function Analyzer* are also present in this environment. While the unary graphical transformations of translation, dilation and contraction, and reflection are powerful and interesting for exploring a wide range of properties of functions, they are not sufficient. This insufficiency stems from the fact that these transformations do not change the nature of the function. A linear function remains linear under these transformations. In the *Function Supposer*: Explorations in Algebra, users are offered the additional capability of performing binary operations on functions. This is an environment whose driving metaphor is that of a calculator that operates on functions using the unary graphical operations and binary symbolic operations. In such an environment it is possible to build any function (that has a Taylor expansion) starting from the constant function and the identity function without ever entering coefficients, variables, exponents, or any of the other symbols that we usually think of as the sine qua non of algebra. As with all of the other environments, all functions are represented symbolically, numerically, and graphically.

THE SYMBIOSIS OF SOFTWARE AND CURRICULUM

In this section, we offer a few examples of how environments of the sort we have been describing can be used in teaching and learning. These examples are presented in three groups: (a) the use of simultaneous linked representations; (b) the power of "driving" each representation; and (c) the manipulation of functions as entities.

We make this categorization for the sake of clarity and in order to illustrate our approach. However, we do not claim that the categories are distinct and disjoint. Each of the examples (many of which have already been used in classrooms) suggests a way to teach and learn both conventional and novel school algebra content in an unconventional manner: creating algebraic understanding based on the knowledge of function and on the use of linked representations.

Using Simultaneous Linked Representations of Functions to Understand Algebraic Entities

We usually teach "number sentences" or expressions by introducing a symbol for a variable and an algebraic expression as a computational process involving numerical substitution of the variable in the expression. Because of the complexity involved in the manipulation of expressions as processes, we soon abandon this approach, which treats expressions as functions that can relate any number in their domain to a corresponding number in their range. Instead, the teaching moves on to equations where the main task is to look for particular number(s) that x represents. Thus, the symbol x changes from one that denotes any number (in the domain) to one that denotes a particular number (or numbers).

We believe that the technology can affect this situation in two ways. First, as a direct result of being able to present multiple linked representations of expressions as functions in three representations (symbolic, numerical, and graphical), we believe that we can help learners think about functions as processes and x as a variable. This kind of software allows users to work in parallel on symbolic and numerical representations as well as graphs. Because numerical and tabular representations make the process interpretation of the function salient, and because graphs make the entity interpretation of the function salient, we believe that working in multiple linked representation environments can broaden and deepen the understanding of function that students develop.

There is a second way in which we believe the technology can affect understanding in algebra. Symbol-manipulating software is becoming increasingly common and inexpensive. We think of such software as the algebraic analog of the hand calculator in arithmetic. The hand calculator relieves us of the need for endless attention to the arithmetic manipulation of numbers and allows us to explore and to do interesting things with numbers and their properties. In algebra, symbolic manipulation software relieves us of the need to spend instructional time on endless symbolic manipulation and allows us to explore and do interesting things with functions and their properties. The following examples illustrate how some of the most basic and traditional concepts of beginning algebra can be presented in an integrated fashion if one adopts a function perspective and works in an environment in which symbolic, graphical, and numerical representations of functions are simultaneously present.

Example 1: Functions as Building Blocks of Algebraic Expressions

Goal 1. To explore the links among the symbolic, numerical, and graphical representations of functions.

Task. To fit a rule to a collection of ordered pairs.

Instruction. Using points, move in the coordinate system and mark points that create a shape.

Watch the Table of Values and write a rule (a function) that describes the shape.

While working on this activity, two seventh graders invented a game. They suggested to one another: Each of us will mark eight points. One will mark them using the table editor and one will mark points on the coordinate system. The winner is the one who first announces a rule that fits.

This suggests a few conjectures: First, they understood the meaning of parallel representations — there is no point in the plane that does not have a numerical representation as an ordered pair, and no ordered pair that does not represent a point in the plane. Second, they understood that each representation has its own strengths. For example, if one thinks about a rule as a numerical process (such as, all pairs in which y is larger than x by 2), then editing a table of values is a more efficient procedure. However, if one thinks of a rule as an entity represented by a visual image, such as a straight line, then moving in the coordinate system is the better procedure. In each case, our goal is to link these approaches, and the use of the *Analyzer* seems to help do that.

Goal 2. To explore connections between geometric transformations and algebraic operations.

Task. Have the students invent transformations and analyze them graphically, numerically, and symbolically. Ask them to compare various moves that produce the same transformation. Ask them to conjecture about the look and the critical points and behavior of the result of the operation or the composition. The role of the activity is to allow students to make connections between functions and operations on functions while emphasizing two aspects of the function: the function as a computational process (e.g., watch the changes in the Table of Values as the function is transformed) and the function as an entity (e.g., watch changes in the shape and/or position of the graph as the function is transformed).

Instruction. Part (a). There are two fundamental functions in algebra. The first is the rule that maps any number into the number 1 (the constant function). The second basic function is the rule that maps any number into itself [the identity function $f(x) = x$].

We introduce the horizontal and vertical translations as primitive graphical operations on functions. The corresponding numerical operations on functions are sliding the $f(x)$ column in a table up or down with respect to the x column and adding a constant to every number in the $f(x)$ column.

Use the *Function Supposer: Explorations in Algebra* to figure out which of the following transformations can be expressed using these two functions and the two transformations. In each case conjecture about the results, then give a visual description of the transformation you made and explain the change in the table of values:

$f(x) = 7$
$f(x) = x+1$
$f(x) = x+x$
$f(x) = 7x$
$f(x) = x(x+1)$
$f(x) = (x/2) - 4$

Part (b). Now start from the same basic functions but add two transformations: that is, vertical and horizontal stretch (contraction and dilation). Can you describe how you would make the functions listed above, starting with the two basic functions?

Part (c). If you were not able to make all of the functions listed above using the transformations we have introduced so far, can you suggest other transformations that will help you complete the task?

Here the student might invent a transformation that would transform each x into $x*x$.

Example 2: Devising Criteria for Comparing Two Functions or Two Expressions

Goal. In the current curriculum, we spend a great deal of time doing procedures that transform expressions into a canonical form, but very little time comparing the internal structure of one expression to another. In contrast with experts, algebra students are not able to identify meaningful components of expressions (e.g., the highest power degree of the variable, or possible canceled terms) in order to reduce manipulation errors. Toward this end, Thompson (1989) suggested presenting expressions as tree structures rather than as strings, as they are usually presented. Thompson found that students who experienced work with the software *Expressions,* which emphasizes the structural features of algebra syntax, found expression trees to be quite intuitive, and that students were more alert to the structural properties of any expression even when dealing with arithmetic expressions.

Another type of representation, one that we have tried, is to present any single variable algebraic expression as a function. Then, once the learner is familiar with the various representations of functions, we can devise several meaningful ways of making comparisons among the algebraic expressions (the functions). These comparisons include:

1. Comparison of critical behavior of the two expressions, such as asymptotic behavior, or location of the intersections of the functions.

2. Evaluation of the graph of the difference of the two functions (difference graphs).
3. Evaluation of the graph of the ratio of the two functions (ratio graphs).
4. Devising a sequence of transformations (or compositions) that will transform one of the functions into the other.

One of our projects involved an experiment with teachers. In this study, elementary school mathematics teachers were asked to compare two expressions and explain why they are different and what might be done in order to make them congruent. During the experiment, the participants worked with the *Function Analyzer,* the *Function Supposer: Symbols & Graphs,* and the *Function Supposer: Explorations in Algebra.* They learned to analyze the impact of various algebraic transformations on the graph of the expressions. They practiced identifying incorrect transformations (i.e., transformations that do not preserve the equivalence of the function), while working both with and without interactive graphs and difference graphs. After the intervention period, they took a test and were interviewed. The results of the experiment are summarized next.

Task. While transforming the given expression, your student made a mistake and reached the following expression. Find the mistake(s) and correct the transformation.

Problem 1 (without graphs)

GIVEN: $-x(5-3x)-4-2x(3-4x(5-x))$

RESULT: $8x^3 + 43x^2 - 11x - 4$

Problem 2 (with graphs)

GIVEN: $-x(7-2x)-2-3x(5-2x(3-x))$

RESULT: $6x^3 + 20x^2 - 22x - 2$

(Another mistaken transformation is shown in Fig. 3.1.)

Seventy-five percent of the teachers ($n=28$) identified the incorrect term using information from the graphs of the functions. In their arguments they made use of both the general shape and the detailed position of the graphs as well as the difference graph. Fifty-seven percent of the teachers actually used the information gathered from the graphs to repair the incorrect transformation. The rest started the problem over, transformed it in their own way, and reached a correct simplified expression.

3. MAKING ALGEBRA INTERESTING   57

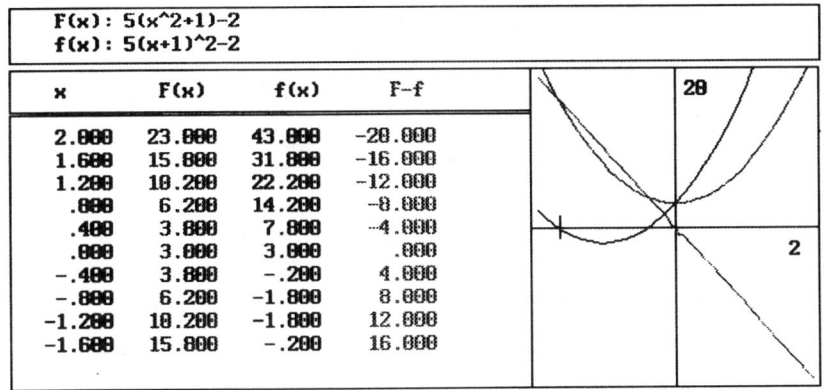

FIG. 3.1. A mistaken transformation.

Differences were found among ability levels: All of the upper ability students used the graph to directly access incorrect steps. The arguments of those who were not able to use the information indicated obstacles rooted in the technical format of the traditional transformational activities. For example, some of the statements offered by teachers in this group were, "It is safer to work from the beginning" and "I am confused when I have to compare the two expressions so I start over."

Measurements of the performance time and the number of steps taken to reach a correct expression show a reduction in time required to complete the problem and a reduction in the number of steps required to complete the problem for those who used the graph. Protocols from the interview and recording of discussions between pairs of students suggest that their analysis of simplifying expressions as a set of function transformations helped them to focus on the false procedure. They used function terminology extensively in their arguments (positive and negative inclination, intercepts, S-shaped and U-shaped forms of "expression," and others).

Example 3: Solving Equations and Inequalities

Perhaps the most common algebraic act we ever commit that is built around the concept of comparing functions is the solving of equations and inequalities. For the most part, solving equations and inequalities is taught as a set of seemingly arbitrary rules that govern allowed and disallowed actions. Thus, for example, it is permissible to bring a term involving x to the other side of the equation (or inequality) provided the sign of the term is changed in so doing. Schoenfeld (1987) argued that "what the mathematician knows about solving equations is not only more than but different from a collection of techniques mastered in individual domains" (p. 16).

Wenger (1987) argued that when students have to select the methods to manipulate expressions or relations, they have great difficulty. They do not seem to "see" the right things in the algebraic expression and, seemingly, often choose their next move almost randomly, without a specific purpose in mind. Which representation could help them develop those skills? We argue that the presentation of any algebraic relation as a comparison between two functions is not only consistent with our approach to algebraic expressions, but also provides us with all of the representations of functions we need to work with while solving equations and inequalities.

Solving equations and inequalities using graphical techniques is presented in some algebra curricula as a procedure to be invoked when no analytic techniques are available—for example, consider the equation $x = \tan(x)$. For the most part, however, this method of solving equations is absent from the algebra curriculum as it is normally taught. We suspect that this is so because in the absence of a technology that makes the plotting of the graphs of functions easy and casual, this technique is not particularly attractive or effective. On the other hand, when the technology permits the easy plotting of functions, this becomes an attractive as well as efficient technique both for solving equations and inequalities and for understanding the essential nature of these constructs.

We have tried this approach in an Algebra I course using the COMPARATOR section of the *Function Supposer: Symbols & Graphs*. The group had previously used the *Function Analyzer* in learning about polynomials and had practiced traditional methods of solving equations. The purpose of the activity was to help students analyze the procedural rules they knew for solving equations and inequalities while at the same time thinking about these constructs as comparisons of functions.

After a short exposure to the representation of equations as $f(x) = g(x)$, the students raised the following points:

1. When you don't touch the x terms and you only subtract a constant from each side you actually change the function but keep the slopes constant, so the x value of the intersection point has to remain the same.
2. When one changes the coefficient of the x terms, the slopes of the lines on both sides of the equation are different and the lines are not parallel to the previous pair but the intersection point stays at the same x value.
3. When I had $-3x = -12$, I could have done lots of operations, but I preferred to add 12 and plot the lines $-3x+12$ and 0 because the intersection point is on the axis and I can read the value $x = 4$.
4. I wonder what would the equation $x = x$ will look like? Where the intersection point will be?

5. I got the following drawing [see Fig. 3.2]; does that mean that the equation does not have any solution?
6. The teacher never suggested that we divide or multiply each side by an expression. I think that if we do that, the whole thing will look very different and probably it will disturb everything because when you multiply by x, you get a different polynomial.

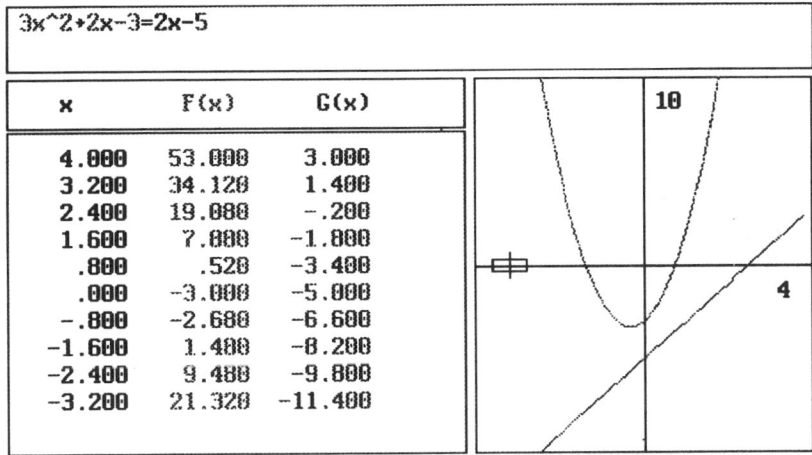

FIG. 3.2. Where is $f(x) = g(x)$?

All of this happened during a single class period followed by homework and discussion. As Kieran (1989) pointed out, students face enormous confusion and difficulties in understanding the equivalence of the equations that are generated in the process of finding a solution set. We agree. Nonetheless, we think these results are sufficiently encouraging to warrant further careful study and development.

The examples we have discussed up to this point have been algebraic procedures that traditionally are not integrated with the learning of functions. In fact, these activities are normally taught and carried out without any mention of function. The examples that follow are all from the traditional teaching of algebraic functions. They deal with such basic concepts of functions and their graphs as the slope of graphs, and the graph of a function as a member of a family of functions, and suggest activities that focus on connected misconceptions.

Example 4: Driving Each Representation

Slope. Students are introduced to the concept of slope in the context of linear functions. After that, the concept is not encountered again until the student begins to learn calculus. There have been several studies (Orton, 1983; Tall, 1987) that suggested that this oversimplified notion of slope present in the algebra curriculum is the source of subsequent student misconceptions in calculus. The traditional introduction of slope argues that the coefficient of the leading term in the expression is the parameter governing the slope. This fact, which is correct in the case of linear functions, is often generalized incorrectly to parabolas and higher order polynomials. This contradiction is rarely resolved and students learn mechanically about derivatives and slopes, without connecting what they do in calculus to what they have already been introduced to in algebra.

A particularly interesting illustration of what is possible comes from an Algebra II class that was being taught by a colleague of ours. Students, who had been assigned the task of characterizing functions by the visual appearance of their graphs, had decided that the asymptotic behavior, number of extrema, and number of roots were interesting categories to use for the classification of functions. There was some attempt to make use of an incompletely formed notion of slope, but most students had difficulty articulating what they wanted to say. Andy, however, volunteered that he had thought about the matter and that he had a way of thinking about it: "The trouble with slope, if the function is not a straight line, is that the slope keeps changing from place to place. That means that the slope has to have an x in it. Suppose you want the slope of the function x^3. If you look at the function 'close-up,' the function looks like a straight line. So you have to figure out how to write x^3 in the form of mx + b. To do this you write $x^2 * x + 0$. Then we see that the x^2 plays the role of the slope."

Families. Any function one chooses to explore in the *Function Analyzer* is, of necessity, a particular function, and both its graph and its symbolic expression are peculiar to it. Let us consider, however, the set of symbolic and graphical manipulations of the function that are available to the user of the environment.

The *Analyzer* allows users to manipulate the symbolic representation of the function by positioning a cursor on any numerical parameter in the expression and sequentially incrementing that parameter with an increment of any size. Each variation of the symbolic representation of the function is plotted and a family of graphs representing the series of modified symbolic expressions is displayed. Clearly each such modified expression is a different particular function, although each belongs to a common family of functions.

The *Analyzer* also allows users to manipulate the graphical representation of functions by horizontal and vertical translations, horizontal and vertical stretching and squeezing, and reflections in the horizontal and vertical axes.

Each variation of the graphical representation of the function is plotted and the corresponding symbolic representation of the modified function is displayed. It is similarly clear that each such graph represents a different particular member of a family of functions.

In each case, a modification of the function creates a new function that belongs to the same family, provided that family of functions is suitably defined. Consider the following examples:

$$f(x) = x^2 - 4$$

Modifying the symbolic expression clearly demonstrates that this function is a member of at least the following three one-parameter families of functions:

$$f(x;A) = Ax^2 - 4$$

$$f(x;B) = x^B - 4$$

$$f(x;C) = x^2 - C$$

and of the three-parameter family

$$f(x;A,B,C) = Ax^B - C$$

as well as of countless other families of functions of even greater generality.

Similarly, modifying the graphical form of the function clearly demonstrates that this function belongs to the following families of functions, among others:

$$f(x;A) = (x-A)^2 - 4$$

$$f(x;B) = x^2 - 4 + B$$

$$f(x;C) = (Cx)^2 - 4$$

and to the three-parameter family

$$f(x;A,B,C) = (Cx-A)^2 - 4 + B$$

as well as to countless other families of functions of even greater generality.

Starting from these observations, we may pursue a different pedagogic approach to the learning and teaching of algebra as a formal mathematical system. Rather than starting, as we normally do, with particular functions expressed symbolically, and then, almost as an afterthought, turning to the graphs of those functions, we turn from the outset to particular functions expressed both symbolically and graphically. Because we have independent ways of manipulating both the symbolic representation and the graphical representation of a function, and because the manipulations that we perform of necessity generate other functions that belong to the same families of functions as the original function, it becomes feasible to ask questions about the invariant properties of those families. The investigation of the invariant properties of a family of mathematical objects, in this case families of functions, is a far more sophisticated and potentially far more rewarding activity for algebra students and teachers than many of the activities one now observes in the algebra classroom.

Exploring particular functions as a way of allowing students to generate conjectures about families of functions turns the learning and teaching of algebra into a different sort of activity. As in the case of the *Geometric Supposer,* no amount of exploration of particular cases is a substitute for formal proof. On the other hand, we expect that as in geometry, the ability to explore particular cases will lead students to demand "ways of knowing" that go beyond demonstration in particular cases.

MANIPULATING FUNCTIONS AS ENTITIES

The discussion of the preceding section was directed to the problem of helping students develop a rich sense of the concept of function. It is at this point that most conventional curricula stop. Rather than accept this position, we see students at this point as situated on the threshold of being able to expand their concept of function beyond that of process to include the concept of function as entity, without losing the process interpretation of function.

We feel that the entity interpretation of function is best developed in an environment in which it is possible to manipulate functions as entities and to combine them in various ways so as to make new entities. In what follows, we offer some examples. Because we have had relatively less experience in exploring these issues in classrooms with students, our examples in this section are sketchier and more suggestive in character than those put forward earlier in the chapter.

Example 5: The Never-Ending Process of Making Entities

Consider any two-parameter family of functions, such as, $\sin(Ax+B)$. This function may be plotted in the x–y plane, and as A and/or B is varied, a family of functions will be displayed in the *x-y* plane. Our initial function, however, with its particular value of *A* and *B* (A_0 and B_0, for example), can be represented as a point in the *A-B* plane. Moreover, the family of functions generated by varying the parameters is represented in the *A-B* plane as a trajectory traced out by the point A_0, B_0. Figure 3.3 shows a particular example.

This example has a virtue that goes beyond focusing attention on invariants of functions of families. Part of our pedagogic agenda is to have students come to think of functions as entities. In this case, each function is represented by a point in the parameter plane. This can only serve to heighten the sense that the function can be treated as an entity. Moreover, although it has not happened at the time of this writing, we fully anticipate that some day a student will observe that if the trajectory in the parameter plane can be parametrized with two parameters, then that trajectory can be represented in yet another parameter plane as a point. Not only is a function an entity; a family of functions can be an entity.

Functions Whose Input Is a Function (Composition)

Previous studies suggest (E. Dubinsky, personal communication) that the most complicated situation involving functions that we ask students to understand is the one in which the input to the function is a function. The ability to understand the composition of functions requires an under-

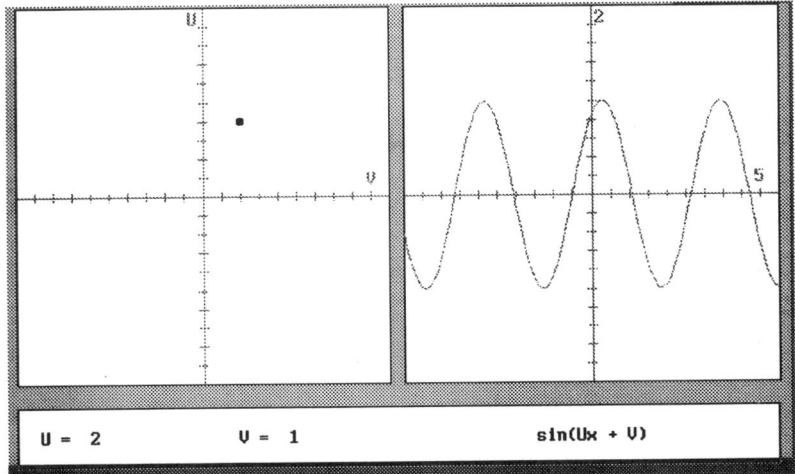

FIG. 3.3. The function sin(2*x* + 1) represented by the point (2,1) in the *U,V* parameter plane.

standing of a function as an entity that operates on another entity. We speculate that questions and problems that press the student into viewing the graph as a visual entity should help students to grasp this difficult concept. The following are examples that can be readily investigated using the *Function Analyzer* or the *Function Supposer: Explorations in Algebra*.

1. Is it possible to compose a polynomial of degree N with a polynomial of degree M to get a polynomial of the degree $(N+M)$? Try a few special cases, generalize your conjecture, and prove it.
    A student who already has had some experience with translations and stretches (which are equivalent to compositions of a function with the linear functions $x+k$ and kx, respectively) will probably recognize the fact that those transformations do not change the "nature" of the function being transformed. This would be an interesting first step toward a broader generalization.
2. If $f(x)$ has no roots, is it possible to find a $g(x)$ such that $f(g(x))$ will have roots? If the functions $f(x)$ and $g(x)$ have the same set of roots, what can be said about the roots of $f(g(x))$?
3. Observe, conjecture about, and prove your conjectures concerning symmetry and asymmetry of compositions. Which properties of functions seem to be affected by the order of a series of compositions and which seem not to be? (Consider properties such as roots, degree, symmetry, odd/even.)

Difference Equations and the Meaning of Differentials

At the time of this writing, there is a great deal of discussion in the world of mathematics education about the nature of the introductory calculus course. Much of this attention focuses on calculus instruction at the secondary level. However widely disparate the opinions voiced in this discussion are, there seems to be universal agreement about the centrality of the concepts of rate of change and of limit.

To deal with the difficulty that many students have with these concepts, Orton suggested that "A first approach to differentiation may be very informal and may be based largely on numerical and graphical explorations assisted by an electronic calculator" (1983, p. 244). Other educators who have studied this problem point to the difficulty of understanding the difference between a difference and a ratio of differences.

Here is an activity that should help students develop a grasp of the derivative function (as opposed to the slope at a point) and its relation to the concept of limit:

Consider a function $f(x)$. Imagine it translated in x by an amount s. Conjecture about the function $H(x) = f(x+s) - f(x)$.

Use the *Function Supposer* to form $H(X) = f(x+s) - f(x)$.

What seems to be true about $H(X)$ as you vary s?

Can you devise a way to stop the function $H(x)$ from vanishing everywhere as you make s smaller and smaller?

Repeat this procedure for several different functions. Can you make any general conjectures?

THE IMPLICATIONS FOR RESEARCH

The approach we are taking to the reformulating of the algebra curriculum raises many issues that ultimately must be answered by empirical research. The reader should not assume that the enthusiasm we have expressed in this chapter for the merits of our approach blinds us to the need for a good deal of hard work that lies ahead in finding out just how to do what we have said we would like to do.

Specifically, several different sorts of research questions must be explored. To begin with, there are questions that relate to individual learning. In contrast to current practice, we feel no particular need to concentrate in the early stages on linear functions. We believe that the tradition of doing so derives from the computational simplicity of evaluating them. On the other hand, Goldenberg (1987) showed that there are some special confusions that obtain when one deals with linear functions in graphical environments.

We need to explore in greater depth and detail our conjecture that the symbolic representation is relatively more powerful in portraying the function as a process and that the graphical representation is relatively more powerful in portraying the function as an entity. Evidence from past studies on related issues is mixed (see Dubinsky, 1988).

Beyond the myriad of research issues that pertain to individual learning, there is an entirely different set of issues that pertain to problems of developing, disseminating, and implementing new curricula, particularly when such curricula may be heavily dependent on technology. We do not minimize the importance of answering such questions thoughtfully and wisely. We are painfully and equally cognizant of the extremists who see the cathode ray tube as the "new papyrus" as well as those who would have our youngsters' mathematical education be slavish rehearsals of their own mathematical educations.

SOME CLOSING REMARKS

For the most part, the mathematics we teach in our primary and secondary schools is the mathematics already made by other people. Were we in the United States to teach language in this fashion, we might ask the students to learn a play by O'Neill, an essay by Emerson, a short story by Hemingway, but we would never ask them to write prose of their own.

We believe that students should be challenged to create in every subject matter we ask them to learn. Let us put aside our philosophical and ideological reasons for this belief. We think this position is an important one for perfectly pragmatic societal reasons. Except for a small percentage of the population at the top level of ability, no contemporary society has succeeded in the serious mathematical education of its young. We believe that no contemporary industrial society can continue to afford the squandering of human resources that results from a wasted mathematical education and a quantitatively illiterate electorate.

From our perspective, a pedagogy of inquiry, with rich opportunities for conjecturing and creating, is the essence of helping students to a better and deeper understanding of mathematics. The possibility of such instructional activities is not new. Why have they not been a serious part of mathematics teaching and learning in the schools? We believe it is because this sort of conjecturing activity is often difficult to carry out without suitable tools and that it is only very recently that such tools have begun to come into widespread use. If people do not have adequate tools to make the effort associated with the creating and exploring of conjectures manageable, they are simply not likely to be interested in doing so. Thus, while the possibility is not new, in a very real sense the opportunity is.

The central proposition of this chapter has been that appropriately designed microcomputer software can restructure the possible content and curricular emphasis in a domain—in this case, in algebra—and provide environments in which users, be they students or teachers, can explore that intellectual domain and their own understanding of it. Making it possible for students and teachers to approach mathematics as a subject to which they can contribute is, in our view, the best possible strategy we can avail ourselves of in the struggle to educate a public that thinks critically about quantitative and other matters.

REFERENCES

Arcavi, A., Tirosh, D., & Nachmias, R. (1990). *The effects of exploring a new representation on prospective mathematics teachers' conception of functions (Report No. 47)*. Tel Aviv: The Knowledge Technology Lab, Tel Aviv University, School of Education.

Dreyfus, T., & Eisenberg, T. (1987). On the deep structure of functions. In J. C. Bergeron, N.

3. MAKING ALGEBRA INTERESTING 67

Herscovics, & C. Kieran (Eds.), *Proceedings of the 11th International Conference for the Psychology of Mathematics Education* (Vol. 1, pp. 190–196). Montreal.

Dreyfus, T., & Vinner, S. (1989). Images and definitions for the concept of function. *Journal for Research in Mathematics Education, 20*(4), 356–366.

Dubinsky, E. (1988). On helping students construct the concept of quantification. In A. Borbas (Ed.), *Proceedings of the 12th Annual Conference of the International Group for the Psychology of Mathematics Education* (Vol. 1, pp. 225–262). Veszprem, Hungary: Ferenc Genzwein OOK.

Dufour-Janvier, B., Bednarz, N., & Belanger, M. (1987). Pedagogical considerations concerning the problem of representation. In C. Janvier (Ed.), *Problems of representation in the teaching and learning of mathematics* (pp. 125–148). Hillsdale, NJ: Lawrence Erlbaum Associates.

Dugdale, S. (1989). Building a qualitative perspective before formalizing procedures: Graphical representations as a foundation for trigonometric identities. In C. A. Maher, G. A. Goldin, & R. B. Davis (Eds.), *Proceedings of the 11th International Conference for the Psychology of Mathematics Education-North American Chapter* (pp. 249–255). New Brunswick, NJ: Rutgers-The State University of New Jersey.

Goldenberg, E. P. (1987). Believing is seeing: How preconceptions influence the perception of graphs. In J. C. Bergeron, N. Herscovics, & C. Kieran (Eds.), *Proceedings of the 11th International Conference for the Psychology of Mathematics Education* (pp. 197–203). Montreal.

Goldin, G. A. (1987). Cognitive representational systems for mathematical problem solving. In C. Janvier (Ed.), *Problems of representation in the teaching and learning of mathematics* (pp. 125–148). Hillsdale, NJ: Lawrence Erlbaum Associates.

Heid, M. K. (1988). *The impact of computing on school algebra: Two case studies using graphical, numerical, and symbolic tools.* Budapest: ICME-6, Theme Group 2, Working Group 2.3.

Herscovics, N. (1989). Cognitive obstacles encountered in the learning of algebra. In S. Wagner & C. Kieran (Eds.), *Research issues in the learning and teaching of algebra* (pp. 60–86). Reston, VA: National Council of Teachers of Mathematics; Hillsdale, NJ: Lawrence Erlbaum Associates.

Janvier, C. (1987). Representation and understanding: The notion of function as an example. In C. Janvier (Ed.), *Problems of representation in the teaching and learning of mathematics* (pp. 67–72). Hillsdale, NJ: Lawrence Erlbaum Associates.

Kerslake, D. (1981). Graphs. In K. M. Hart (Ed.), *Children's understanding of mathematics: 11-16* (pp. 102–119). London: John Murray.

Kieran, C. (1989). The early learning of algebra: A structural perspective. In S. Wagner & C. Kieran (Eds.), *Research issues in the learning and teaching of algebra* (pp. 33–56). Reston, VA: National Council of Teachers of Mathematics; Hillsdale, NJ: Lawrence Erlbaum Associates.

Kieran, C. (1990). Cognitive processes involved in learning school algebra. In P. Nesher & J. Kilpatrick (Eds.), *Mathematics and cognition: A research synthesis by the International Group for the Psychology of Mathematics Education* (pp. 96–112). Cambridge: Cambridge University Press.

Lesh, R. (1987). The evolution of problem representation in the presence of powerful conceptual amplifiers. In C. Janvier (Ed.), *Problems of representation in the teaching and learning of mathematics* (pp. 197–206). Hillsdale, NJ: Lawrence Erlbaum Associates.

Orton, A. (1983). Students' understanding of differentiation. *Educational Studies in Mathematics, 14*(3).

Schoenfeld, A. H. (1987). Cognitive science and mathematics education: An overview. In A. H. Schoenfeld (Ed.), *Cognitive science and mathematics education* (pp. 1–32). Hillsdale, NJ: Lawrence Erlbaum Associates.

Schwartz, J. L., & Yerushalmy, M. (1983, 1985, 1988, 1990). *The geometric supposer* [Computer software and teachers' guide]. Pleasantville, NY: Sunburst Communications.

Schwartz, J. L., Yerushalmy, M., & Education Development Center. (1988). *Visualizing algebra: The function analyzer* [Computer software and teachers' guide]. Pleasantville, NY: Sunburst Communications.

Schwartz, J. L., Yerushalmy, M., & Education Development Center. (1990). *The function analyzer* [Computer software and teachers' guide]. Pleasantville, NY: Sunburst Communications.

Tall, D. (1987). Constructing the concept image of a tangent. In J. C. Bergeron, N. Herskovics, & C. Kieran (Eds.), *Proceedings of the 11th International Conference for the Psychology of Mathematics Education* (Vol. 3, pp. 69–75). Montreal.

Tall, D. (1989). Different cognitive obstacles in a technological paradigm. In S. Wagner & C. Kieran (Eds.), *Research issues in the teaching and learning of algebra* (pp. 87–92). Reston, VA: National Council of Teachers of Mathematics; Hillsdale, NJ: Lawrence Erlbaum Associates.

Thompson, P. W. (1989). Artificial intelligence, advanced technologies, and learning and teaching algebra. In S. Wagner & C. Kieran, *Research issues in the learning and teaching of algebra* (pp. 135–161). Reston, VA: National Council of Teachers of Mathematics; Hillsdale, NJ: Lawrence Erlbaum Associates.

Vinner, S. (1981). The nature of geometrical objects as conceived by teachers and prospective teachers. In C. Comite & G. Vergnaud (Eds.), *Proceedings of the Fifth International Conference for the Psychology of Mathematics Education* (pp. 375–380). Grenoble.

Vinner, S., & Hershkowitz, R. (1983). On concept formation geometry. *Zentralblatt fur Didaktik der Matimatik, 15*, 15–20.

Wenger, R. H. (1987). Cognitive science and algebra learning. In A. H. Schoenfeld (Ed.), *Cognitive science and mathematics education* (pp. 217–252). Hillsdale, NJ: Lawrence Erlbaum Associates.

Yerushalmy, M. (1991). Effects of computerized feedback on performing and debugging algebraic transformations. *Journal of Educational Computing Research, 7*(3), 309–330.

Yerushalmy, M., & Chazan, D. (1990). Overcoming visual obstacles with the aid of the SUPPOSER. *Educational Studies in Mathematics, 21*, 199–219.

4
Aspects of Understanding: On Multiple Perspectives and Representations of Linear Relations and Connections Among Them*

Judit Moschkovich
Alan H. Schoenfeld
University of California-Berkeley

Abraham Arcavi
Weizmann Institute of Science-Rehovot

INTRODUCTION

The algebra standard for middle school mathematics in the NCTM *Curriculum and Evaluation Standards* (1989, p. 102) includes the following statement:

In Grades 5-8, the mathematics curriculum should include explorations of algebraic concepts and processes so that students can

- understand the concepts of variable, expression, and equation;
- represent situations and number patterns with tables, graphs, . . .
- develop confidence in solving linear equations using concrete, informal, and formal methods.

The Standards for Grades 9-12 extend the desired competencies in significant ways. All students are expected to:

- use tables and graphs as tools to interpret expressions . . . (Standard 5, Algebra, p. 150);
- represent and analyze relationships using tables, verbal rules, equations, and graphs (Standard 6, Functions, p. 154);
- translate among tabular, symbolic, and graphical representations of functions (Standard 6, Functions, p. 154);
- analyze the effects of parameter changes on the graphs of functions (Standard 6, Functions, p. 154).

---
*This chapter was a collaborative effort. The order in which the authors are listed was determined by a random choice procedure.

These are statements of expected performance. In contrast, the reality of current student achievement is indicated by the following quotation from *The STATE of Mathematics Achievement* (Executive Summary), a distillation of results from the 1990 National Assessment of Educational Progress:

> When the mathematics became at all complicated, performance fell off dramatically, even for twelfth graders. For example, high school seniors had considerable difficulty with the following set of questions.

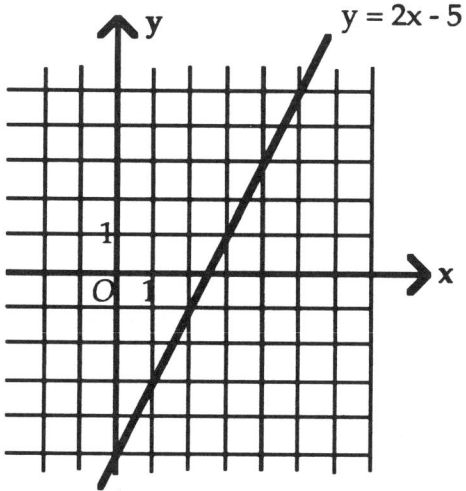

a. On the axes above, *draw* a line parallel to $y = 2x - 5$ that goes through the origin O.

b. On the line below, *write* an equation of the new line.

Equation: _____

Only 32% of the high school seniors drew the new parallel line on the graph, when a correct response essentially required the ability to find the origin O on the graph, the ability to find the existing line on the graph, and an understanding of the term "parallel." Sixteen percent of the twelfth graders answered both parts of this question correctly. (Mullis et al., 1991, p. 11)

Because there is significant variation in state mathematics requirements across the country, some of the students who took the test may not have had formal instruction in the mathematics of the Cartesian plane. However, given the large number of students who have studied the relevant material and the relative simplicity of the question, the fact that only 16% of the high-school seniors who took the test were able to answer both parts (a) and (b) correctly is distressing. From the perspective of people who are comfortable with the properties of algebraic functions and their graphs, the question is straightforward. The skills required to answer part (a) correctly have already been noted. A correct though telegraphic explanation of (b) is as follows: "The desired line has the same slope as the line $y = 2x - 5$, so its slope is 2; it passes through the origin, so its y intercept is 0; hence its equation is $y = 2x$."

The level of complexity of the NAEP task is increased substantially (and the task becomes more interesting and more mathematically important) if in its statement "the origin" is replaced by any other point in the plane. One would expect that the percentage of graduating seniors who could solve Problem 1, which follows, would be quite small.

Problem 1. Determine an equation of the line that is parallel to $y = 2x - 5$ and that goes through the point (1,4).

This chapter explores the complexity of Problem 1 and of a family of related problems concerned with different symbolic representations (algebraic expressions, tabular representations, and graphs) of linear relations. Our analysis indicates that there is more complexity to the domain than would appear at the surface level: Consistent with other research in the domain (see later discussion), we indicate that students must come to grips with connections across representations (e.g., the meanings of algebraic parameters in a geometric context) and, depending on context or interpretation, with different perspectives regarding the functions themselves. Those perspectives are as follows.

From the *process perspective*, a function is perceived of as linking x and y values: For each value of x, the function has a corresponding y value.[1] From the *object perspective*, a function or relation and any of its representations are thought of as entities — for example, algebraically as members of parametrized classes, or in the plane as graphs that, in colloquial language, are thought of as being "picked up whole" and rotated or translated.

The recent literature (see, e.g., Even, 1990; Schwarz & Yerushalmy, 1992; Sfard, 1992) makes it clear that coming to grips with both the object and the process perspectives is an essential part of learning about functions and graphs. The following quote indicates aspects of the process-object distinction.

Consider now the two functions

$x + 3$ and $4 + x - 1$.

---

[1] We thank Ed Dubinsky for pointing out that much of our discussion applies to *relations* (in which an x value is not necessarily linked with a *unique* y value) as well as to functions.

From the point of view of the process that is carried out with the recipe, these are two different recipes. If, however, one were to plot the output of each of these recipes against its input on a Cartesian plane then the two recipes would be indistinguishable. We see that the symbolic representation of function makes its process nature salient, while the graphical representation suppresses the process nature of the function and thus helps to make the function more entity-like. A proper understanding of algebra requires that students be comfortable with both of these aspects of function. (Schwartz & Yerushalmy, 1992, p. 265)

Breidenbach, Dubinsky, Nichols, and Hawks (1992) described the two perspectives as follows:

A process conception of function involves a dynamic transformation of quantities according to some repeatable means that, given the same original quantity, will always produce the same transformed quantity. The subject is able to think about the transformation as a complete activity beginning with objects of some kind, doing something to these objects, and obtaining new objects as a result of what was done. . . . A function is conceived of as an *object* if it is possible to perform actions on it, in general actions that transform it. (Breidenbach et al., p. 263)

Both the process and object perspectives shed light on the behavior of functions, in every representation, but the perspectives are differentially useful, in that one perspective may be usefully invoked in some problem contexts and not in others. In part, we argue that developing competency with linear relations means learning which perspectives and representations can be profitably employed in which contexts, and being able to select and move fluently among them to achieve one's desired ends. We illustrate these perspectives with the analysis of one approach to solving Problem 1. Note that many of the processes that experts use to solve such problems are automatic. We do not claim that people who solve Problem 1 consciously invoke all of the information described, but that such knowledge and perspectives do underlie a competent solution.

To begin, one knows that Problem 1 can be solved because any two independent pieces of information (e.g., two points on the line, the slope and the value of the y intercept, and so on) are enough to determine a line, in any of its representations. This general knowledge cuts across representations and perspectives. However, since the problem asks for an equation of a line parallel to one expressed in the form $y = mx + b$, it seems reasonable to use that form. One expects to write the equation of the line (call it L) in the form $y = mx + b$, where m and b are parameters whose values must be determined.

The object perspective is natural for determining m. One attribute of a line as a whole is its slope. Parallel lines have the same slope, so L has the same slope as the line whose equation is $y = 2x - 5$. One can read the slope of that line directly off its equation, as the coefficient of x. Hence $m = 2$, and the equation of L is given by $y = 2x + b$. Now the value of b needs to be determined.

Here a change of perspective is in order. The second piece of information in the problem statement is that the graph of L passes through the point (1,4). Exploiting this information depends on using the crucial information contained in the (deceptively simple) statement of the *Cartesian Connection* "A point is on the graph of the line L if and only if its coordinates satisfy the equation of L." From the process perspective, (1,4) lies on the graph of L, so the equation for L must produce the y value of 4 when the corresponding x value is 1. Hence $4 = 2(1) + b$, and $b = 2$. Thus the equation of L is $y = 2x + 2$.

We make some preliminary comments. First, the preceding analysis may seem like an exercise in overkill. One might have described the solution to Problem 1 in just a few lines. Is all of the complexity described in the previous two paragraphs really necessary? We argue that it is, especially to capture (and facilitate!) the learning process. By way of crude analogy, consider all the things one must learn when first learning to ride a bicycle. From the perspective of someone who, having had much practice, simply hops on a bike and rides off, bike riding could hardly seem simpler. But watch a child first struggling to master a two-wheeler, training wheels and all, and the full complexity of the domain is revealed. Skills, connections, and coordinations that are quite difficult to develop may seem trivial once they have been mastered. Similarly, unraveling the complexity of the domain serves a useful pedagogical function. On the basis of prior research (e.g., Schoenfeld, Smith, & Arcavi, in press), we can assert that some aspects of the domain that we take to be trivial are major stumbling blocks for students. And, knowing what the underlying skills and perspectives actually are can serve as a guide to developing curricula.

In a narrow sense, then, this chapter seeks to elaborate the theme announced in its title—to elaborate aspects of an understanding of linear relations that correspond to the ability to move flexibly between the process and object perspectives in a variety of representations (our focus here being on algebraic, tabular, and graphical representations[2]). In a broader sense, we view this effort as part of a research and development program whose

---

[2]We note that in invoking these three representations we are not necessarily invoking objects that are well defined or well understood. Dan Chazan has remarked to us, for example, that equations have different entailments depending on the ways we think of them (e.g., as "number recipes" or transformations); our descriptions of the objects within the representational categories must become more nuanced to be fully useful.

intention is to map out understandings of complex domains and to construct curricula that help students come to grips with that complexity. We place our main efforts much more along the lines of "seeing and exploiting connections" than on procedural knowledge. The next section of this chapter briefly provides background and context for the body of research that led to this particular study and for the research perspective that was sketched in this section. We focus on a 2 by 3 matrix or framework (the two perspectives and the three representations just discussed) and illustrate how the framework can be used to help construct or assess curricula. The remainder of the chapter is devoted to two "data stories" taken from a series of ongoing tutoring studies, where the curriculum for the studies was informed by our preliminary ideas about the Cartesian Connection.

We hope that the tight focus of this chapter helps to cast some issues in high relief. The elaboration of the process and object perspectives in the case of linear relations points to some loci of conceptual difficulty for students, as seen in our data stories. Generally speaking, our notion of domain competence includes having the ability to use mathematical ideas to deal with somewhat novel and complex problems—problems much more complex and mathematically interesting than the NAEP example that started this chapter. Our analysis of this domain indicates that such competence rests in part on the ability to know which representations and perspectives are likely to be useful in particular problem contexts, and to switch flexibly among representations and perspectives as seems appropriate. The framework delineated in this chapter, suitably expanded, can highlight the kinds of connections one needs to make. Such a detailed delineation can serve both as a means of assessing curricula and as a heuristic frame for curriculum construction. The framework can also serve as a guide for interpreting and understanding students' solutions to problems in this domain.

Two caveats are appropriate here. The first is that we make no claims for the completeness of the framework. This chapter does not discuss verbal representations, for example, or the construction of mathematical models (in any representation) that capture the great variety of real-world situations that embody linear relations. Nor does it discuss other central mathematical ideas (e.g., proportionality) on which an understanding of linearity depends. Here we elaborate on what it means to understand part of a domain—certainly not all of it. However, one can easily envision extensions of the framework that deal with such issues. Second, our data stories are intended to be illustrative and suggestive, and are rather sketchy in consequence. The reader will not find here the welter of detail that typifies our cognitive analyses (see Schoenfeld et al., 1993; Schoenfeld et al., in press).

BACKGROUND

Our intention in this section is to provide the reader with some perspective on the task in which we are engaged and on the issues we consider important. The Functions Group at Berkeley has been engaged since 1985 in a series of studies related to students' understanding of functions and graphs. Early on, we constructed a computer-based microworld called GRAPHER (Schoenfeld, 1990), designed to help students come to grips with aspects of the domain. Ultimately the research group spent a year and a half engaged in the very fine-grained analysis of 7 hours of videotape of one student working with GRAPHER—the goal of that analysis being to understand precisely how her understanding of functions and graphs changed over the period she worked in our lab (Schoenfeld et al., in press). The analysis resulted in our description of the Cartesian Connection, a characterization of the understandings possessed by people who are knowledgeable in the domain. The analysis also indicated that students who appear to be competent in the domain can, indeed, miss fundamental connections. For example, students can treat the algebraic and graphical representational domains as though they are essentially independent. Although the m in the equation form $y = mx + b$ is typically referred to as the slope and a student may refer to it as such, the student may not attribute any slope-related graphical properties to equations that have differing values of m. Similarly, the student may refer to the b value of the equation as the y intercept but may not know that the point $(0,b)$ lies on the graph of the equation—even though the student uses the term y intercept when referring to properties of the graphs. Or the student may not realize that the parameters m and b in the form $y = mx + b$ are independent—that is, that one can change one of the parameters while leaving the other constant, and generate a family of lines with specific properties (see also Moschkovich, 1989, 1990).

Subsequently, the research group organized a collection of problems designed to focus on aspects of the Cartesian Connection and to serve as the basis for a curriculum introducing students to linear functions and graphs. We are now engaged in the extended analyses of videotapes of students and tutors working through that tutoring curriculum, with the goals of (a) extending the cognitive analyses in Schoenfeld, Smith, and Arcavi (in press), (b) delineating the complexities of the tutoring process and constructing a tutor model (Arcavi & Schoenfeld, 1993; Schoenfeld et al., 1993), and (c) constructing a trial curriculum for technology-based classroom instruction on linear functions.

One result of the research has been to delineate the complexity of what, unexamined, might appear to be absolutely straightforward. For example, many educational researchers and developers seem to have the belief that "once things are shown clearly on the computer screen (as opposed to the crude drawings we produce by hand) then students will understand." The work of Goldenberg (1988), discussed elsewhere in this volume, showed all too clearly that such assumptions are unwarranted. Here is another example that points to the dangers of naive curricular assumptions.

As part of a curriculum development project, Magidson (1989) developed a guided discovery unit in which student volunteers were introduced to linear equations and their graphs. Early in the unit, pairs of students who worked on the curriculum were instructed to use the available graphing software (*Green Globs* on an Apple II) to work the following problem.

Clear the screen and type in these equations, one at a time:

$y = 2x + 1$

$y = 3x + 1$

$y = 4x + 1$.

What do you notice?

How are these lines similar?

How are they different?

What do you think will happen if you type in $y = 5x+1$? Sketch your prediction on this empty graph [which was provided on a work sheet] and then try it on the computer.

What happened?

The intent of the problem should be obvious, as were Magidson's expectations of what the students should see: All of the lines pass through the point (0,1), and the larger the coefficient of x, the steeper the line (see Fig. 4.1) But then again, we know what to look for. Here are the responses from one pair of students, reproduced verbatim:

What do you notice? *The higher the number you are multiplying by x the more upright the line.*

How are these lines similar? *All practically the same angle.*

How are they different? *They're not the same angle.*

What do you think will happen if you type in $y = 5x+1$? Sketch your prediction on this empty graph and then try it on the computer. (The students' sketch is given in Fig. 4.2.)

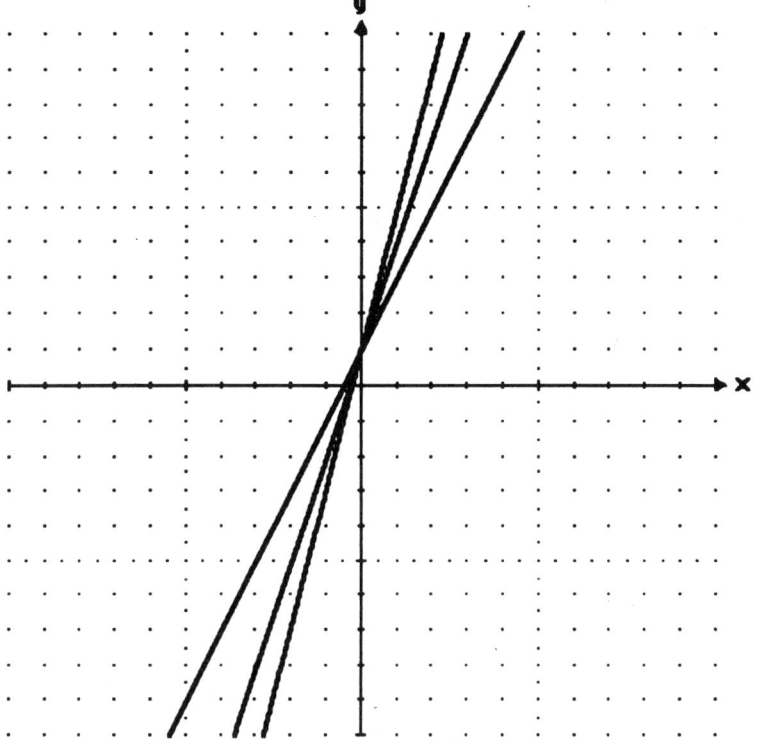

FIG. 4.1. The graphs of $y = 2x+1$, $y = 3x+1$, $y = 4x+1$.

What happened? *It went more upright than the other lines but less upright than we thought it would be.*

Note that the students' sketch passes through the point (0,5). The students had failed to notice that all the lines graphed by the computer had passed through (0,1)! At least, these students did notice the relationship between the coefficient of x and the steepness of the line. In general the students who worked the problem made two kinds of observations, neither of which had anything to do with the common point of intersection. The first kind was related to the position of the line. The observations sometimes dealt with steepness, as above. But they frequently dealt with the manner in which the lines appeared on the screen. The software produces any graph in order of increasing x values. Thus when they appeared, all of the lines "started" at the bottom of the screen, moving upwards and to the right. Many of the students made only the following type of observation: "As the numbers get bigger, the lines start further to the right."

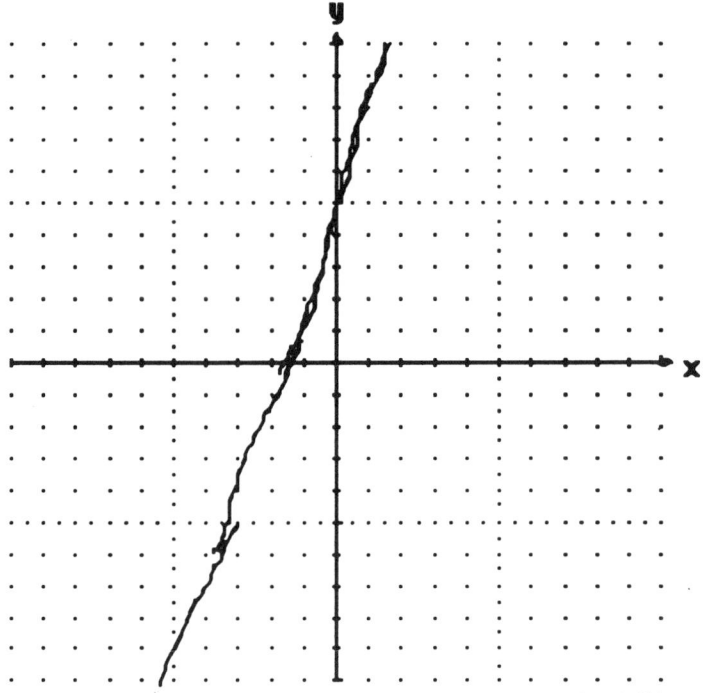

FIG. 4.2. The students' sketch of $y = 5x+1$.

The other kind of observation made by the students with whom Magidson worked had to do with the physical appearance of the lines graphed on the screen. As illustrated in Fig. 4.1, each graph is composed of a series of pixels (small rectangular shapes) on the screen. This produces an approximation, but not a perfectly smooth one, to the line being represented. Different patterns of pixels, corresponding to different slopes, produce different degrees of jaggedness on the graphs. Many of the students noted this and discussed it at length. In essence,[3] their response to the question "What do you notice?" was "The lines get more jagged as the number in front of the x gets bigger."

To sum up the moral of these examples in brief, what we perceive is shaped by what we know. People familiar with the domain know what to ignore and what to focus on, while newcomers to it do not. As a result, we see Fig. 4.1 differently than students do. What is obvious to us may not even be apparent to them. The large part of our task is to learn to see the

---

[3]We say "in essence" because we have paraphrased the students' responses. One of the clearer student statements, verbatim, was: "When you try an equation with smaller numbers the line gets straighter; when you type higher numbers the line gets thicker."

domain as students do, and to help them develop the understandings that allow them to perceive and understand the objects in it in a manner consistent with our perceptions and understandings.

THE FRAMEWORK

As indicated in the introduction, working competently in this domain involves thinking along at least two dimensions. One dimension refers to available means of representing linear functions (our focus here being on the three most common symbolic representations, algebraic, tabular, and graphical); the second refers to the perspective from which a linear function is seen or operated on. These two aspects are represented in Table 4.1. The solutions to tasks or problems may reside solely within one cell of the table; they may move across representations within one perspective or across perspectives within one representation; or they may move across both dimensions. As will be seen, the main curricular and mathematical interest is in tasks whose solutions call for moving both horizontally and vertically across the cells of Table 4.1. We begin with simple illustrations, building up in complexity.

TABLE 4.1. A Schematic Characterization of the Framework, Indicating Alternate Perspectives of Functions in Typical Representations.

Perspective	Representation		
	Tabular	Algebraic	Graphical
Process			
Object			

The Process Perspective

In the process perspective one's attention is directed to the relationship between the x and y values of a linear relation. The relation itself may be represented in tabular form [a list of (x,y) pairs], as an algebraic equation, or as a graph (where, thanks to the Cartesian Connection, the x and y coordinates of points on the graph are seen as "satisfying" the equation and corresponding to possible entries in the tabular form). The focus is on the x and y values and the relationship between them, on the variables in an equation that stand for those numbers, or on the sets of individual points in the Cartesian plane that, collectively, constitute lines. To begin with a trivial example, the solution to

Problem 2. Given the equation $y = 3x + 2$, find y when $x = 5$.

involves only the process perspective within the algebraic representation. It is easy, though boring, to find tasks whose solutions reside within just one cell of Table 4.1; we will give no further examples. Here is a slightly more interesting problem.

> Problem 3. Given the following table of values, find the corresponding linear equation.
>
> | x | y |
> |---|---|
> | 0 | 0 |
> | 1 | 3 |
> | 2 | 6 |
> | 3 | 9 |
> | 4 | 12 |

Solving this problem or its dynamic version, Guess My Rule, is a bit more complicated. Obtaining a solution involves making a connection between two representations, the tabular and the algebraic. Typically, students do so entirely within the process perspective: The goal is seen as finding, by guesswork or some previously developed algorithm, the algebraic expression with the property that when the x values in the table are input into the function, the corresponding y values are produced. Although it certainly is possible to think about this problem from the object perspective, we suggest that students do not.[4]

Similarly, the following two problems call for aspects of the Cartesian Connection linking the algebraic and graphical representations at the process level.

---

[4]There is a subtle distinction to be made here. One might think that in saying "the desired function has the form $mx + b$" and guessing at the values of m and b, the student is necessarily invoking the class of linear functions with m and b as parameters—hence invoking the object perspective. In fact, in many situations where students *appear* to be dealing with variables or parameters, they are not. There is an extensive literature (see, e.g., Wagner & Kieran, 1989) indicating that in introductory algebra classes, students perceive the task in the problem "solve the equation $x + 3 = 5$" to be that of "determining an unknown" (i.e., a single, predetermined value) and do not conceive of the x in the equation as a variable. Likewise, determining which rule of the form $y = mx + b$ will generate the given table can be seen as a task of determining particular, predetermined values of m and b: m and b themselves may not be perceived as parameters, and the function as an as-yet-not-specified process.

Problem 4. Why is the number 4 in the equation $y = 3x + 4$ the y intercept of its graph?

Problem 5. Find the x intercept of the graph of the equation $y = 2x - 2$ and explain how you did so.

The Object Perspective

The situation becomes more complex when one considers the following:

Problem 6. The table on the left represents specific values of the function $f(x) = x^3 - 3x^2 + 2x$. Fill in the table on the right, which represents the function $y = x^3 - 3x^2 + 2x + 1$.

x	y
0	0
1	0
2	0
3	6
4	24

$y = x^3 - 3x^2 + 2x$

x	y
0	
1	
2	
3	
4	

$y = x^3 - 3x^2 + 2x + 1$

There are at least two ways to approach this problem. One could complete it without referring at all to the table on the left. The function under consideration is $y = x^3 - 3x^2 + 2x + 1$, and one can simply calculate the y values that correspond to $x = 0, 1, 2, 3$, and 4. That, of course, is treating the function as a process. But it's also a lot of work. People with more mathematical sophistication are likely to approach the problem by saying (at least implicitly): "If I call the two functions $f(x)$ and $g(x)$ respectively, then $g(x) = f(x) + 1$. I can complete the table for $g(x)$ by simply adding 1 to each of the entries for $f(x)$." Here the two functions are treated as objects, and the transformation "adding 1" as an operator that converts one object to the other. The values of g are determined without direct computation.

Two more obvious examples at the object level are the following:

Problem 7. What are possible values for the slope of a line that lies entirely in the shaded region of the figure below?

Problem 8. Write an equation that characterizes an arbitrarily chosen line from those illustrated in the figure below.

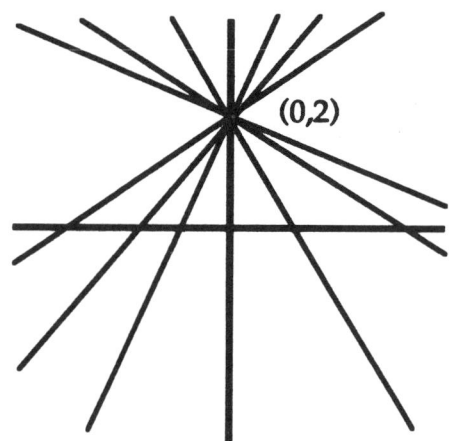

In both of these problems, candidate lines are thought of as being indexed by their slope. In Problem 5, the lines are of the form $y = mx$; the region is bounded by the x axis and the line $y = x$, which have slope 0 and 1 respectively; and, because lines of increasing slope increase in steepness, the slopes must lie between 0 and 1. For Problem 6, one notes that the family suggested by the figure is "all of the lines passing through (0,2)" or perhaps "the set of lines you get rotating a line through (0,2)" and that the corresponding algebraic form is $\{y = mx + 2: m \in R\}$.

Connections

Connections between perspectives allow for the possibility of flexibly switching from viewing a line (or an equation) as an object that can be manipulated as a whole, to viewing a line (or an equation) as made up of individual points (ordered pairs). This is an especially crucial flexibility when explaining why objects in the domain behave in the manner that they do. For example, consider the following question:

> Problem 9. Why is the graph of $y = 3x$ steeper than the graph of $y = 2x$? What about $y = 4x$, $y = 5x$, $y = 10x$?

One first answer could be "because it has a larger slope." However, this still requires an explanation of *why* a larger slope number corresponds to a steeper line. There are various ways in which one can answer this question, but no matter how one goes about it, the answer involves making connections across representations and perspectives. As in the previous two examples, each of the lines is indexed by its slope; the value of the slope parameter determines which lines are being considered. Here the lines are considered as entities, and the object perspective is employed in the algebraic and graphical representations. But to explain how the values of m correspond to the steepness of the individual lines requires the process perspective. Here is the standard way:

Consider any line L: $y = mx + b$ in the plane. Let (x_1, y_1) and (x_2, y_2) be any two distinct points on L. Some algebraic hocus-pocus results in the following:

$$y_2 - y_1 = (mx_2 + b) - (mx_1 + b) = m(x_2 - x_1) \text{ so that } m = \frac{y_2 - y_1}{x_2 - x_1}.$$

The key to interpreting this ratio for m graphically lies in the Cartesian Connection. By convention the expressions $(y_2 - y_1)$ and $(x_2 - x_1)$ represent directed line segments in the plane. Those line segments have magnitude [the lengths $|y_2 - y_1|$ and $|x_2 - x_1|$, respectively] and direction [up or down, respectively, if $(y_2 - y_1)$ is positive or negative, and to the right or left, respectively, if $(x_2 - x_1)$ is positive or negative]. It follows from these conventions that m is positive if and only if $x_2 > x_1$ implies $y_2 > y_1$.

One way to determine the steepness of the line L is as follows. Take any two points on L whose x coordinates differ by 1. Then $(y_2 - y_1) = m(x_2 - x_1) = m(1)$, so $y_2 = y_1 + m$. If m is positive, the graph of L rises m units for each unit change in x. Thus larger (positive) m corresponds to steeper slope.

As noted above, there are a number of ways to solve Problem 9. Here is another, briefer solution. All lines of the form $y = mx$ pass through the origin. The lines L_1: $y = m_1x$ and L_2: $y = m_2x$ pass respectively through the points $(1,m_1)$ and $(1,m_2)$; hence if $m_2 > m_1$, L_2 rises more steeply (see Fig. 4.3). Even in this condensed solution, however, one can see aspects of both the process and object perspectives, in the algebraic and graphical representations. From the object perspective, the individual equations and lines are considered as members of the parametric family $\{y = mx: m \in R\}$. But using points on the graphs, and determining their coordinates using the equations of the lines, employs the process perspective.

A third solution, suggested by Ed Dubinsky, is as follows. In $y = mx$, a change of value in x (run) causes a change in y (rise). The larger the value of m, the larger will be the rise for the given run. This solution might be considered the compiled, process version of the first "standard" solution.

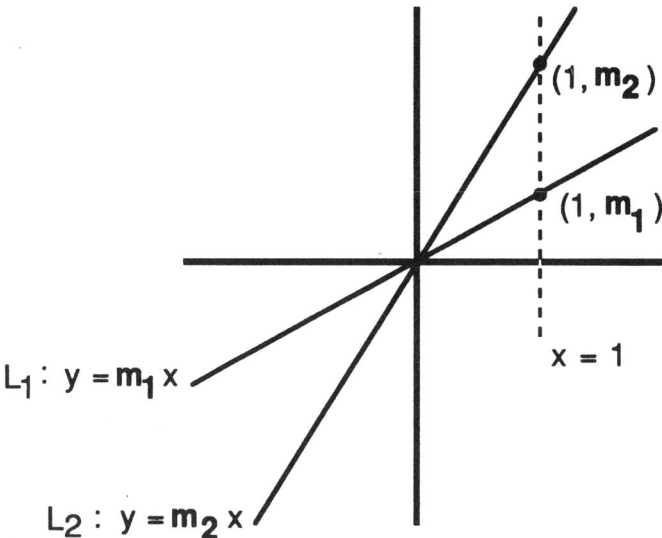

FIG. 4.3. A graphical illustration of why $y = m_2x$ is steeper than $y = m_1x$ if $m_2 > m_1$.

A Particularly Rich Curricular Example

The following problem is adapted from Resnick (1987, p. 158). To solve it calls for moving back and forth between graphs and equations using the Cartesian Connection, and both the process and object perspectives. We find the problem particularly rich in connections across cells in the framework, and think it provides a rich context for assessing the degree to which students have made such connections (as well as for inducing them).

Problem 10.

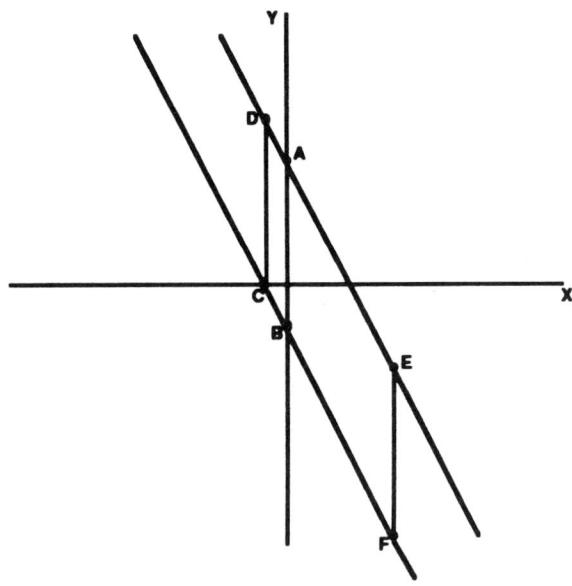

(a) What can you say about the slopes of these two lines?

(b) What can you say about the y intercepts of these two lines?

(c) The following list includes the equations of the two lines. Match each line with its equation.

$y = 2x + 6$ $\quad y = 2x - 2$ $\quad y = -2x - 2$ $\quad y = -2x + 6$

(d) Find the coordinates of points A, B, C, and D, knowing that the line segments CD and EF are parallel to the y axis.

(e) If the x coordinate of point E is 5, find its y coordinate and the coordinates of point F. (Does it look right on the graph?)

(f) Find the lengths of segments EF, CD, and AB. Does your result make sense? Why?

(g) Draw another segment connecting the two lines which is parallel to the y axis. Can you predict its length without knowing the coordinates of the end points? Explain. Would the equation help in this task? Why?

Having presented solutions to other problems in rather fine-grained detail, we forego a similar analysis here. However, we suggest that the reader work through the problem, noting the ways in which the perspectives and representations come into play. It is also worth noting how many different ways there are to solve the problem. Examples of student work on it are discussed in the next section.

NAVIGATING THROUGH THE CURRICULUM: TWO DATA STORIES

As explained above, part of our research and development efforts involved the creation of a tutoring curriculum designed to introduce students to the ideas represented in the framework. Here we describe some of our rationale for curricular construction, and the work of some students who worked through the tutoring curriculum. Our intention is to show how this framework (and more generally, how a detailed analysis of what it means to understand a domain) can be used as a heuristic guide to curriculum development, and also to illustrate, with samples of student work, the complexity of knowledge development within the domain. We have two main story lines. The first illustrates a sequence of activities designed to have students become familiar with the object perspective. The second illustrates our emphasis on "flexible competence" rather than procedural mastery as an indicator of what it means to understand a domain. Here we paint with a broad brush. Detailed analyses of the student work (replications and extensions of Schoenfeld et al., in press) and of the character of the tutoring interactions (Arcavi & Schoenfeld, 1993; Schoenfeld et al., 1992) are in progress.

Data Story 1. Coming to Grips With the Object Perspective

It may appear that there is a natural progression from the process perspective to the object perspective, that one must first learn to "put a function together" and see how it works at the process level before one is capable of thinking of the function as an object. That is, how could one think about the graph of a function as an object until one could produce it—and how could one produce it except as a rule (in algebraic representation) or by plotting points (in graphical representation), using the process perspective? Indeed, absent the current technology, that may have been the only learning path available to students. Schoenfeld (1990, pp. 285-286) wrote autobiographically as follows:

> In school, I learned to draw graphs (for concreteness, say the graph of $y = x^2 + x - 3$) by calculating the y-values for different x-values (usually $x = 0, \pm 1, \pm 2, \pm 3$, etc.; more points if more detail was necessary), making

a table of values, plotting the points from the table, and joining the plotted points with somewhat curvilinear segments. Early on the "overhead" for all the subsidiary operations was tremendous; after the trouble of calculating, plotting, and drawing, there was hardly the focus to reflect on the curves or their properties. . . . Over time and with extensive experience, I came to abstract the idealized mathematical curve from the empirical procedure. . . . Even so, this was only the first step in a long progression.

Consider what I saw when I compared the graphs of $y = x^2 + x - 3$ and $y = x^2 + x - 7$. The two looked similar. Moreover, having composed the tables for both, I know that the latter had the property that each of its y-values was precisely four units below the corresponding y value for the former. Nonetheless, this was a "point by point" comparison. . . . The mathematician also thinks of the two curves as representing the graphs of *entities*, where $y = x^2 + x - 7$ is the function (note the singular; it is one object!) obtained by "subtracting four" from the function $y = x^2 + x - 3$, and the graph of $y = x^2 + x - 7$ (again, singular; it is perceived of as a whole, single entity) is obtained by *shifting* the graph of $y = x^2 + x - 3$ four units downward in what is technically called vertical translation. At some point, I developed this understanding based on my ability to think of the functions and graphs themselves as *conceptual entities*—objects that, despite their complexity, I could think of as single, concrete manipulable objects. Later, I could begin to imagine the transformations as dynamic manipulations. Having graphed $y = \mathbf{a}x^2$ for many values of **a**, often on the same sheet, I could eventually "see" the idealized family of parabolas $y = \mathbf{a}x^2$ vary continuously as **a** varied.

This description of conceptual development may make it seem as though there is a natural progression from process to object perspective, in which (a) one must master the former before grappling with the latter, and (b) the object perspective ultimately supersedes the process perspective, and mathematical "adepts" will work solely at the object level. We believe that neither assertion is correct. Regarding (b), we point to the solution of Problem 1 discussed in the introduction to this paper, noting that both the process and object perspectives were components of its solution. On that point, Kieran (1991, p. 252) wrote as follows:

> The acquisition of structural conceptions by which expressions, equations, and functions are conceived as objects and are operated on as objects does not eliminate the continued need for the procedural conception . . . both play important roles in mathematical activity. However, very few studies have addressed the issue of the role and interaction of both conceptions in doing algebra. . . . The challenge to classroom instruction is to . . . develop the abilities to move back and forth between the procedural and structural conceptions and to see the advantages of being able to choose one perspective or the other—depending on the task at hand.

Regarding (a), a main point in the design of GRAPHER was that with the help of the technology, students no longer had to follow the path just described.

> The path we wish to make smoother with the help of the technology is a "learning trajectory" that results in the kinds of competence held by people who are fluent in the domain. This learning trajectory *via* the technology need not recapitulate the learning trajectory taken by those without it. With the help of the technology, students might (a) have different experiences with the mathematics, and (b) be able to deal with it in a different order or in different ways. For example, one need not start by having students graph very simple functions, if the computer will display complex functions that the students can analyze. Different sequencing made possible by the technology may allow for different kinds, and orders, of "scaffolding." (Schoenfeld, 1990, p. 285)

Our tutoring curriculum was designed to provide students with a set of experiences, early on, in which they manipulate graphs as objects. As such, it was intended to allow students to develop an intuitive feel for such manipulations—to give them an intuitive base for understanding slope as a parameter, and the correlation between values of m and the orientation of lines of slope m. Later, when students encountered the formal definition of slope, that definition might help explain phenomena about which the students had an intuitive grasp.

As we see in the discussion of student work, coming to grips with the object perspective takes time. Being able to think of functions of the form $y = mx + b$ as members of a two-parameter family in which the parameters m and b (a) are independent, (b) determine the position of a line graphically, and (c) each "move" the graph of a line in particular ways as they are varied, requires making a lot of connections. There are two "morals" in the data story that follows. First, giving students access to the object perspective in the ways suggested in the previous paragraph seems to be of some use. Second, it is easy to read either too much or too little into student actions as one observes them working in the domain. Saying when a student actually "has" the object perspective is not a simple matter. It is not a yes/no kind of knowledge, but one of degrees, and the process of learning is not one of simple monotonic growth, but one that includes a fair amount of oscillation.

In the first unit of our tutoring curriculum, students were introduced to standard conventions of the Cartesian plane. They were then given a number of point-plotting exercises that resulted in the graphs of straight lines, such as "plot five points for which the y coordinate is twice the x coordinate plus 1. What do you notice?" These were followed by exercises that made similar links between tables and graphs (given a table, abstract the rule; write the equation, graph the function), tables to algebraic expressions (Guess My Rule), graphs to tables and equations (given a graph, make a table and find the equation), etc. The overall lesson of that first unit was this: "Certain tables, verbal statements of relations between x and y values, verbal and algebraic expressions of the form $y = $ (something)$x + $ (something), and linear graphs in the plane, are different ways of representing the same things."

The second unit of the curriculum begins with the Starburst problem (Magidson, 1993). Students are presented with Fig. 4.4. Their task is to reproduce the Starburst. They can type in equations of the form $y = mx$, and GRAPHER will produce the graphs of those equations. Note that this activity is used before students have seen the formal definition of slope. It is motivational and engages students for long periods of time; although it provides little formal structure, it allows students to make numerous observations about the relationship between the value of m and the properties of the line.

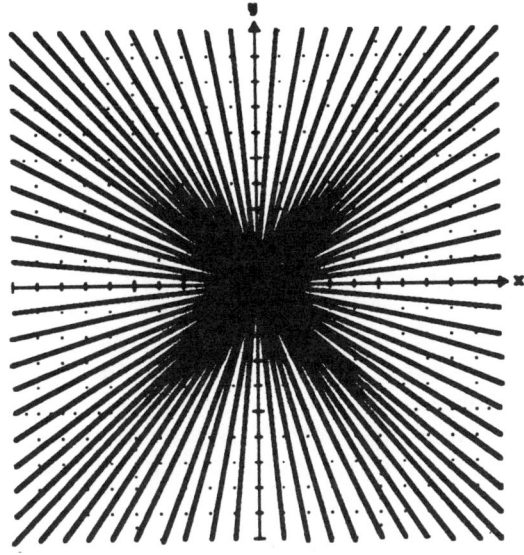

FIG. 4.4. A Starburst.

Almost all students observe that larger positive values of m produce steeper lines, but that the resulting spacing for $m = 1, 2, 3, 4, 5, \ldots$ is not uniform. Many observe that m must be between 0 and 1 to produce a line between 0° and 45°; many observe that the line $y = -mx$ produces the vertical reflection of $y = mx$, so that solving "half the problem" (getting even spacing for lines that have positive slope) will solve the whole problem. And they can do all this without knowing what slope is—thus laying an informal, empirical foundation for the formal definition.

As an example, AK, an eighth-grade student for whom our curriculum was the first introduction to the Cartesian plane, spent almost an hour working on Starburst. With the guidance of his tutor CK, he became aware of all of the issues just mentioned: that integer values of m produce nonuniform spacing, that values of m between 0 and 1 are required to produce lines that make angles between 0° and 45° with the positive x axis, and that lines of slope $-m$ are the vertical "mirror images" of lines with slope m. Asked "What does the number before the x do?" he wrote the following:

> The number determines where the line goes. The smaller the number gets the closer the line is to the x-axis (when positive). And when it's negative the bigger the number is the closer the line gets to the x-axis. When big and positive [it] gets closer to y-axis. When smaller and negative gets closer to y-axis.

Once one understands AK's language use (note that his use of "larger" and "smaller" for negative numbers does not refer to absolute value but to position on the number line; he correctly refers to -0.1 as a "large" negative number and -10 as a "small" one), it is clear that AK's statement is correct. And, AK has not only located the position of various lines as determined by their slopes, but he has used the language of change; it certainly appears that he has a sense of parametric variation. Hence, there is a strong temptation to say that AK has—at a correlational level rather than one of mechanism, because his conclusion is based on pattern recognition and he does not have any idea why particular values of m correspond to particular orientations—a good sense of the role of m as a parameter.

That was the goal of the Starburst task, of course. However, one must be careful in what knowledge one attributes to a student at any particular time. For example, the best case attribution for AK is that he understands one-parameter variation. As it happens, that attribution is too generous in some ways. A task given shortly after Starburst asked AK to "predict what the equations $y = 10x$ and $y = -10x$ would look like." His sketch appears in Fig. 4.5.

Note that in labeling the point (1,10), AK used his knowledge of the process perspective: the equation $y = 10x$ produces a y value of 10 for $x = 1$. The same holds for his graph of $y = -10x$, which unambiguously passes through $(1, -10)$.

The next part of the problem asked: "Now graph the lines $y = 10x$ and $y = -10x$ using the computer. Was your prediction correct? Do you have any ideas that explain what happened?" AK's response, "10 is the opposite of -10 so they slant down in opposite directions," indicates clearly that he

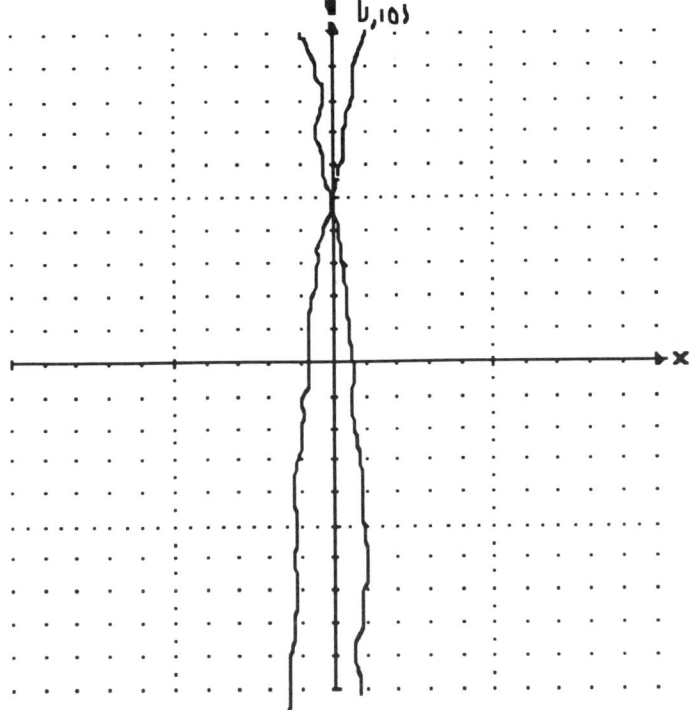

FIG. 4.5. AK's sketch of $y = 10x$ and $y = -10x$.

had observed that the two lines, of opposite slope, are symmetric with regard to the y axis. Yet his graph, and his response, miss what to adepts in the domain may be the most salient feature of Starburst—the fact that lines of the form $y = mx$ must pass through the origin! We stress that Starburst did a good deal of what it was intended to do in helping AK develop an intuitive sense of m as a parameter, but also that the observer must be cautious, for it is easy to read more than is warranted into student performance. It is most accurate to say, here, that AK had taken some important steps toward the development of an understanding of one-parameter variation.

We now fast-forward through two more units of the curriculum. In those two units, AK explored the properties of slope and y intercept. He encountered the informal definition of slope as "rise over run" and learned to use the slope formula $m = \dfrac{(y_2 - y_1)}{(x_2 - x_1)}$; he learned to "read" the y intercept of a graph from the equation, and also from a table of values (that is, he understood that the y intercept occurred at the point on the graph when

$x = 0$, which corresponds to the tabular entry when $x = 0$); and he learned that parallel lines have the same slope. Toward the end of his sixth tutoring session, AK encountered Problem 11:

Problem 11. Draw different lines which cross exactly 2 quadrants.
- What can you say about the slope of all these lines?
- What can you say about the y intercept of all these lines?

The intent of the curriculum designers was to have the student conceive of the answer in terms of parametric families. Three families of lines satisfy the given condition: all vertical lines except for the y axis, all horizontal lines except for the x axis, and all lines except the two axes that pass through the origin. Algebraically, those families are:

$$x = a \leftarrow (a \neq 0)$$

$$y = b \leftarrow (b \neq 0)$$

$$y = mx \leftarrow (m \neq 0)$$

There is a subtle distinction to be made in understanding the solution to this problem, one that parallels the distinction between "variable as specific but unknown quantity" and "variable as quantity that takes on a range of values" discussed in footnote 4. It is one thing to characterize a given set of lines by referring to its parameter values; it is yet another to invoke a class of lines by referring to a range of parameter values. That is, in working Problem 7, students may observe that every specific line they generate that lies within the region has a slope value between 0 and 1, and say "all the lines have slope between 0 and 1." Although this statement is true, it does not necessarily invoke the family of lines of the form $y = mx$ (where $0 < m < 1$). What we see in AK's work on Problem 11, discussed briefly below, is the way in which he comes to realize that a set of parameter values can invoke a family of lines.

AK began Problem 11 by interpreting it in a way that is typical of student approaches, illustrating the problem's tacit use of the conventional mathematics register (Pimm, 1987). Although the lines the student is told to draw are intended to be general (in that the student is intended to draw general conclusions from them), the problem statement permits a literal interpretation that is not general. Hence AK answered the first part of the problem by drawing $y = x$ and $y = -x$, and said their slopes were plus and minus 1 respectively. He noted that the y intercept of both lines is 0. AK then reread the last part of the problem, focusing on the world "all." After saying

"What can you say about the y intercept of all these lines?," he realized that vertical lines (except the y axis) also pass through precisely two quadrants. He realized that he was dealing with more than one family of graphs, and that caused some difficulty, as indicated by the following segment of dialogue with his tutor, CK:

> CK: If the line is straight up and down what can you say about the y intercept?
>
> AK: It doesn't have one.
>
> CK: What if the line slants?
>
> AK: It has one.
>
> CK: Aha, so what kind of y intercept is it?
>
> AK: It's a sometimes y intercept . . . it intercepts sometimes.

One can only speculate about mechanism, for there is scant evidence on the videotape, but it appears that the observation about vertical lines (all vertical lines save for the y axis meet the condition of the problem) led AK to realize that the two lines $y = \pm x$ were only two examples of a class of lines that meet the two-quadrant criterion. From this point on he appears to be referring to the families of lines as partial solutions of the problem. And, he seeks a way to characterize those families. He and CK focus on nonhorizontal lines. When reminded of the problem condition, that the lines must pass through precisely two quadrants, he says:

> So it's always going to be—two of them [that is, two of the lines that meet the desired criterion] are only gonna be $y = -x$ and $y = x$ . . . and you can't get it any other way . . . y equals something without a plus or minus, you can't have a plus or minus. So $y = 2x$ will go through the origin and work.

Here AK has come up with a verbal formulation of the relevant algebraic class: Implicitly he refers to equations without a constant term ("you can't have a plus or minus"), and in which the coefficient of x can vary ($y = x$ and $y = -x$ are prototypes, but "$y = 2x$ will go through the origin and work"). Having classified the set of lines that pass through the origin, AK turns his attention to a classification of vertical lines. He has some difficulty, and then says:

> When it's going vertical . . . any number going vertical will not touch the y axis except when it's $x = 0$, so . . . x can be anything, it's x equals anything [writes "$x = $ anything" on the work sheet], any equation like that. Let's say

x equals *n* [writes "*x* = *n*" on the work sheet]. So *x* = *n*, it will go through two quadrants.

In this segment of dialogue we see AK developing the understanding that the family of lines can be invoked by its parametric description. He is still a long way from understanding the family $\{y = mx + b: m, b \in R\}$ as a two-parameter family, where *m* and *b* are independent, but he has made tremendous strides. We pursue the issue of AK's learning trajectory—more generally, the issue of what it means to come to grips with subtle notions such as parametric representations—in the concluding discussion.

Data Story 2. On Flexible Understanding: The Importance of "Making Connections" Above and Beyond "Learning Procedures"

Here we discuss two students' work on Problem 10, which was introduced in the section entitled The Framework. We wish to highlight the facts that (a) there are multiple ways to solve the problem, and (b) solving the problem calls for making connections across representations and for employing both the process and object perspectives. These are two of the main reasons that the task is mathematically rich and interesting. Problem 10 begins the fifth unit of our tutoring curriculum. AK, with whom the reader is familiar from the discussion in Data Story 1, worked the problem at the beginning of his sixth tutoring session. CK was his tutor.

Absent the equations for the two slanted lines in Problem 10, parts (a) and (b) can only be answered in qualitative terms: The two lines have (the same) negative slope; one has a negative *y* intercept, the other a positive *y* intercept. On the basis of that information AK was able to make the proper selection from the candidates given in part (c) of the problem. He identified the lines correctly as $L_1: y = -2x - 2$ and $L_2: y = -2x + 6$.

It is interesting to note that AK then had some trouble with part (d). Having dealt with the lines wholistically as objects, he found himself at a loss when asked to determine the coordinates of point A—even though he had just determined the equation of L_2. Looking at the picture, he said:

AK: It's zero something, it's (0,*x*), but there's no way you can tell what A is.

AK's problem was not caused by a lack of knowledge, but rather by a difficulty in seeing and pulling together the relevant information. When CK reminded him that he had determined the equations of the lines:

CK: Well . . . you know the equation for these lines, right? . . . 'cause you just did that.

AK's response was almost immediate:

AK: Oh, so it's six . . . so this is zero, six [writes down coordinates for A].

AK then estimated the x coordinate of point D as -1 ("because it looks that way") and computed its y coordinate as 8. He then turned to compute the coordinates of point B. Despite having just gone through the process, AK found himself at a loss once again: He needed to be reminded that he had determined the equation of L_1, and that he knew one of the coordinates of point B. Things progressed more smoothly for points C and D. AK computed the x coordinate of C by solving the equation $-2x - 2 = 0$, and substituted that value into L_2 to obtain the y coordinate of D. Moreover, he commented that it was sure to be right, as opposed to his earlier guess.

What we find notable here is that AK had all of the relevant information at his disposal, but that he found this part of the problem difficult. We believe that it is because the diagram leads one to focus on features of the lines as objects, while determining the coordinates of points A, B, C, and D requires a switch to the algebraic representation of the function, and to the process as well as object perspective.

AK began part (e) of Problem 10 by substituting $x = 5$ into the equation of L_2. Then, after having some difficulties with the coordinates of the various points he had labeled in the figure, he reverted to geometric reasoning. Observing that the vertical distance between the points C ($-1,0$) and D ($-1,8$) is 8, he drew several segments parallel to EF, saying:

AK: Eight . . . eight . . . eight . . . once I knew that this one [CD] was gonna be eight, I knew the other one [EF] was gonna be eight.

His tutor, pushing for him to make the connection to the equations, asked:

CK: Could you see it [the distance of 8] from the equations?

and AK responded as follows.

AK: Look, they are 8 apart [points to the two equations] they are 8 apart because 2 and -6 is 8 apart.

In brief: Although AK had much of the "within representation" and "within perspective" knowledge required for this problem, solving it calls for moving flexibly across representations and perspectives where necessary or appropriate. Such flexibility does not come easily. Much of AK's difficulty came when he had to shift gears—or more precisely, when he needed to shift from one perspective or representation to another.

We turn now to a second student-tutor pair. AU was an eighth grader, the one student who entered the curriculum with some prior knowledge of graphs of linear functions (she often worked mathematics problems with her father in the evenings). Her tutor was AS, one of the authors of this paper.

AU moved smoothly through parts (a) through (c) of Problem 10. She, like AK, was at first stymied by part (d). An interjection by the tutor pointed her to the equation:

> AS: Let's take one point at a time. Given what you know, which is the equation of this line [points to L_1] and that line [points to L_2], can you tell me what the value of A is?
>
> AU: It's 6. I get it now. This [the y coordinate of A] is 6, so this [points at D] would be . . . well . . .
>
> AS: Let's take them one at a time. A was 6. The next one is B . . .
>
> AU: Negative 2.
>
> AS: How about C?
>
> AU: Negative 1 I think.
>
> AS: How did you get that?
>
> AU: Negative 2x negative 1 is 3 minus 2 equals . . . Wait, no. I mean if you substitute x for negative 1 [in $y = -2x - 2$] that equals 3 — Wait . . . equals 2 minus 2, equals zero.[5]
>
> AS: So that's (−1,0) [pointing to C on the work sheet].

AU took a different approach when identifying the coordinates of point D. She observed that the vertical distance from C to D is the same as the vertical distance from A to B (which is 8); hence, the y coordinate of D is $(0 + 8) = 8$, and the coordinates of D are (−1,8). And she employed yet another method for part (e) of the problem, using the slope formula (!) to obtain the y coordinate of the point $(5,y)$ on L_2. [She knew the slope of L_2 to be −2; since the points (0,6) and $(5,y)$ lie on L_2, it follows that $-2 = (y - 6)/5$, and that $y = -4$.] Parts (f) and (g) of Problem 10 were solved using geometric reasoning.

We make two observations. First, AU's difficulty came at a point in the problem that called for changing perspectives: She had dealt with parts (a) through (c) of the problem using the object perspective, but she needed to

---

[5]As far as we can tell, AU guessed/intuited the solution after she examined the equation $-2x - 2 = 0$, and she then checked her conjectured solution by seeing if $x = -1$ satisfied the equation.

move to the process perspective (substituting into the equations) to solve part (d). AU, like AK, was stymied at a point that called for a shift in perspective. Second, a major cause of AU's overall success was that, in general, she managed to move flexibly between the process and object perspectives. She solved or substituted into equations (process perspective) where necessary to determine the coordinates of particular points, but exploited global properties of the lines (object perspective) when considering their slope, or the distance between them. As discussed next, such flexibility is a hallmark of competence.

CONCLUDING DISCUSSION

Our first major goal in writing this chapter was to introduce and elaborate the framework for understanding functions that was outlined in schematic form in Table 4.1. There we pointed to two ways of viewing functions (the process and object perspectives) and the three most prominent representations of functions (in tabular, graphical, and algebraic form).[6] We hope to have indicated that competence in the domain consists of being able to move flexibly across representations and perspectives, where warranted: to be able to "see" lines in the plane, in their algebraic form, or in tabular form, as objects when any of those perspectives is useful, but also to switch to the process perspective (in which an x value of the function "produces" a y value), where that perspective is appropriate. As the data indicate, developing such flexibility is difficult: AK and AU, good students both, faltered at points in Problem 10 where a change in perspective was called for. Again, we stress that Table 4.1 is incomplete; there is much that it leaves out. Despite this limitation, however, the table has much to offer. For us, now augmented by columns representing "real world contexts" and "verbal representations," it serves both as a heuristic guide to curriculum development (Does any curriculum we propose make adequate connections across representations and perspectives? If not, it had better be revised) and for understanding and assessing student learning (Can the student move flexibly across representations and perspectives when the task warrants it?).

We suggest that the approach illustrated here—seeing understanding as making connections, and analyzing content domains to see what kinds of connections competent practitioners make—will be a profitable approach for both curriculum development and (student and curriculum) assessment. Indeed, our second major goal has been to focus the reader's view on this version of understanding and away from more procedurally oriented

---

[6]Note that although our discussion has focused primarily on linear functions, everything suggested in Table 4.1 applies to functions that can be expressed in closed form.

definitions such as "The student understands linear functions when she/he has mastered and can use the point-slope formula, the two-point formula (etc.) when appropriate." One of the points we hope emerged from the discussion of AK's and AU's work is that they produced delightfully different solutions to parts of Problem 10, and that such variation should be treasured and encouraged. Procedural competence is a component of understanding, but it should not be mistaken for the real thing. As a case in point, we highlight recent work by Dowker (1991, 1992). Dowker asked 44 mathematicians to perform a range of numerical estimation tasks. Not only did she discover a huge amount of variation in the procedures employed by the mathematicians—as many as 23 different strategies for one problem!— but she found that the experts were not consistent in the ways they approached the problems. Eighteen of the mathematicians were tested on the same problems a few months later. The mathematicians frequently used different strategies to solve the same problems: a minimum of 9 alternate strategies on the 20 problems, a maximum of 17 out of 20. In short, real competence consists of being able to get the job done easily, not in doing it the same way over and over again. We should teach for connections and understanding, not merely for procedural skills.

A third goal was to remind the reader of the subtlety and complexity of the learning process. The work here, both in research methodology and in its view of the domain, is grounded in the data provided in Schoenfeld et al. (in press). In particular, we believe that the growth and change of knowledge (i.e., learning) is a slow and complex affair. Coming to grips with parametrization, for example—recall Data Story 1—is not easy regardless of the visual assistance that computer-based technology can offer. Further exegeses of the learning process in this domain, which were only hinted at in Data Story 1, are in progress.

A fourth and final top-level goal was to try to place the use of instructional graphing software in a reasonable perspective. It should be clear that we have an investment in such media, and believe they can help change things for the better: GRAPHER as a tool, and tasks such as Starburst, offer students the opportunity to deal with some aspects of functions, and develop some intuitions, in ways (and sequences) that were simply inaccessible prior to the availability of computer-based tools. Specifically, the software allows students to operate on equations and graphs as objects, and that may facilitate the development of the object perspective in ways not possible before the existence of such technologies. Having said this, however, we must repeat the caution discussed in the background section: The action is not what takes place on the screen, but is what the individual student puts together for herself or himself. And coming to grips with complex mathematical notions will take time and experience. In short, we are somewhat technophilic but hope not to be naive technophiles. We hope to develop, in a principled way, a deeper understanding of what it means to understand complex mathematical domains, and to exploit available resources to help students develop such understandings. Perhaps the work described here will be a step in that direction.

ACKNOWLEDGMENTS

This research was supported by the National Science Foundation through grant MDR-8955387. The NSF support does not necessarily imply endorsement of the ideas expressed in this chapter. The tutoring curriculum was developed by Abraham Arcavi, Miriam Gamoran, Judit Moschkovich, and Ching-Fen Yang, and it drew on materials generated by various people including Sue Magidson. A very rough draft of this paper profited from the critical comments of Dan Chazan, Tommy Dreyfus, Ed Dubinsky, Michael Leonard, Sue Magidson, Henri Picciotto, Tom Romberg, Anna Sfard, Jack Smith, and members of the Learning subgroup of the Functions Group. The paper as a whole profited from Cathy Kessel's comments and editing. Our thanks to all for the improvements that resulted.

REFERENCES

Arcavi, A. A., & Schoenfeld, A. H. (1993). Mathematics tutoring through a constructivist lens: The challenges of sense-making. *Journal of Mathematical Behavior*, 11 (4).
Breidenbach, D., Dubinsky, E., Nichols, D., & Hawks, J. (1992). Development of the process conception of function. *Educational Studies in Mathematics*, 23, 247–285.
Dowker, A. (1991). *Mathematicians' estimation strategies: Some cognitive implications.* Manuscript available from author, Experimental Psychology, University of Oxford, England.
Dowker, A. (1992). Computational estimation strategies of professional mathematicians. *Journal for Research in Mathematics Education, 23*(1), 45–55.
Even, R. (1990). Subject matter knowledge for teaching and the case of functions. *Educational Studies in Mathematics, 21,* 521–544.
Goldenberg, E. P. (1988). Mathematics, metaphors, and human factors: Mathematical, technical, and pedagogical challenges in the educational use of graphical representation of functions. *Journal of Mathematical Behavior, 7,* 135–173.
Kieran, C. (1991). A procedural-structural perspective on algebra research. In F. Furinghetti (Ed.), *Proceedings of the Fifteenth PME Conference* (Vol. 2, pp. 245–253). Assisi, Italy: International Group for the Psychology of Mathematics Education.
Magidson, S. (1989). *Revolving lines: Naive theory building in a guided discovery setting.* Unpublished manuscript, EMST, School of Education, University of California, Berkeley, CA 94720.
Magidson, S. (1993). *From the laboratory to the classroom: A technology-intensive curriculum for functions and graphs.* Paper presented at the Annual Meeting of the American Educational Research Association, San Francisco, CA.

Moschkovich, J. (1989, April). *Constructing a problem space through appropriation: A case study of tutoring during computer exploration.* Paper presented at the Annual Meeting of the American Educational Research Association, San Francisco, CA.

Moschkovich, J. (1990, July). *Students' interpretations of linear equations and their graphs.* Paper presented at the XIV Annual Meeting of the International Group for the Psychology of Mathematics Education, Mexico.

Mullis, I. V., Dossey, J. A., Owen, E. H., & Phillips, G. W. (1991, June). *The STATE of Mathematics Achievement* (Executive Summary). Washington, DC: National Center for Educational Statistics.

National Council of Teachers of Mathematics. (1989). *Curriculum and evaluation standards for school mathematics.* Reston, VA: Author.

Pimm, D. (1987). *Speaking mathematically: Communication in mathematics classrooms.* London: Routledge and Kegan Paul.

Resnick, Z. (1987). *Functions* (Textbook from the Mathematical Chapters Series for 9th grade. In Hebrew). Rehovot, Israel: Department of Science Teaching, The Weizmann Institute of Science.

Schoenfeld, A. H. (1990). GRAPHER: A case study in educational technology, research, and development. In A. diSessa, M. Gardner, J. Greeno, F. Reif, A. H. Schoenfeld, & E. Stage (Eds.), *Toward a scientific practice of science education* (pp. 281-300). Hillsdale, NJ: Lawrence Erlbaum Associates.

Schoenfeld, A. H. (1992). On paradigms and methods: What do you do when the ones you know don't do what you want them to? *Journal of the Learning Sciences, 2*(2), 179-214.

Schoenfeld, A. H., Gamoran, M., Kessel, C., Leonard, M., Orbach, R., & Arcavi, A. (1993). *Toward a comprehensive model of human tutoring in complex subject matter domains.* Paper presented at the Annual Meeting of the American Educational Research Association, San Francisco, CA.

Schoenfeld, A. H., Smith, J., & Arcavi, A. (in press). Learning: The microgenetic analysis of one student's understanding of a complex subject matter domain. In R. Glaser (Ed.), *Advances in instructional psychology* (Vol. 4). Hillsdale, NJ: Lawrence Erlbaum Associates.

Schwartz, J., & Yerushalmy, M. (1992). Getting students to function in and with algebra. In G. Harel & E. Dubinsky (Eds.), *The concept of function: Aspects of epistemology and pedagogy* (MAA Notes, Vol. 25, pp. 261-289). Washington, DC: Mathematical Association of America.

Sfard, A. (1992). Operational origins of mathematical objects and the quandary of reification—the case of function. In G. Harel & E. Dubinsky (Eds.), *The concept of function: Aspects of epistemology and pedagogy* (MAA Notes, Vol. 25, pp. 59-84). Washington, DC: Mathematical Association of America.

Wagner, S., & Kieran, C. (1989). *Research issues in the learning and teaching of algebra.* Hillsdale, NJ: Lawrence Erlbaum Associates.

11 STUDENT THINKING

5 Functions and Graphs — Perspectives on Student Thinking

Sharon Dugdale
University of California-Davis

Recent efforts to improve students' conceptualizations of functional relationships as described by graphs have emphasized the need to move beyond plotting and reading points to interpreting the global meaning of a graph and the functional relationship that it describes. Some instructional approaches show promise for improving students' qualitative understanding of graphs that describe aspects of familiar physical phenomena. The availability of function-plotting software has raised the possibility of visual representations of algebraic functions playing a more important role in mathematical reasoning, but new instructional models are needed to encourage graphical reasoning. Educators should be aware of, and attempt to minimize, common misconceptions among students using function-plotting tools. Given that changing perceptions and evolving ideas are a normal part of learning, students also need to develop a capacity for recognizing and resolving apparent contradictions and refining mathematical reasoning through experience.

This chapter addresses three interrelated aspects of student thinking about functions and graphs:

1. Students' conceptualizations of functional relationships in a global sense — interpreting qualitative features of graphs.
2. Students' perceptions of functional relationships as facilitated by computer function-graphing software.
3. Students' misconceptions of functional relationships as revealed through computer graphing activities.

The first section, Qualitative Interpretation of Graphs, deals primarily with students' ideas about graphs that describe aspects of familiar physical phenomena—in particular, functional relationships that are often not easily described by algebraic equations. These experiences with graphs are considered important because of their potential for providing a qualitative basis for students' conceptualizations of graphs that describe functional relationships between variables. Qualitative aspects of graphs can be approached early in the curriculum, using familiar situations that are not necessarily described by algebraic equations. Algebraic graphs tend to have fewer obvious interesting features for a novice to talk about than some other graphs that represent sequences of events.

The second section, The Impact of Function-Plotting Tools, addresses the potential for graphing software to enhance students' understanding of functional relationships. Function-plotting tools have facilitated the movement away from a focus on calculating values and plotting points toward a more global emphasis on the behavior of entire functions, and even families of functions. Easy manipulation of graphical representations has raised the possibility of visual representations of functions playing a more important role in mathematical reasoning, investigation, and argument. However, new instructional models are needed to ensure students' involvement in graphical reasoning.

The third section, Students' Misconceptions, discusses students' misinterpretations of functional relationships as presented by graphing software. Researchers have noted some common difficulties among students using function-plotting tools. One response to these misconceptions is to analyze their causes and revise the tools and teaching methods in an attempt to avoid the occurrence of the misconceptions noted. While attending to this matter, the opportunity to address a larger issue should not be neglected. Changing perceptions and evolving ideas are a normal and unavoidable part of learning. Working within the mathematical environment of a function-plotting tool, misconceptions may provide opportunities to help students develop a habit of recognizing and resolving apparent contradictions and refining mathematical reasoning through experience.

QUALITATIVE INTERPRETATION OF GRAPHS

Students' Difficulties in Interpreting Graphs

Graphing of functions has long been a part of the study of mathematics. Traditional emphasis has been on procedures, such as computing function values and plotting points. Developing a qualitative perspective of the graphs of functional relationships in general has been a relatively recent curricular concern. To support this concern, there is ample evidence of students' difficulties in conceptualizing functional relationships and making qualitative interpretations of graphs. For example, Karplus (1979) studied secondary school students' conceptualizations of functional relationships in which continuous variables such as time, frequency, distance, and temperature depend on one another. Karplus reported:

> It is clear that few students had become aware of continuous functional relationships prior to taking a trigonometry course. Some evidence of this shortcoming has been reported previously (Karplus et al., 1977), where spontaneous student investigations of variables in a chemical experiment were described. The students' approach emphasized dichotomous rather than continuous descriptions of the variables, as in their references to hot versus normal, or cold versus normal temperature, rather than a serial ordering of cold-normal-hot in qualitative or quantitative terms.

Given data pairs describing a continuous functional relationship, students were most likely to process the data in a mechanical way, without evidence of reasoning about the physical context in which the number pairs were introduced. As Karplus summarized:

> Most serious among the problems we have uncovered is many students' failure to conceptualize a functional relationship that is given quantitative significance by the data. Instead, they concentrated on numerical properties of the data and applied an arithmetic or graphical algorithm.

To avoid this difficulty, Karplus suggested that science and mathematics teaching should include the extensive qualitative representation of functions by means of graphs, with attention shifted from numerical values to the overall relationship between the variables represented.

Similar concerns were raised in a study involving the introductory honors physics course at the University of Washington (Peters, 1982). Students were shown an apparatus designed to produce a natural motion in one dimension. The apparatus used an incline, a magnet, and a spring to control the motion of an object, causing it to speed up, slow down, and reverse directions. In conjunction with 20 minutes of repeated demonstrations of this motion, students were asked to sketch the position-versus-time and velocity-versus-time graphs for the motion. The results were disappointing. To quote from the report:

> Overall, out of 66 responses, only 30% of the students represented the motion reasonably accurately. Another 45% had one or more serious defects in their graphs, showing major points of confusion. The remaining 25% were so in

error in one or the other graphs as to indicate a complete lack of understanding of position or velocity or both.

Peters was careful to point out that these were honors students, comprising approximately the top 3% of the students taking introductory physics.

Peters concluded that such conceptual difficulties must be even more serious among students from the general population, but that because most texts (and presumably most lecturers) do not explicitly address these problems, most authors and instructors may not be aware of the basic conceptual difficulties that impede the progress of large numbers of students.

Peters also reported that although many of the honors students entered the class expecting to discuss esoteric topics in physics, most of them found the pursuit of understanding the fundamental concepts an intellectually satisfying objective.

Efforts to Improve Students' Graph Interpretation

Graphs That Tell a Story

Various recent efforts to improve students' understanding of graphs have emphasized the need to move beyond plotting and reading points to interpreting the global meaning of a graph and the functional relationship that it describes. Researchers at the Shell Centre for Mathematical Education at the University of Nottingham have been active in exploring students' perceptions of graphs and designing activities to develop qualitative understanding of graphs. Using the microcomputer program *Eureka* (Phillips, Burkhardt, & Swan, 1982), researchers have involved groups of students in:

- Manipulating a familiar physical situation and seeing the resulting changes in an associated graph.
- Examining a graph and describing various events that might have produced the graph.
- Producing their own scenarios and corresponding graphs, then comparing their scenarios with classmates' interpretations of the same graphs.

The *Eureka* program displays a bathtub and allows the user four options:

P: Put the plug in or out.

T: Turn the taps on or off.

M: Put a man into the tub or take him out.

S: Have the man sing or stop singing.

As the picture changes, the program graphs the water level over time. Figure 5.1 shows a typical sequence of displays. The first three options (plug, taps, and man) can affect the water level, but will not always do so. For example, turning on the taps will not affect the water level unless the plug is in or there is water already in the tub. The fourth option (sing) illustrates that the variable being graphed (water level) is dependent on only certain things, and singing is not among those.

The model is simplified so that the features addressed are readily apparent from the graph. For example, the water level rises at a constant rate, even though the profile of the man sitting in the tub suggests that we might expect the water level to rise more rapidly when filling around the wider parts of the man's body than when filling at the level of, say, his neck. This simplification does not appear to be a problem in children's interpretations of the model, although it might be an interesting issue to raise in some circumstances.

Shell Centre researchers used various approaches to guide children to think about what the graph of water level reveals about the situation (Phillips, 1986). For example, children were asked to create their own scenarios, to draw graphs that describe the water level throughout their scenarios, and to ask classmates to interpret the graphs to tell what might have happened. *Eureka* also provides stored examples for which the graphs may be viewed without the pictorial story with the bathtub. Children were asked to describe what could have happened to produce a given graph. They could then view the pictorial story that produced the graph. Of course, one child's scenario will not be the same as another's, nor will it necessarily match the Eureka program's story that produced the graph.

Rather than simply stating that the man got out of the tub and then got back in and then turned on the taps, children were encouraged to say why the particular events may have happened. For example, a child might suggest that the telephone rang, so the man had to get out of the tub to answer the telephone, and when he got back, the water had gotten cold, so he added some hot water, and so on.

Some of the examples stored by the program are simple and straightforward, whereas others can be quite difficult to explain. Figure 5.2 shows a difficult example graph stored by the program. The story produced by one child and reported by Phillips follows:

Our house has caught fire. I go into the bathroom, put in the plug and turn on the taps. I know the bath is going to overflow, but it doesn't matter because it's an emergency. I go and look for a bucket, and fill this from the bath, and throw the water at the fire. I fill the bucket again and throw the water at the fire. The third bucket of water puts the fire out. I rush back into the bathroom and pull out the plug to stop it overflowing. Then I turn off the taps.

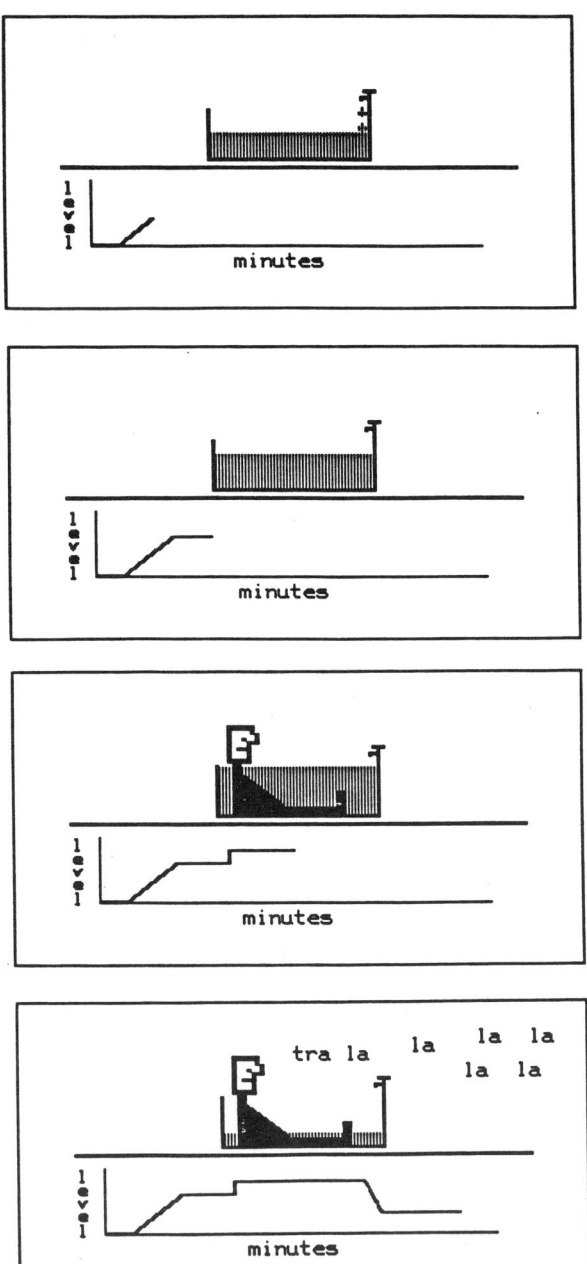

FIG. 5.1. A sequence of displays from *Eureka*, by Richard Phillips, Hugh Burkhardt, and Malcolm Swan. Copyright © 1982 by UK Crown. Reprinted by permission.

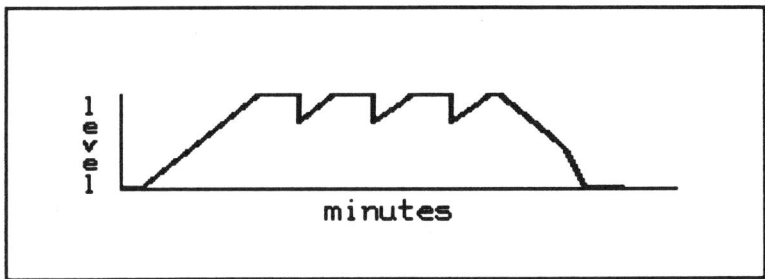

FIG. 5.2. An example graph stored by *Eureka*, by Richard Phillips, Hugh Burkhardt, and Malcolm Swan. Copyright © 1982 by UK Crown. Reprinted by permission. Children are asked to tell a story about what might have happened to produce the graph.

Phillips pointed out the importance of such elaboration:

> Consider the real life skill of interpreting a graph. We open a newspaper and see a graph of unemployment statistics. The unemployment rate goes up and down at different times of year. To understand this graph, it's not sufficient to read off values, or to note which month unemployment is highest. To make sense of it, we have to put a story to it, even though we cannot be certain that our story is true. We might say "It's low there because there are more jobs in the summer," or "It's high there because of school leavers." This is real *graph interpretation*.

Further, if we think of typical uses of data in scientific investigation, we can see the need to consider various possible interpretations, then to consider what further investigation might be useful for supporting or ruling out some of the possible interpretations.

The approach described by Phillips exhibits important characteristics of what Brown, Collins, and Duguid (1989) advocated as honoring "the situated nature of knowledge" and what Davis (1989) referred to as "experiential education." The ideas of functions and graphs are developed and used in the context of a familiar activity, the emphasis is on relating the new representational method to students' everyday experience, and students learn the power and limitations of the representational method through creating and comparing their own scenarios and graphs.

A further project at the Shell Centre involved recording the electrical consumption of a typical household over a 2-week period and presenting the data graphically (Phillips, 1988). Unlike the graphs presented in the *Eureka* program, the electrical consumption graph is recorded from a real household, and it is not simplified to facilitate interpretation.

As shown in Fig. 5.3, in viewing the graph for a single day or part of a

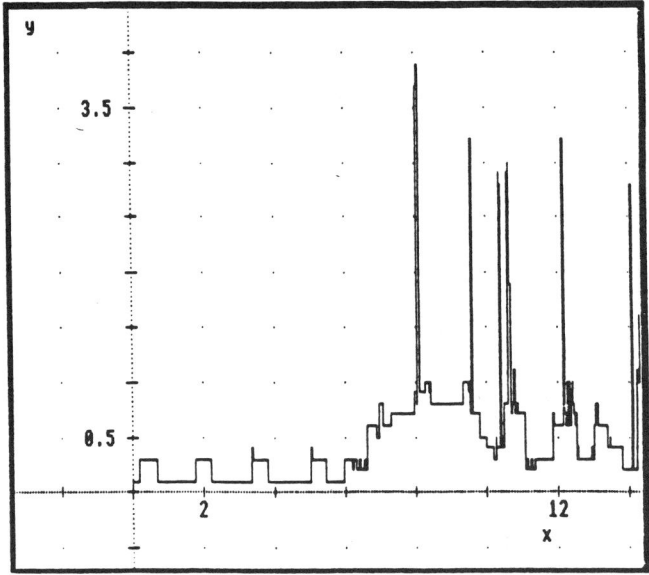

FIG. 5.3. A graph of household electrical consumption (in kilowatts) from midnight until about 2:00 p.m. The illustration is from *Mouse Plotter*, by Richard Phillips. Copyright © 1988 by the Shell Centre for Mathematical Education. Reprinted by permission.

day, many features are available as clues to household activities. Although little has been written about students' interpretation of the electrical consumption graph, the questions suggested by Phillips (1988, p.71) illustrate the potential for providing interesting discussion and enhancing students' understanding of graphical representations:

- At what time in the morning do people get up?
- Is this summer or winter?
- What causes the repeating pattern on the graph during the night-time?
- What would a similar graph look like based on where you live?
- Could you write a story about the people living in this house and what they did over this time?

Like *Eureka*, the work with graphs of electrical usage focuses on a single physical situation, with in-depth discussion of one type of graph. Other efforts have looked at students' skills in interpreting graphs of a variety of situations involving different variables, so that the labels on the axes must be carefully considered in interpreting each graph.

In work with about 10 classes of high school students (Dugdale, 1984), students were given descriptions of physical situations and were asked to choose an appropriate graph from three given graphs. As with the bathtub stories in *Eureka*, the graphs were of familiar circumstances that required no particular content background. One frequently observed student error was to ignore the labels on the axes and simply look for a graph that resembles a "picture" of the event described. For example, given the item in Fig. 5.4, students would often choose the graph that looks like a hill. This graph might be appropriate if the axes had been labeled elevation versus distance traveled, but it is not a likely choice for speed versus time. (This graph-as-picture confusion has been noted also in other studies, such as that of Mokros and Tinker [1987], which is discussed later.)

Students worked in pairs and were encouraged to discuss their reasons for choosing particular graphs. This interaction helped students clarify their ideas about the graphs and also helped observers understand what the students were thinking. Students were free to ask for help, and observers sometimes interjected comments without being asked. Typical comments from an observer might be, "Look at the graph you chose. Where does it show the bicyclist's speed increasing?" "Now think about riding a bicycle over a hill. Where do you really speed up?" This sort of guidance (having

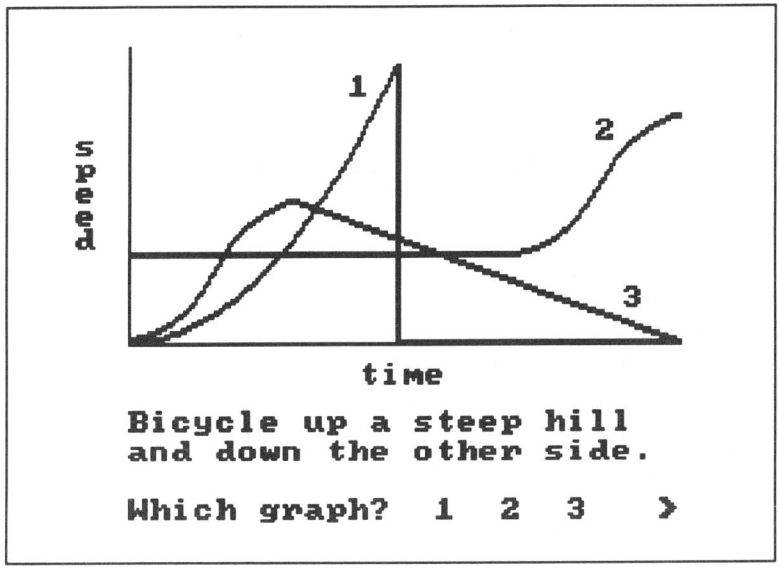

FIG. 5.4. A display from *Relating Graphs to Events* (Dugdale and Kibbey, 1983b, 1986b). Students are asked to choose an appropriate graph for the event described.

students relate the incorrect graph choice to the labels on the axes, then contrast it with the given situation) was usually sufficient to reorient students' thinking from the graph-as-picture notion to a focus on the behavior of the variables being graphed.

Observation of the students' approach to the task and their discussions with their companions indicated that they were not accustomed to making qualitative interpretations of graphs of physical events—this was a new idea to them. However, through discussion and some coaching, the students became oriented to the task and developed proficiency at choosing appropriate graphical representations for familiar situations.

Interpreting Simultaneous Graphs

A later study (Dugdale, 1986-1987) involved high-school students in interpreting simultaneous graphs of three different aspects of a situation in order to reconstruct the events described by the graphs collectively.

The context of the graph interpretation activity was a microcomputer environment, *Pathfinder* (Dugdale & Kibbey, 1987), in which a ball's movement around the computer screen is controlled by the placement of "tokens" on the screen. Six types of tokens control three aspects of the ball's movement:

- TURN LEFT and TURN RIGHT tokens control the direction of the ball, turning it to its left or its right.
- GET BIGGER and GET SMALLER tokens control the size of the ball, causing it to increase or decrease in size.
- SPEED UP and SLOW DOWN tokens control the speed of the ball.

Each time the ball encounters a token, the ball changes direction, size, or speed, as specified by the token encountered. As the ball moves around the screen, responding to the tokens it encounters, the ball's direction, size, and speed are graphed at the bottom of the screen.

As with the *Eureka* example, students began by manipulating the model—constructing courses to route the ball, sometimes in quite intricate patterns. Courses that moved the ball in a closed path, without leaving the screen or going outside of the allowed ranges of size and speed, could be stored as challenges for classmates to reconstruct by interpreting the graphs of the ball's direction, size, and speed.

After some experience with the model, students moved on to the task of using the graphs produced by a given course to reconstruct the course. Although stored problems were provided by the program, the students showed a marked preference for working on the challenges saved by their classmates. Most students worked in pairs, sometimes with an extra classmate watching and contributing comments. Discussion among the students revealed their strategies and insights to the observer as well as to each other.

The Three Graphs. Students were given no instruction about the properties of the graphs and how they interrelate. The three graphs, recording direction, size, and speed of the ball, address quite different aspects of the situation. While students worked on reconstructing classmates' courses, their comments, ideas, and strategies were recorded as they dealt with these aspects:

Direction. The direction graph is fundamental to the overall layout of the course. It is possible to sketch the approximate shape of the entire course using only the graph of the ball's direction. This is not possible with either the size or speed graph.

Size. The size graph relates to the other two graphs in a simple, straightforward way. The ball's size changes take place at certain points relative to the other graphs, but placement of size tokens has no effect on the other graphs. For this reason it is possible to add the size tokens last, after the course has been laid out and the other two graphs have been completely matched.

Speed. The speed graph, however, interacts with the other graphs in a more subtle and complicated way. Changes in speed produce corresponding changes in the other two graphs. For example, if the ball speeds up, it takes less time to traverse a given distance, and hence it shows a shorter segment in the direction graph. Although the basic layout of a course can be determined from the direction graph alone, accurate placement of the direction tokens depends on the placement of the speed tokens. For this reason the speed graph, unlike the size graph, cannot be ignored while matching the other two graphs.

Hence, the activity is not an exercise in interpreting three graphs with essentially equivalent properties and significance, but rather an exploration of relationships among three fundamentally different graphs and the situation that they collectively describe.

Students' Strategies. Students' strategies were observed and recorded, and changes in strategies were apparent as students gained more experience in interpreting the graphs. The initial approach of most students was to lay out the entire course from the graphs, then begin debugging. After some experience, students shifted to a more interactive method of building a small section, checking it against the graphs, debugging it, adding a few more tokens, checking against the graphs again, debugging the new section, and so on. From students' discussion during the activity and their later responses to questions about their strategies, it appeared that they found building and debugging a series of small sections more rewarding because it provided early and frequent feedback and transformed the larger problem into a series of more easily addressed smaller problems. This provided a series of successes as each problem was worked out.

Besides the change of overall strategy, students' use of the three graphs changed as they gained experience. It was generally understood quite early that the direction graph was special, in that it defined the overall layout of the course. However, for most students, the fundamental difference between the size and speed graphs was recognized much later, and this was often a dramatic revelation. Students who figured out the effect of the speed graph on the other two graphs helped their classmates understand the idea, and by the end of the study 15 of the 23 students had been observed to remark about the interaction of the speed graph with the other graphs.

In sum, beginning students were unlikely to make any distinction between the size and speed graphs or to recognize that a repeating path would produce periodic graphs. Experienced students, however, were more likely to recognize the different properties of the speed and size graphs, using the speed graph along with the direction graph, and saving the size graph until last. They were also more likely to use the periodic nature of the graphs to get an overview of the situation before starting to place tokens.

Although the construction of courses as challenges for classmates to reconstruct does not directly involve graph interpretation, observation of students in this activity revealed a marked difference in thinking between students taking general mathematics classes and those enrolled in college preparatory classes. Constructing a course requires the placement of direction, size, and speed tokens to keep the ball moving in a closed path. If the ball leaves the screen or encounters a token that would put it outside of the allowed sizes or speeds, students must change one or more tokens to fix the problem. Most students, both general mathematics and college preparatory, were observed to trace the ball path in order to fix direction problems. Beyond this, there was little overlap between strategies used by the two groups of students.

The general mathematics students used local strategies, attending only to the token where the difficulty became apparent. These students might run the ball through their courses until it hit a bug, then simply reverse the "offending" token. Nearly 80% of the general mathematics students were observed to use this technique at least once. For example, if the ball left the screen after encountering a "turn right," the "turn right" was changed to a "turn left," as if the problem were the single token, rather than the way it related to the rest of the course. If reversing the token did not solve the problem, the token might then be changed to an unrelated token (for example, a "get bigger" token) or deleted entirely. This sort of strategy was sometimes successful, and after repeated trial and error the students could end up with a running course.

In contrast, the college preparatory students made more use of global strategies, showing more awareness of the total picture. Their discussions and changes nearly always took into account more than the immediate trouble spot.

Experiences with Probeware

Recent work with "probeware" has documented students' experience with the real-time generation of graphs related to experiments and activities conducted by the students. Probes attached to a microcomputer sense physical phenomena such as temperature, light, and motion, and the microcomputer produces graphs as the data are gathered.

One prominent effort of this type is the Microcomputer-Based Labs (MBL) materials developed at the Technical Education Research Centers (TERC). As explained by Mokros and Tinker (1987),

> In developing the MBL materials, it was decided to use graphs as a central means of communication with students. Data are reported to the students in the form of graphs that evolve as the experiment is underway. The students predict results in terms of graphs, and if there is a discrepancy between the graphs of the observations and the predictions, students must recognize this and make the necessary corrections. (pp. 369-370)

In using MBL with middle-school students, Barclay (1985) reported that students were generally able to correctly answer questions involving direct reading of values from the axes of a graph, whereas they were confused by questions that required consideration of larger sections of a graph. For example, questions involving the slope of the graph or the significance of positive and negative values, and the relating of these features to their meaning in terms of the physical events being graphed, were more difficult. Barclay observed that these more difficult aspects of graph interpretation may require explicit discussion and/or acting out of the event described by the graph. For example, students can be given a graph of a person's motion and asked to reproduce the graph themselves, using the MBL motion detector. Barclay suggested that some important attributes of students' learning experiences through MBL are:

- The grounding of the graphical representation in the concrete actions of the students.
- The inclusion of different ways of experiencing the material: visual, kinesthetic, and analytic.

- The fast feedback that allows students to immediately relate the graph to the event.

In a more formal study to determine the effect of three months of MBL instruction on middle-school students' graph interpretation skills, Mokros and Tinker (1987) reported a significant change in students' ability to interpret and use graphs, with greatest gains on items where the mental image of the phenomenon and the graph of the phenomenon were discrepant. For example, when asked about the speed of a ball rolling freely over a given terrain, students who used MBL became more likely to choose an appropriate graph rather than a graph that resembled a picture of the terrain.

Particular attention was given to graphs that double back in time—that is, after proceeding forward on the time axis for a while, the graphs reverse horizontal direction and go backward. Students' responses to pretest and posttest questions indicated a significant improvement in their ability to tell that there is a problem with such a graph and to state the nature of the problem.

THE IMPACT OF FUNCTION-PLOTTING TOOLS

Perhaps the most dramatic and widespread influence on students' work with functions and graphs has been the recent proliferation of computer function-plotting tools. Such tools have facilitated the movement away from a focus on calculating values and plotting points toward a more global emphasis on the behavior of entire functions, and even families of functions.

Consider, for example, the student's work shown in Fig. 5.5. In a game of *Green Globs* (Dugdale & Kibbey, 1983a, 1986a), the student has constructed a parabola to hit several green globs, and has then added $1/(x-3.5)$ to the equation. The effect of this extra term is negligible for all values of x except those close to 3.5, so the resulting graph is nearly the specified parabola except around $x = 3.5$. As x gets close to 3.5, the denominator of the extra term approaches zero, making a vertical asymptote. Using this technique, the student has caused the graph to leave the parabolic path briefly to hit three more globs.

It is more typical for algebra students to encounter expressions with a variable in the denominator in textbook exercises similar to those shown in Fig. 5.6. Students are asked to indicate for what value of the variable each expression is undefined. Following the example provided, students figure out for what value of the variable each denominator is zero and record that value as the answer.

5. PERSPECTIVES ON STUDENT THINKING 115

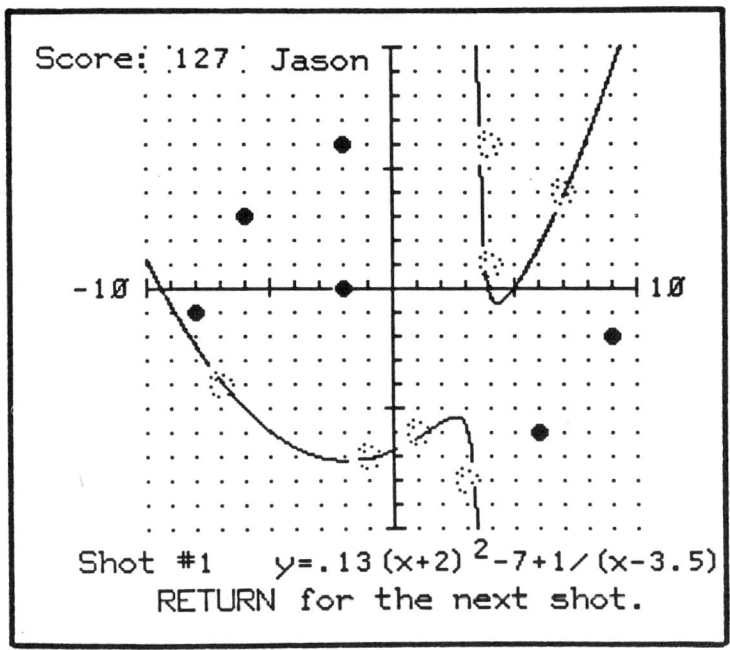

FIG. 5.5. A display from *Green Globs*. The program displays coordinate axes with 13 green globs scattered to appear randomly placed. The object of the game is to hit the globs with graphs specified by entering equations. When a glob is hit, it explodes and disappears. Here, a game from the Records Section shows a student's "shot." The student's name is displayed at the top of the screen.

Contrast the student's experience in Fig. 5.6 with that in Fig. 5.5. The mathematical content is similar—when the denominator is zero, the function is undefined. But instead of being concerned only with one particular x value (where the function is undefined), the student in Fig. 5.5 is using the behavior of the entire function, with particular attention to the function near the undefined value. This puts the discontinuity into a meaningful context, because the function gets very large in a positive direction on one side of the particular x value, and very large in a negative direction on the other side of that x value.

The easy manipulation of graphical representations allowed by current function-plotting tools has raised the possibility of visual representations of functions playing a more important role in mathematical reasoning, investigation, and argument. Relationships among functions can be readily observed, conjectures can be made and tested, and reasoning can be refined through graphical investigation. Eisenberg and Dreyfus (1989) suggested that students need a repertoire of basic functions, such as $\sin(x)$ and $\log(x)$,

> **EXERCISES**
>
> State the value of the variable for which each expression is undefined.
>
> Example: $\dfrac{3y}{18+3y}$
>
> Solution: -6, because when $y=-6$, $18+3y=0$.
>
> 1. $\dfrac{5}{6m}$
> 2. $\dfrac{2}{b-4}$
> 3. $\dfrac{-7}{5t}$
> 4. $\dfrac{2x}{8+2x}$
> 5. $\dfrac{2k}{3k-12}$
> 6. $\dfrac{3n+4}{3n+4}$
> 7. $\dfrac{z+2}{z^2}$
> 8. $\dfrac{4y+10}{5-6y}$
> 9. $\dfrac{4}{x^2-4}$
> 10. $\dfrac{m-3}{5m+10}$
> 11. $\dfrac{2t}{3t^2-5t-2}$
> 12. $\dfrac{7a}{a^2+8a+12}$

FIG. 5.6. Typical textbook exercises. Given a worked-out example to follow, students are asked to tell for what value of the variable each expression is undefined.

for which the graphical representation is as familiar as the analytic, and that students should feel comfortable manipulating the graphical representations.

In a study intended to analyze how students develop a sense for function transformations, Eisenberg and Dreyfus focused on students' use of visual and analytic thinking and their connections between the visual and analytic representations. Through a combination of instructional sessions and computer time with the *Green Globs* software, students were introduced to various classes of functions and transformations to which the functions could be subjected. The goal of the treatment was to help students gain a sense for visualizing function transformations. Students' performance on posttest questions indicated "a readiness . . . to approach problems by visual means, to use qualitative arguments, and to process information visually." In 57% of their responses to posttest questions, students' solutions included visual arguments. Further, 71% of the problems that were processed visually were answered correctly, compared to only 22% of those that were processed nonvisually.

Eisenberg and Dreyfus concluded that working with function transformations in a graphical context helped students understand the inherent connection between the algebraic-symbolic description of a function and its graphical representation: "They [the students] did not, as we felt in the beginning, always look at the graph as an extra load the function must carry." Through individual interviews, the researchers found that "students who viewed function transformations visually were more competent on non-standard transformation questions than those who viewed them algebraically."

Eisenberg and Dreyfus expressed concern that students tended to describe transformations as sequences of two static states, rather than as a dynamic process. They suggested that movable transparencies, movie films, or computer software showing transformations dynamically could be used to emphasize the process of transformation from one graph to another.

Encouraging Students to Think About Functions and Graphs

Although there is clear potential for graphing software to support and encourage mathematical reasoning, use of plotting tools does not in itself guarantee emphasis on mathematical reasoning. New instructional models are necessary to foster a more thoughtful approach to functions and graphs. This was apparent in a recent study comparing two approaches to incorporating graphing software into an instructional unit on trigonometric identities (Dugdale, 1989, 1990). One group of students experienced a traditional treatment of trigonometric identities supplemented with graphing activities, whereas the other group experienced graphical representations as the foundation for trigonometric identities.

The Supplemented Traditional Treatment focused on the algebraic-symbolic proof of identities, with graphs as an additional representation. Students in this treatment plotted many graphs (more than students in the other treatment) and related the graphs to the identities being proved. However, these students were not required to engage in any graphical reasoning tasks.

The Graphical Foundations Treatment introduced identities graphically, and the usual symbol manipulations were used to justify the relationships evidenced in the graphing activities. Students in this treatment were engaged in graphical reasoning tasks, such as:

- Analyzing graphic feedback and revising functions to change their graphs.
- Using graphs of functions to predict graphically the shapes of other, related, functions before plotting. For example, from graphs of $y = \sin(x)$ and $y = \sec(x)$, students figured out where $y = \sin(x) \cdot \sec(x)$ would have zeros, asymptotes, and positive and negative values, then predicted the general shape of the graph. After recognizing that the resulting graph looked very much like

$y = \tan(x)$, students were asked to construct an algebraic-symbolic argument to verify the apparent equivalence.

In the graphical reasoning tasks, interaction with classmates was a valuable source of insights. Students used and shared a variety of ideas. For example, in using two graphs to predict the shape of a related function, students made observations such as:

- When one graph has a function value of 1, the product is the corresponding point on the other graph. (And similarly, when one graph has a function value of -1, the product is the opposite of the corresponding point on the other graph.)
- When, within an interval, one graph has function values very near 1, the product within that interval is approximated by the other graph.
- Where two graphs cross (that is, their function values are equal), the quotient is 1.

Useful observations were easy to make, because the basic trigonometric functions are periodic, with frequent zeros, asymptotes, and relative maxima and minima of values 1 and -1.

Beyond using graphical representations as the foundation for trigonometric identities, the Graphical Foundations Treatment was intended to involve students in:

- Experiencing active participation in the development of mathematical ideas.
- Building a qualitative perspective before formalizing procedures.
- Applying previous knowledge and skills to a current problem without being told what, in particular, to do.

In addition to showing superior posttest performance in relating functions to their graphical representations, students in the Graphical Foundations Treatment Group exhibited more variety and personal involvement in their approaches to the standard content of proving identities. For example, given the goal of constructing a convincing algebraic-symbolic argument to justify an observed graphical equivalence, one student in the Graphical Foundations Treatment used the definitions of trigonometric functions in terms of a right triangle with sides a, b, c, using $\sin(x) = a/c$, $\cos(x) = b/c$, etc., although the two classmates with whom she was working chose a more standard approach to proving the identity. By the end of the second class session, this student had verified to herself that her method was essentially equivalent to what her classmates were doing and that their approach was probably less cumbersome. Although she abandoned her initial method, it provided some synthesis between the current topic and earlier material, and the ownership she felt for her method was clearly important to her.

A Question of Goals

In the trigonometric identities study already described, one group of students was provided examples and procedures to follow, whereas the other group was expected to draw from their experience and decide what they already knew that could be used to construct a justification for their observations. These two instructional methods illustrate two contrasting goals, and the choice between these two instructional goals is fundamental in deciding what kind of thinking students engage in. The two contrasting goals could be summarized as:

- To develop proficiency in solving specific types of problems.
- To encourage mathematical reasoning and investigation and establish strategies for thinking about mathematics.

Clearly there is merit in each of these goals and, in practice, each goal can support the other. Students who have learned a repertoire of specific techniques and can recognize standard situations that call for those techniques may be better prepared to apply a variety of techniques in investigating new problems. Also, students who are experienced in mathematical investigation and reasoning may be better equipped to assimilate new techniques, adapting and combining those techniques as necessary in approaching new situations.

Textbooks are typically organized around the first goal: to develop proficiency in solving specific types of problems. Problems are presented as a follow-up to the introduction of applicable solution techniques, so that students have opportunities to apply the newly learned methods. Of course this is important, but other types of experiences are necessary to address the second goal: to encourage mathematical reasoning and investigation and establish strategies for thinking about mathematics.

The context in which a task is presented can have a substantial influence on the types of thinking in which students engage. Presenting a task outside of the context of its most efficient solution can sometimes provide a richer experience. For example, consider the task in Fig. 5.7. Given a polynomial function and its graph on the left-hand grid, the task is to find a polynomial function that will produce the graph shown on the right-hand grid — that is, a graph that looks identical to the one on the left-hand grid, even though the scale on the right-hand grid is different. Further, if such a polynomial function is found, can a general solution be formulated to produce the same effect on any polynomial?

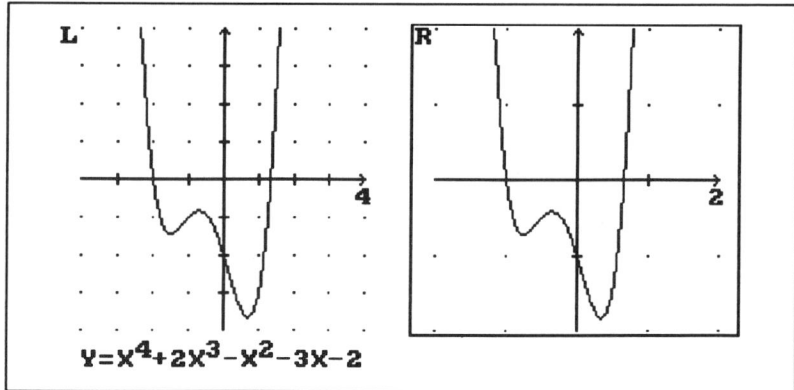

FIG. 5.7. Given the polynomial function and its graph on the left-hand grid, the task is to find a polynomial function that will produce the graph shown on the right-hand grid—that is, a graph that looks identical to the one on the left-hand grid, even though the two grids are scaled differently.

Presented as a follow-up to a unit on function transformations, this task might be simply routine. But consider the same task in the context of a unit on graphs of polynomial functions—viewing the graph of a polynomial function as the sum of the graphs of its terms, and recognizing the features of each term in the graph of the sum (Dugdale, Wagner, & Kibbey, 1992). In this context, it is natural for students to consider investigating each term individually, whether they decide this on their own or as a result of teacher suggestion. Figure 5.8 shows one possible solution sequence.

In Fig. 5.8, the requested function is built from the given function. The given polynomial function is graphed on the left-hand grid, adding one term at a time, and at each step a matching graph is constructed on the right-hand grid by altering the coefficient of the newest term. Thus, proceeding term by term, the requested function becomes evident. Comparing the coefficients of the two polynomials and making a table, it appears that a general solution can be formulated:

General Solution I. The desired function can be constructed from the given function by multiplying the coefficient of each term as indicated in Table 5.1.

Applying the transformation specified by Table 5.1 to produce the same effect on a few other polynomial functions, it seems that this may be, indeed, a general solution.

Allowing students to pursue a task of this sort can result in a variety of methods that can be compared and contrasted. Some students may be content with a conversion table as in General Solution I, and others may want to find a more general rule for the table of values.

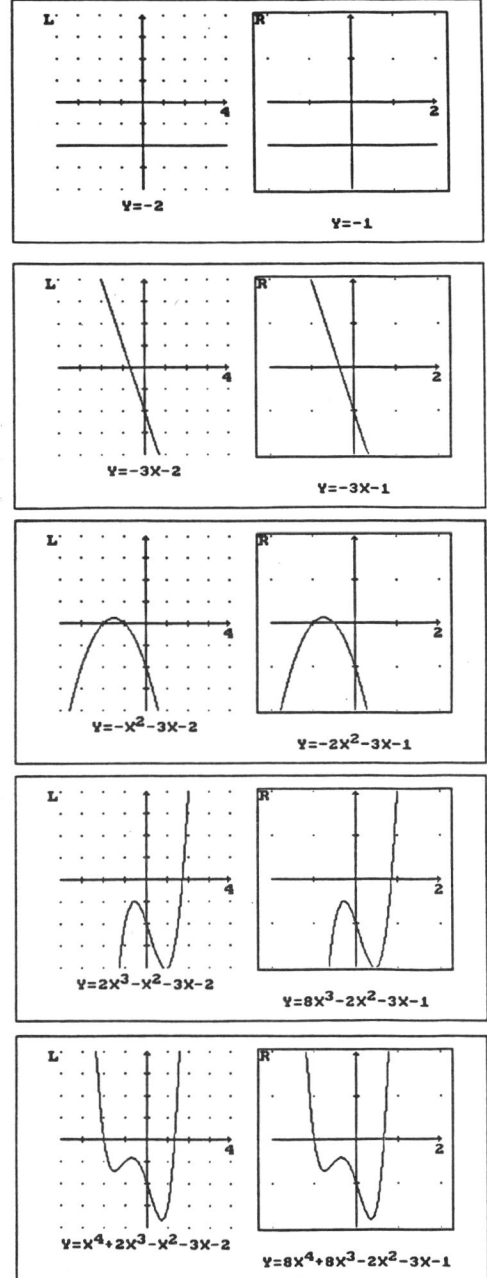

FIG. 5.8. One possible solution sequence to the task presented in Fig. 5.7. The given function is graphed on the left-hand grid, adding one term at a time, and at each step a matching graph is constructed on the right-hand grid by altering the coefficient of the newest term.

TABLE 5.1. A Solution to the Problem Posed in Fig. 5.7.

Term	Multiply Coefficient By:
x^4	8
x^3	4
x^2	2
x^1	1
x^0	0.5

A pattern relating the two columns in Table 5.1 above is not so readily apparent. This is one of those "Guessing Functions" games (Davis, 1967) that require some fiddling with (a technique that deserves more respect than the name seems to imply). Doubling the second column of Table 5.1 reveals a more recognizable pattern, as shown in Table 5.2: For each term x^n, the third column is 2^n. Hence, the second column is $2^n/2$, and the transformation can be stated more succinctly, as in General Solution II.

General Solution II. Replace each term ax^n of the polynomial with $(2^n/2)ax^n$, which simplifies to $a(2x)^n/2$. That is, replace x with $2x$ and divide each term by 2.

This is certainly more concise than General Solution I, but a further simplification produces:

General Solution III. Replace x with $2x$, and replace y with $2y$.

From here it is easy to make and test conjectures about the effect of replacing x with Ax and y with Ay for any value of A (or even more generally, replacing x with Ax and y with By for any values of A and B). Of course, how any such activity develops depends heavily upon the ideas of the participants. Students well versed in a transformations approach to functions (and with a healthy habit of relating current problems to techniques not just covered) may quickly recognize the problem as a "size change." This can provide a comparison for other students' solution efforts.

TABLE 5.2. Adding a Third Column Reveals a Pattern.

Term	Multiply Coefficient By:	Double Second Column
x^4	8	16
x^3	4	8
x^2	2	4
x^1	1	2
x^0	0.5	1
x^n	$2^n/2$	2^n

For example, this may be just the motivation needed to proceed from a tabular solution such as General Solution I, to find out how it relates to a more concise solution.

Although the problem could have been quickly dispensed with in the context of function transformations, approaching it in a less direct context provides a ready opportunity for experience with:

- Investigation of a nonstandard problem.
- Generalization of a solution by seeking patterns.
- Comparison of solutions and methods.

General Solutions I, II, and III, are all correct, and they produce equivalent transformations on polynomial functions. However, there are various characteristics to consider in choosing a solution method to remember and use. These might include economy of statement, efficiency of use, and generality—does the method work for functions other than polynomials? Experiences and issues such as these can be helpful in encouraging mathematical reasoning and in establishing appropriate ways of thinking about mathematics.

STUDENTS' MISCONCEPTIONS

With the proliferation of function-plotting tools, researchers have noted some common difficulties and misconceptions among students using these tools. As Goldenberg (1988) noted, "Our earliest experiments showed that students often made significant misinterpretations of what they saw in graphic representations of function. Left alone to experiment, they could induce rules that were misleading or downright wrong (p. 137)." Goldenberg described several visual illusions that can confuse students. For example:

- In viewing the graphs of linear functions, $y = ax + b$, varying b can appear to move the graph either horizontally or vertically, depending on the shape of the viewing window and the angle that the graph makes with that window.
- A parabola graphed on different scales in the same size window appears to change shape, and two congruent parabolas positioned at different heights in the same viewing window appear to have different shapes.
- Although it is apparent from the symbolic representation of a parabola that any value can be substituted for x, and hence the domain is unlimited, students' visual impression of the graph of a

parabola often leads them to reason as if the domain of the function is bounded by values within the viewing window.

These illusions may be easily ignored by anyone with sufficient algebraic sophistication, but beginners tend to find them misleading. Goldenberg suggested that students who experience some graphing-by-hand establish the connection between the analytical representation and the graphical representation of a function more effectively than students who experience only computer graphing. Further, he suggested that graphing software should mark one or more points on graphs and show the translations of the marked points along with translations of the graphs.

From a detailed study of the mathematical understandings of one student during her exploration of the graphs of simple algebraic functions over seven weeks, Schoenfeld, Smith, and Arcavi (in press) noted the instability of this student's notions of the slope and y intercept of a linear graph. We often observe such changing perceptions among students as they develop and refine new concepts. For example, during classroom testing of recently developed materials (Dugdale et al., 1992), a pair of students seemed to be progressing very well in determining equations for polynomial graphs. In constructing equations for the given graphs, these students were using a combination of techniques, sometimes starting with the leading term, and at other times building from the constant term up to the leading term. They were good at recognizing the effects of the various terms in a graph, and they had worked out on their own the idea that the constant term is the y intercept.

After successfully matching a variety of graphs of polynomials of degree 6 or less, this pair of students encountered the parabola shown in Fig. 5.9. Going first for the constant term, one of the students looked startled and asked, "What's going on here? Is it a decimal?" Although they had already solved several problems of this type, something had slipped. Through some examples of polynomials for which the y intercept was the visual "center" of the graph, their understanding that the constant term defines the y intercept had apparently shifted to the misconception that the constant term is the function value of the "center," which for a parabola is obviously the vertex. The vertex of the given graph had a y coordinate between -1 and -2. Because no previous examples had required noninteger coefficients, the students suspected a problem. Presumably, had the function value at the vertex been an integer, the students would have proceeded until they encountered some other apparent contradiction. Changing their focus to the leading term, the students easily found the quadratic term, then the linear term, and ended with the constant term (an integer, as expected). Having encountered an apparent discrepancy, the students chose an alternate solution method and successfully completed the task.

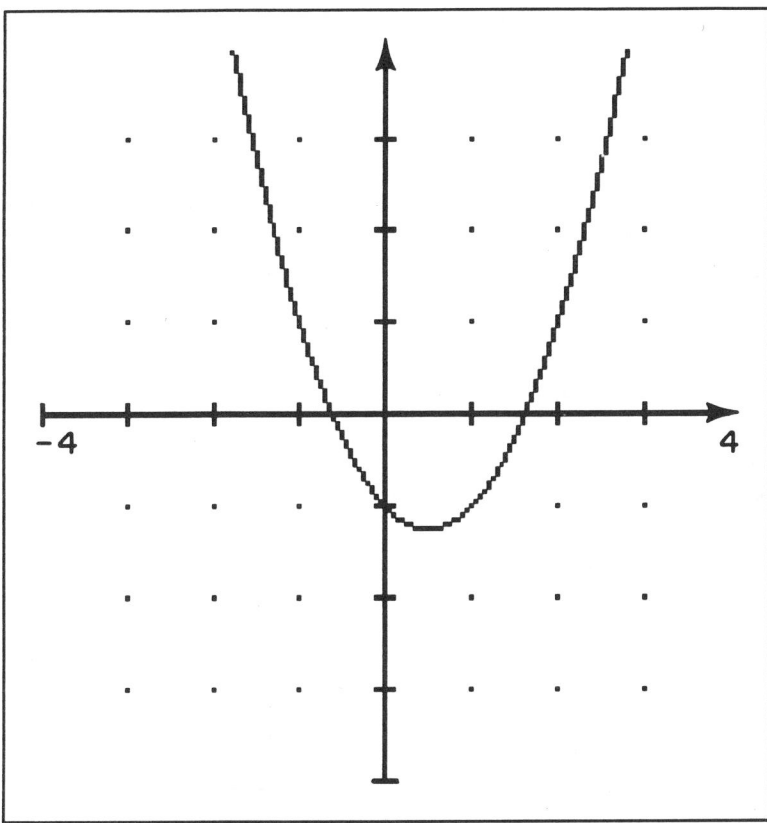

FIG. 5.9. By successive graphing of terms, students were asked to construct a polynomial function to match the given graph.

Educators should certainly be aware of likely misconceptions and attempt to minimize them. However, concepts are not static. Rather, they evolve in a context of growing and changing understandings. Hence, it is also important to equip students with a capacity for recognizing their own misconceptions, or drifting conceptions, and for learning how to recover from them. We need to develop instructional methods and learning tools that improve students' skills in using multiple approaches, doing parallel checking, and resolving apparent contradictions. In the example just given, the students took an alternate approach, but there was no overt indication that they returned to the unsuccessful approach and resolved the apparent contradiction before proceeding to the next problem. The disappointment in this case is not that the students made a common error, but that they evidently did not see a need to resolve the contradiction.

Learning is an iterative process. We need to be aware of likely misconceptions, but perhaps more importantly, we need to provide opportunities for students to recognize discrepancies and refine their thinking accordingly.

In the trigonometric identities study described earlier (Dugdale, 1989, 1990), for example, one of the graphical reasoning tasks was to use the graphs of two functions to predict the shape of a related function. Two of the most useful and easily noted features for this task are zeros and asymptotes, and in predicting the product of two functions, it is common for students to assume that:

- Where either function has a zero, the product has a zero.
- Where either function has an asymptote, the product has an asymptote.

These assumptions are useful and often true. However, the limitations of this reasoning can become apparent when other cases are encountered. Consider, for example, $\sin(x) \bullet \cot(x)$, where one factor is zero at the same x values where the other factor is undefined. Hence, one factor, $\sin(x)$, suggests zeros at the same x values where the other factor, $\cot(x)$ suggests asymptotes. Such discrepancies can introduce the necessity of refining mathematical reasoning appropriately as new cases are encountered. Although it is reasonable to suspect asymptotes where a function is undefined, additional checking near discontinuities is necessary to be sure. Carefully chosen cases, encountered at appropriate times, can provide opportunities for students to use mathematical reasoning and refine that reasoning through experience.

It is not uncommon to observe shifts in concepts as students reorganize their ideas to accommodate new information, apply previous ideas in different contexts, and establish interconnections. This seems a normal part of learning: a process of changing perceptions and evolving ideas. As Schoenfeld et al. (in press) commented, "Errors, bugs, misconceptions — call them what you will — are a natural and inevitable consequence of attempts to come to grips with complex domains." We need to regard changing perceptions as a normal part of learning and to develop in students a habit of recognizing and resolving apparent contradictions and refining mathematical reasoning through experience.

SUMMARY

This chapter has addressed three interrelated aspects of student thinking about functions and graphs:

- Students' conceptualizations of functional relationships in a global sense—interpreting qualitative features of graphs.
- Students' perceptions of functional relationships as facilitated by computer function-graphing software.
- Students' misconceptions of functional relationships as revealed through computer graphing activities.

Graphing of functions has long been a part of the study of mathematics. Traditional emphasis has been on procedures, such as computing function values and plotting points. Developing a more qualitative perspective of the graphs of functional relationships in general has been a relatively recent curricular concern. To justify this concern, there is ample evidence of students' difficulties in conceptualizing functional relationships and making qualitative interpretations of graphs. For example:

- Students who are able to correctly answer questions involving direct reading of values from a graph are often confused by questions that require consideration of changes over larger sections of a graph (Barclay, 1985).
- In choosing or constructing a graph to describe a given aspect of a familiar physical situation, students frequently favor a graph that resembles a "picture" of the event described, rather than a graph that adequately illustrates the behavior of the variables involved (Barclay, 1985; Dugdale, 1984; Mokros & Tinker, 1987).
- In interpreting multiple graphs that describe different aspects of the same sequence of events, students often lack a sense of the relationships among the graphs (Dugdale, 1986–1987).

Various recent efforts to improve students' understanding of graphs have emphasized the need to move beyond plotting and reading points to interpreting the global meaning of a graph and the functional relationship that it describes. To this end, the ideas of functions and graphs are developed and used in the context of familiar activities, with emphasis on relating graphical representations to students' everyday experiences. These efforts have involved students in manipulating familiar physical models and seeing the resulting changes in related graphs, examining graphs and describing (or acting out) events that could have produced the graphs, and constructing their own scenarios and corresponding graphs.

Such activities have shed light on students' perceptions of functional relationships as described by graphs and have improved students' abilities to interpret and use graphs. For example:

- Phillips (1986) recounted students' reasoning about the events evidenced by particular features of graphs, and also about the possible causes of those events.

- Mokros and Tinker (1987) reported a significant change in students' abilities to interpret and use graphs, with greatest gains on items where the mental image (the "picture" of the situation) and the appropriate graph were discrepant.
- Dugdale (1986–1987) described changes in students' graph interpretation strategies as students gained more experience interpreting simultaneous graphs that describe different aspects of the same sequence of events.

Other efforts have focused on students' understanding of graphs of algebraic functions. Perhaps the most dramatic and widespread influence on students' work with functions and graphs has been the recent proliferation of computer function-plotting tools. The easy manipulation of graphical representations allowed by current function-plotting tools has raised the possibility of visual representations of functions playing a more important role in mathematical reasoning, investigation, and argument. Relationships among functions can be readily observed, conjectures can be made and tested, and reasoning can be refined through graphical investigation.

Based on a study of students' use of visual and analytic thinking and their connections between the visual and analytic representations, Eisenberg and Dreyfus (1989) concluded that working with function transformations in a graphical context helped students understand the inherent connection between the algebraic-symbolic description of a function and its graphical representation. Further, the researchers found that students who viewed function transformations graphically performed better on nonstandard transformation questions than those who viewed them only algebraic-symbolically.

Although there is clear potential for graphing software to support and encourage mathematical reasoning, new instructional models are also necessary to foster a more thoughtful approach to functions and graphs. In a study comparing two approaches to incorporating graphical representations into a unit on trigonometric identities (Dugdale, 1989, 1990), one group of students was given a traditional algebraic-symbolic treatment of trigonometric identities, supplemented with related graphing activities, and the other group was involved in:

- Using graphical reasoning techniques.
- Experiencing active participation in the development of mathematical ideas.
- Building a qualitative perspective before formalizing procedures.
- Applying previous knowledge and skills to a current problem without being told what, in particular, to do.

In addition to showing superior posttest performance in relating functions to their graphical representations, students in the second treatment group exhibited more variety and personal involvement in their approaches to the standard content of proving identities.

Beyond developing proficiency in solving specific types of problems, we need to encourage mathematical reasoning and investigation and establish appropriate ways of thinking about mathematics.

Computer function-plotting tools offer unprecedented opportunities for exploration of functions and graphs. However, with the proliferation of function-plotting tools, researchers have noted some common difficulties and misconceptions among students using these tools. For example, Goldenberg (1988) described several visual illusions that can confuse students. He suggested that students who experience some graphing-by-hand may establish the connection between the analytical representation and the graphical representation of a function more effectively than students who experience only computer graphing. Schoenfeld et al. (in press) noted the instability of a student's notions of the slope and y intercept of a linear graph, as developed through computer graphing experiences. Classroom testing of materials addressing polynomial graphs (Dugdale et al., 1992) also revealed shifts in students' perceptions of the relationships between graphs and equations.

Educators should be aware of likely misconceptions and attempt to minimize them. However, it is not uncommon to observe shifts in concepts as students reorganize their ideas to accommodate new information, apply previous ideas in different contexts, and establish interconnections. This seems a normal part of learning: a process of changing perceptions and evolving ideas. Hence, it is also important to equip students with a capacity to recognize their own misconceptions, or drifting conceptions, and learn how to recover from them. We need to develop instructional methods and learning tools that improve students' skills in using multiple approaches, resolving apparent contradictions, and refining mathematical reasoning through experience.

REFERENCES

Barclay, W. L. (1985, November). Graphing misconceptions and possible remedies using microcomputer-based labs. *Proceedings of the National Educational Computer Conference* (NECC), Technical Education Research Center (TERC) (Tech. Rep. 85-5). Cambridge, MA: TERC. ERIC.

Brown, J. S., Collins, A., & Duguid, P. (1989). Situated cognition and the culture learning. *Educational Researcher, 18*(1), 16–25.

Davis, R. B. (1967). *Explorations in mathematics: A text for teachers.* Palo Alto, CA: Addison-Wesley.

Davis, R. B. (1989). The culture of mathematics and the culture of schools. *Journal of Mathematical Behavior, 8*(2), 143-160.

Dugdale, S. (1984). Some computer applications for the pre-college mathematics and science curriculum. *Technology in education and training: Planning and management* (pp. 163-174). Silver Spring, MD: Information Dynamics Inc.

Dugdale, S. (1986-1987). Pathfinder: A microcomputer experience in interpreting graphs. *Journal of Educational Technology Systems, 15*(3), 259-280.

Dugdale, S. (1989). Building a qualitative perspective before formalizing procedures: Graphical representations as a foundation for trigonometric identities. In C. Maher, G. Goldin, & R. Davis (Eds.), *Proceedings of the Eleventh Annual Meeting of the North American Chapter of the International Group for the Psychology of Mathematics Education* (pp. 249-255). New Brunswick, NJ: Rutgers-The State University.

Dugdale, S. (1990). Beyond the evident content goals, Part III—An undercurrent-enhanced approach to trigonometric identities. *Journal of Mathematical Behavior, 9*,(3) 233-287.

Dugdale, S., & Kibbey, D. (1983a). *Graphing equations* [Computer-based instructional package]. Iowa City, IA: Conduit.

Dugdale, S., & Kibbey, D. (1983b). *Interpreting graphs* [Computer-based instructional package]. Iowa City, IA: Conduit.

Dugdale, S., & Kibbey, D. (1986a). *Green globs and graphing equations* [Computer-based instructional package]. Pleasantville, NY: Sunburst Communications.

Dugdale, S., & Kibbey, D. (1986b). *Interpreting graphs* [Computer-based instructional package]. Pleasantville, NY: Sunburst Communications.

Dugdale, S., & Kibbey, D. (1987). *Pathfinder* [Computer-based instructional package]. Pleasantville, NY: Sunburst Communications.

Dugdale, S., Wagner, L. J., & Kibbey, D. (1992). Visualizing polynomial functions: New insights from an old method in a new medium. *Journal of Computers in Mathematics and Science Teaching, 11*(2), 123-141.

Eisenberg, T., & Dreyfus, T. (1989). *On visualizing function transformations* (Tech. Rep.). Beer Sheva, Israel: Ben Gurion University.

Goldenberg, E. P. (1988). Mathematics, metaphors, and human factors: Mathematical, technical, and pedagogical challenges in the educational use of graphical representation of functions. *Journal of Mathematical Behavior, 7*(2), 135-173.

Karplus, R. (1979, October-December). Continuous functions: Students' viewpoints. *European Journal of Science Education*, 397-415.

Karplus, R., Lawson, A. E., Wollman, W. T., Appel, M., Bernoff, R., Howe, A., Rusch, J. J., & Sullivan, F. (1977). *Science teaching and the development of reasoning*. Berkeley, CA: University of California.

Mokros, J. R., & Tinker, R. F. (1987). The impact of microcomputer-based labs on children's ability to interpret graphs. *Journal of Research in Science Teaching, 24*(4), 369-383.

Peters, P. C. (1982). Even honors students have conceptual difficulties with physics. *American Journal of Physics, 50*(6).

Phillips, R. J. (1986). Micro primer maths—There's more than meets the eye. *Mathematics in Schools, 15*(2), 29-32.

Phillips, R. J. (1988). *Mouse plotter.* [Computer-based software]. Nottingham, UK: Shell Centre for Mathematical Education.

Phillips, R. J., Burkhardt, H., & Swan, M. (1982). *Eureka* [Computer-based software]. UK: Crown.

Schoenfeld, A. H., Smith, J. P., & Arcavi, A. (in press). Learning: The microgenetic analysis of one student's evolving understanding of a complex subject matter domain. In R. Glaser (Ed.), *Advances in instructional psychology* (Vol. 4). Hillsdale, NJ: Lawrence Erlbaum Associates.

III TEACHER THINKING

6 Teachers' Thinking About Functions: Historical and Research Perspectives

Thomas J. Cooney
University of Georgia

Melvin R. Wilson
University of Michigan

> Research themes on teachers' thinking about mathematical functions are explored by considering the historical development of functions in school mathematics and research related to teachers' thinking, knowledge, and beliefs. The historical account traces the mathematical development of functions and related pedagogical issues. Research on teachers' thinking, knowledge, and beliefs and theoretical models of teachers' beliefs are considered as they relate to developing these research themes. These themes suggest that future research should be sensitive to the interplay between teachers' thinking specific to functions and the contextual nature of their thoughts, decisions, and actions while teaching.

Research in the field of mathematics education has not focused to any great extent on teachers' thought processes and has only recently begun to consider the beliefs mathematics teachers hold about mathematics, the teaching of mathematics, and how students learn mathematics. Nevertheless, increased attention is now being given to the importance of considering the teacher's role in the classroom and to research related to understanding various aspects of that role. Presently, much of that research is descriptive and focuses on the meanings teachers ascribe to classroom events and, in some cases, to their own learning as well (Brown, Cooney, & Jones, 1990; Thompson, 1992). With the exception of some work on addition and subtraction (Carpenter, Fennema, Peterson, & Carey, 1988) and some on division and rational numbers (Ball, 1990, 1991), research on teachers' thinking about specific content areas is virtually nonexistent.

If the pattern of research on teaching in general or mathematics teaching in particular over the past 30 years is reviewed, it becomes obvious that research has moved in the direction of trying to understand the complexities of classroom life and the cognitive processes of the participants, that is, of teachers and students. As noted by Clark and Peterson (1986), the second *Handbook of Research on Teaching* focused on teachers' thought processes for the first time in the 20-year history of the handbooks. Similarly, researchers are now focusing on the beliefs of mathematics teachers with an intensity and vigor that was unforeseen 20 years ago (Cooney, Grouws, & Jones, 1988). Thus, it might well be that subsequent research in the 1990s on mathematics teachers' thinking and beliefs will continue this trend but also extend it by concentrating the research on specific content areas such as functions.

One of the difficulties associated with conducting research on teachers' thinking about functions is determining what domain(s) of research can provide a foundation for conceptualizing the research. The study of teachers' thought processes, knowledge, and beliefs can provide at least a partial basis for developing a research framework. Given the specificity of functions, it seems reasonable, however, that the framework should include an analysis of the content domain as well. In keeping with this perspective, we have considered historical perspectives of functions as defined by mathematicians, as presented in school textbooks, and as discussed relative to pedagogical concerns. In the absence of research on teachers' thinking about functions, our best guess as to what teachers think about this concept is based on how the mathematics and mathematics education communities discuss the role of functions in school mathematics and on how functions are presented in school textbooks. The way in which teachers teach functions is undoubtedly related to what they know about functions and what they value and consider important for their students to know, not only about functions but also about mathematics in general. Hence, investigation of the relationship between functions in school mathematics and what teachers know and believe about functions and about mathematics is relevant to the development of a research agenda on teachers' thinking about functions.

THE FUNCTION CONCEPT IN SCHOOL MATHEMATICS

The most significant evolution in the mathematical definition of *function* occurred before the 20th century. This pre-20th century evolution involved what Kleiner (1989) referred to as "a tug of war between two elements, two mental images: the geometric (expressed in the form of a curve) and the algebraic (expressed as a formula)" (p. 282). In describing the continued but less dramatic evolution of the definition of *function* during the 20th century, Kleiner (1989) identified a third element, namely, the "logical" definition of function as a correspondence (with the mental image of an input-output machine). In the wake of this development the geometric conception of function is gradually abandoned. A new tug of war ensues (and is, in one form or another, still with us today) between this novel "logical" ("abstract," "synthetic," "postulational") conception of function and the old "algebraic" ("concrete," "analytic," "constructive") conception. (p. 282)

The notion of a function first as a correspondence between variables and then as a mapping between arbitrary sets, rather than as a geometric curve or an algebraic formula, became dominant in the mathematics of the 20th century.

Mathematical discussions about the function concept have produced a gradual evolution in its accepted meaning. Similarly, this century has seen an evolution in pedagogical philosophy regarding the function concept. This pedagogical evolution has included recommendations for changes in emphasis regarding the importance of the function concept to school mathematics as well as changes in how the concept should be taught. The evolution has paralleled, to some extent, the changes in the mathematical definition of function during the 20th century, but has included other themes as well. In this section, we summarize some of the discussions about the function concept and the way it was taught during the course of this recent revolution.

The Mathematical Development of the Function Concept

Following a brief overview of the mathematical development of the function concept, particularly as that development relates to school mathematics, we examine the similarity between the mathematical development of functions and discussions on teaching functions in school mathematics.

The Early Development of the Function Concept. Function-like activities can be traced all the way back to the mathematics of ancient civilizations. For example, around 2000 BC, ancient Babylonians developed tables for finding reciprocals, squares, square roots, cubes, and cube roots (Kline, 1972). During the Middle Ages mathematicians expressed general notions of dependence between varying quantities by using geometric terms or verbal descriptions (Youschkevitch, 1976). While these and other contributions foreshadowed the modern development of function, it was not until the late 16th century that functions were studied as objects in their own right.

The modern function concept did not appear in mathematics by chance.

During the late 16th and early 17th centuries, as scientists such as Galileo and Kepler studied physical problems associated with motion, mathematicians searched for tools to describe and model observed phenomena. The algebraic notation developed by Viéte and Descartes also contributed significantly to the modern development of function as their work provided a concise and powerful way of representing mathematical ideas (Kline, 1972; Malik, 1980; Youschkevitch, 1976).

The birth of calculus is often thought to parallel the beginning of the modern concept of function. Newton wrote about *quantitas correlata* and *quantitas relata* in referring to independent and dependent variables, respectively (Youschkevitch, 1976). Leibniz first used the term function in 1673 in reference to geometric quantities such as subtangents and subnormals of a curve (Youschkevitch, 1976). Between 1694 and 1698 the term was adopted by Leibniz and Bernoulli, as evidenced by their correspondence, when they referred to dependence relations as defined by algebraic expressions. Mathematicians during the 18th century defined functions in terms of expressions. For example, Bernoulli in 1718 defined "function of a certain variable" as "a quantity that is composed in some way from that variable and constants" (Youschkevitch, 1976, p. 60). Euler, a student of Bernoulli, later replaced the term "quantity" with "analytical expression" (Youschkevitch, 1976, p. 61).

During the 19th century the accepted meaning of function evolved to include functions that were not necessarily continuous, differentiable, or defined by analytical expressions. Fourier's work on heat conduction and the ensuing debate concerning his writings spurred this evolution. Dirichlet was another prominent mathematician involved in expanding the idea of function to include arbitrary correspondences in addition to those defined by analytical expressions. We have, for instance, Dirichlet's famous "salt and pepper" function, which pairs the rational numbers with 0 and the irrational numbers with 1 (Kleiner, 1989; Youschkevitch, 1976).

Functions as Correspondences. During the first part of the 20th century most mathematicians thought of functions as dependence relations or correspondences between variables (Kleiner, 1989). A typical definition of function early in the century was,

> If for each value of a variable x there is determined a definite value or set of values of another variable y, then y is called a function of x for those values of x. (Townsend, 1915)

According to this definition, relations pairing values of the second variable with one or several values of the first variable were considered functions. Emphasis was placed on the dependence of the second variable on the first, as evidenced by the use of the terms dependent variable and *independent variable* to describe functions. Although the work of Dirichlet during the middle of the 19th century illustrated the idea that function could include arbitrary correspondences, many prominent mathematicians in the early 1900s (including Lebesgue, Baire, and Borel) supported the requirement of a definite "law" of correspondence in the definition of functions (Kleiner, 1989).

The acceptance of a definition of function that included arbitrary correspondences was gradual. Hamley (1934) referred to functions exclusively in terms of correspondence, but he further noted that the correspondence need not involve numbers but could also involve relationships between other "objects" that vary. He quoted Russell, who stated that "[functions] can be extended to all cases of one-many relations, and the 'father of x' is just as legitimately a function of which x is the argument as is 'the logarithm of x'" (cited in Hamley, 1934, p. 28).

Functions as Sets of Ordered Pairs. The accepted mathematical definition of function during the middle of the century did not differ significantly in substance from that of the early 1900s. That is, for the most part, what most mathematicians considered to be a function in 1900 was still considered a function in 1950. However, the definition was refined somewhat to incorporate an increased emphasis on the concept of set, the gradual acceptance of functions as arbitrary correspondences, and the requirement that each value of the independent variable has a unique image.

The concept of set became a fundamental concept of mathematics and, in fact, was considered by some mathematicians to be more basic than that of number (May & Van Engen, 1959). Whereas discussions of functions had previously been centered on dependence and correspondence, with emphasis on the associated formulas and graphs, later discussions referred to functions as sets of ordered pairs. In 1939, Bourbaki proposed the following definition of function:

> Let E and F be two sets, which may or may not be distinct. A relation between a variable element x of E and a variable element y of F is called a functional relation in y if, for all x in E, there exists a unique y in F which is in the given relation with x.
>
> We give the name of function to the operation which in this way associates with every element x in E the element y in F which is in the given relation with x; y is said to be the value of the function at the element x, and the function is said to be determined by the given functional relation. Two equivalent functional relations determine the same function. (Kleiner, 1989, p. 299)

May and Van Engen (1959) claimed that "contemporary" definitions of relation and function represented a "fundamental change" in the way people were thinking about functions and that defining relation as a set of ordered pairs gave a previously ambiguous term a precise mathematical meaning (p. 65). Similarly, they claimed that the notion of functions as sets of ordered pairs for which certain conditions were satisfied gave new generality and precision to the meaning of function. The authors further argued that although a table, graph, rule, or verbal description might describe a function, it was not logical to consider any of these as being the function. Rather, it was more precise to refer to the function as the set of ordered pairs defined by the table, graph, rule, or description. For example, the function defined by the formula $y = 3x$ would be more precisely referred to as the set of ordered pairs of the form (x, y) such that $y = 3x$, or using abbreviated mathematical notation, as $\{(x, y)| y = 3x\}$.

Because discussions about functions (and relations) centered on sets, it was natural to refer to the set of all first elements of the function (domain) and to the set of all second elements (range). Graphs of relations were defined in terms of Cartesian products of sets. For example, A × B is the set of all pairs that can be formed by selecting a first component from A and a second component from B. The new ordered-pair definitions of function and relation made the definitions of inverse relation and function more precise, as well as those of composition, union, and intersection.

Recent Conceptions of Function. Discussions about how mathematicians should define functions have not changed significantly since the middle of this century. The definition proposed by Bourbaki is still the most generally accepted and used definition in mathematics today. Nevertheless, the issue has not been completely settled. Some recent discussions among mathematicians have attempted to replace set theory with category theory as a foundation for mathematics. In these discussions, the concept of function has played a significant role. Kleiner (1989) stated:

> [Category theory] describes a function as an "association" from an "object" A to another "object" B. The "objects" A and B need not have any elements (that is, they need not be sets in the usual sense). In fact, the arguments A and B can be entirely dispensed with. A category can then be defined as consisting of functions (or maps), *which are taken as undefined (primitive) concepts* satisfying certain relations or axioms. (p. 299)

The notion that categories are more basic to mathematics than sets, and that categories can be defined in terms of functions, has not yet gained wide acceptance among mathematicians. It does suggest, however, that the evolution of the mathematical definition of function is continuing and dynamic.

Perspectives on the Teaching of Functions

Many mathematics educators during the early 20th century believed there was a need for greater emphasis on functional thinking in school mathematics. In 1904, the German mathematician Klein (cited in Hamley, 1934, p. 52) referred to functions as the "soul" of mathematics and opined that "an elementary treatment of the function . . . ought to be in the regular course of all types of high schools." His message influenced mathematics education in the United States at least at the policy level. In 1921, in an effort to unify the secondary mathematics curriculum, the National Committee on Mathematical Requirements of the Mathematical Association of America recommended that functional thinking be the unifying principle of secondary mathematics (Hedrick, 1922).

Recommendations during the 1920s suggested that functional thinking should be emphasized in every area of secondary mathematics. Breslich (1928) claimed that since most of mathematics deals with relationships between quantities, "without functional thinking there can be no real understanding and appreciation of mathematics" (p. 42). He suggested that equations, polynomials, ratio, proportion and variation, relationships stated in words, relationships in tabular representations of numerical facts (including tables used in graphic representation), and relationships represented by formulas all provided opportunities for emphasis in functional thinking. He recommended that the use of graphs could be an excellent tool in helping students understand relationships; he maintained that students should be taught to think graphically about numerical facts. Breslich also contended that functional relationships abound in geometry. For example, he pointed out that

> [T]he pupil learns in geometry that the area of an equilateral triangle is found by means of the formula $A = (a^2/4)(\sqrt{3})$ but remains ignorant of the fact that the area varies directly as the square of the side. (p. 50)

Hedrick (1922) emphasized the same point when he stated that "Functional relations — that is, relations between quantities — will occur on every page of every book on mathematics unless we suppress them. We have been suppressing them" (p. 195).

Another emphasis that gave impetus to the importance of functions in the early part of the 20th century was the notion that functions were prominent in the real world. Hedrick (1922) noted that

> the reason for insisting so strongly upon attention to the idea of relationships between quantities is that such relationships do occur in real life in connection

with practically all the quantities with which we are called upon to deal in practice. . . . [T]here can be no doubt at all of the value to all persons of any increase in their ability to see and to foresee the manner in which related quantities affect each other. (p. 165)

Because of the prevalence of functions in the real world, real-world situations were considered to provide valuable opportunities for students to gain understanding in functional thinking. Schorling (1936) stated that "there is nothing in [the function] concept . . . which prevents the presentation of specific concrete examples and illustrations of dependence even in the early parts of the course" (p. 6).

A major issue in discussions of pedagogy around the middle of the century dealt with the importance of emphasizing mathematical structure in teaching school mathematics. The fundamental concept in school mathematics became that of set. Function was still considered to be a unifying mathematical concept, but the new emphasis on mathematical structure brought radical changes involving the teaching of functions. Recommendations made during the 1950s placed a pedagogical emphasis in school mathematics on the definition of function as a set of ordered pairs, as opposed to previous definitions involving correspondences, rules, or graphs. May and Van Engen (1959) argued that since previous definitions of function were "vague and did not satisfy the requirements for precise statements demanded by the mathematical world," these definitions of function would not "satisfy the requirements of good teaching" either (p. 110). They continued,

> In contrast, the definition based on set considerations is precise and clear. A function or relation is a set of ordered pairs. This is a definite entity; one you can almost put your hands on. This being the case, it would seem logical that it be considered as the basis for instruction in elementary mathematics. (p. 110)

Lovell (1971) made a similar argument in support of teaching functions from an ordered pair perspective when he wrote,

> There is insufficient evidence to suggest good ways of introducing functions that involve proportion, but it would seem inadvisable to study such functions until pupils have considerable experience with the general case outlined by the definition. (p. 21)

However, there was not total agreement upon the merits of emphasizing functions as sets of ordered pairs. Willoughby (1967) stated:

> The ordered-pair definition of function is correct and convenient to use; however, it has serious defects from a pedagogical point of view. The ordered-pair idea gives a static impression to the pupil, where a dynamic impression is far more appropriate. Even though it may not be as elegant, or as formally simple, a dynamic impression of a function will be far more appealing to children, and will put them in a much better position to use their knowledge about functions. (p. 226)

Buck (1970) had similar reservations. He stated that "experience seems to show us that the 'a function is a class of ordered pairs' approach is one that imposes severe limitations upon the student and provides a poor preparation for any further work with functions" (p. 255).

Another change in emphasis around the middle of the century regarding the teaching of functions related to its placement and emphasis in the curriculum. Curricular recommendations of the 1950s, as reflected by the Commission on Mathematics of the College Entrance Examination Board (CEEB), suggested that functions be treated as a topic worthy of study in its own right and that a separate functions course replace the traditional advanced algebra course with the intent of providing a unified treatment of the elementary functions: polynomial functions, rational functions, logarithmic functions, exponential functions, and trigonometric functions (College Entrance Examination Board [CEEB], 1959).

Implementations of Functions in School Mathematics

Although the broader mathematical community was emphasizing the importance of functions at the turn of the century, we question whether that emphasis reached the secondary school level based on our analysis of two algebra texts—although we hasten to add that we do not know whether the texts were representative of other texts in use. One of the texts (Marsh, 1905) made no mention of functions. The other text (Hawkes, Luby, & Touton, 1909) provided the following definition of function:

> An algebraic expression involving one or more letters is a *function* of the letter or letters involved.
>
> Thus $2x + 3$ and $x^2 + 5x - 6$ are functions of one letter, x; $x^2 - 2xy + y^2$ and $x^3 + y^3$ are functions of two letters, x and y. The letters of a function are usually referred to as *variables*. (p. 259)

While the authors provided graphical representations of linear and quadratic functions, the primary emphasis was on solving equations graphically. There were no references to correspondence or to the notion of relationships between dependent and independent variables. Further, no references were made to any linkage between the concept of function and real-world phenomena.

In 1921, the National Committee on Mathematical Requirements of the Mathematical Association of America recommended that the study of functions should be given a central focus in secondary school mathematics. The effect of the recommendation appears to be mixed. Three secondary mathematics methods textbooks (Butler & Wren, 1941; Schorling, 1936; Young, 1927) encouraged teachers to emphasize functions as the unifying principle of school mathematics. Osborne and Crosswhite (1970) found two textbooks published in the 1930s that attempted to unify a course around functions (one by Swenson and one by Betz). However, Breslich (1928) analyzed four secondary textbooks in geometry and algebra and concluded that "[these texts] showed practical disregard of opportunities for training in functions" (p. 43). We analyzed nine textbooks published between 1923 and 1950 and found only one that attempted to make functions the unifying principle of the course (Mallory & Fehr, 1940), although we are aware that other texts provided some increased attention to functions.

Hamley (1934) criticized textbook writers for misinterpreting recommendations and thinking that "the function concept was synonymous with the graphical representation of functions" (p. 79). Breslich (1928) pointed out that many teachers were skeptical about teaching the function concept in secondary-school mathematics. He opined that:

> Some [teachers] insist that [functions] should be saved for higher courses. Other teachers agree as to the importance of functional thinking but misunderstand the method of attaining it. It is not advisable to teach functions as a topic of algebra or geometry. As is the case with all new concepts, the idea of relationship is best developed when it grows out of a child's concrete experiences with numerical relationships, beginning in simple forms as early as possible. (p. 54)

Rosskopf (1970) speculated that the reason for the failure of teachers to implement these recommendations was that they may have been preoccupied with more immediate instructional problems resulting from new compulsory attendance laws.

Nevertheless, we did find evidence that the recommendations for increased attention to functions had an impact on school textbooks. For example, Wells and Hart (1929) introduced what they called "functional relationship" by using a sequence of examples such as the following: "The wages earned by a workman who is being paid 75¢ per hour depends upon the number of hours he works" (p. 59). The authors define *functionally related quantities* as those that exhibit the following characteristics: "Always there is one number or quantity so related to one or more others that any change in the latter causes a change in the former" (p. 59). Subsequent discussions and exercises focused on students noting and writing relationships between the area of a triangle and its base and height. Effects on graphs of linear equations in which the parameters were changed were also presented. In their Preface, Wells and Hart made the following statement:

> Attention is called to the chapter on *Functional Relationships*. The desire to place in the hands of teachers and pupils a satisfactory treatment of this subject, *which has come to be stressed in recent years* [italics added], was one of the chief reasons for writing this new text. The treatment will be found simple and adequate without being verbose or extended unnecessarily.

Apparently the authors took seriously those discussions and recommendations that emphasized the teaching of functions.

A similar emphasis on function was given by Betz (1931). Betz emphasized dependence early in his algebra text ("Whenever two variables are so related that a change in the value of one causes a change in the other, we say that one depends upon the other," p. 26) and later provided the following, more formal, definition of function:

> If two variables, such as x and y, are so related that to each value of x (the independent variable) there corresponds a definite value or set of values of y (the dependent variable), y is called a function of x. (p. 26)

Like Wells and Hart (1929), Betz also emphasized graphical representations and examined the impact on the graphs of changing the parameters that defined the functions.

A review of two Algebra I texts published 20 years later (Schorling, Smith, & Clark, 1949; Welchons & Krickenberger, 1949a) revealed little emphasis on functions. A companion Algebra II text (Welchons & Krickenberger, 1949b) provided a modicum of emphasis on function. The authors offered the following definition: "If two variables are so related that for any value of one there is a value (or values) of the other, then the second variable is a function of the first variable" (p. 157). Subsequent exercises involved evaluation exercises [e.g., find $f(2)$ if $f(x) = x^3$] or "functional change" exercises in which students were asked to consider effects such as that on the area of a triangle if the height is cubed. Exercises involving graphs were also presented. The primary emphasis of the exercises appeared to be skill oriented, however.

In contrast to the uneven or, perhaps, even sparse implementation of recommendations regarding the teaching of functions during the first half of the century, the recommendations of the 1950s that emphasized mathematical structure and defined functions in terms of sets of ordered pairs were implemented quite extensively. A survey of 35 elementary algebra and college algebra textbooks by Kennedy and Ragan (1969) found that the textbooks before 1959 used definitions for function that involved rules or correspondences between variables, whereas most of the later textbooks (after 1959) used a definition involving sets of ordered pairs. Our analysis of 16 high-school textbooks published between 1958 and 1986 indicated that functions were consistently defined in terms of sets, either as sets of ordered pairs or as correspondences between elements of two sets.

The CEEB (1959) recommendation that functions be given increased importance in secondary school mathematics, including its recommendation for a separate course on functions, seemed to generate considerable activity with respect to functions. Between 1960 and 1967, the School Mathematics Study Group (SMSG) produced a series of secondary textbooks designed to implement the 1959 CEEB recommendations, including a textbook designed specifically for a high-school functions course (School Mathematics Study Group [SMSG], 1965). Three of the six textbooks we analyzed that were published during the 1960s referred to CEEB or SMSG recommendations when justifying their structural treatment of functions. All six precalculus textbooks we analyzed that were published between 1958 and 1971 were centered on the function concept. Four of these had titles that included the word function.

Each of the six intermediate algebra texts we analyzed that were published between 1963 and 1972 contained a separate chapter on functions. It is interesting to note that in most of these texts, functions were introduced after the treatment of linear and quadratic equations and their graphs, thus suggesting that function was not cast as a unifying concept — at least not until precalculus mathematics.

Texts developed during the 1960s emphasized the notion of sets and concomitantly the notion of domains and ranges of functions. In the popular Algebra I series by Dolciani, Berman, and Freilich (1962), a function was introduced as a subset of a set constituting a relation: "Thus a *function* is a relation which assigns to each element of the domain one and only one element of the range" (pp. 438-439). Subsequent exercises placed a heavy emphasis on graphs of relations, determining whether the relations were functions, and identifying the domain and range of given relations. The companion Algebra II text (Dolciani, Berman, & Wooton, 1963) provided a similar definition of function and introduced the following notation for a function:

$$f: f = \{(x, y): y = 2x + 1\}.$$

Clearly these texts reflected the emphasis on sets in the mathematical community and the argument that functions can be more precisely presented using sets. The texts also provided extensive visual representation of relations and functions through the use of graphs.

In some texts functions were presented as mappings. The experimental course *Elementary Functions*, developed by SMSG (1965), provided the following definition early in the text:

> Definition 1-1: If with each element of a set A there is associated in some way exactly one element of a set B, then this association is called a *function from A to B*. (p. 2)

Functions were discussed in terms of mapping elements in the domain to elements in the range.

Subsequent texts continued the emphasis on sets in defining function. For example, the Algebra I text by Henderson, Pingry, and Klinger (1968a) defined a function as "A set of ordered pairs which has no members having the same first component and different second components is called a *function*, or a *mapping*." While the authors provided virtually no graphical interpretations, there was a heavy emphasis on sets of ordered pairs and on determining which sets determine a function. The companion Algebra II text (Henderson, Pingry, & Klinger, 1968b) continued this theme but from a more sophisticated perspective. The authors provided the following definition.

> Definition: For every (x, y) and $(u, v) \in A \times B$, if when $x = u$, it follows that $y = v$, then $A \times B$ is called a *function* (p. 163).

Again, there were minimal references to graphs but the notion of an inverse of a function was introduced. Numerous other functions were also discussed.

Dolciani, Wooton, Beckenbach, and Sharron (1983) defined a function as "a set of ordered pairs in which each first component is paired with exactly one second component" (p. 67). Arrow diagrams were used to develop a conceptual model of the pairing between the sets of first and second components. Graphs of functions received little attention. Jacobs (1979) defined a function as a "pairing of two sets of numbers so that to each number in the first set there corresponds exactly one number in the second set" (p. 78). References were made to real-world settings; graphing functions was a central theme.

Foerster (1984a) also placed a heavy emphasis on graphs as a way of introducing functions, which he defined as a "set of ordered pairs (x, y) for which there is never more than one value of y for any one given value of x" (p. 568). Foerster's (1984b) Algebra II text defined a function as a "relation in which there is exactly one value of the dependent variable for each value of the independent variable in the domain" (p. 35). Graphs play a major part in the development, but the orientation toward sets is minimized. A similar development can be seen in an Algebra II text by Smith, Charles, Dossey, Keedy, and Bittinger (1990). An analysis of these recent texts reveals that the treatment of function is still quite formal, although slightly less so than that in the 1960s texts. The more recent texts also place a greater emphasis on graphical representations of functions and on interpreting real-world phenomena using functions.

Presently, discussions about the teaching of functions have incorporated many of the past recommendations. The emphasis on functions as a unifying mathematical concept, as a representation of real-world phenomena, and as an important mathematical structure remains central to contemporary discussions. The following statements from the *Curriculum and Evaluation Standards for School Mathematics* (National Council of Teachers of Mathematics [NCTM], 1989) illustrate this emphasis.

> The concept of function is an important unifying idea in mathematics. Functions, which are special correspondences between the elements of two sets, are common throughout the curriculum. (p. 154)
>
> [T]he study of functions should begin with a sampling of those that exist in the students' world. Students should have the opportunity to appreciate the pervasiveness of functions through such activities as describing real world relationships that can be depicted by graphs. (p. 154)

While the emphasis on sets in defining functions provides a certain clarity and precision, many mathematics educators question whether this clarity and precision enables students to develop a better understanding of functions. It may well be that the dependence conception is less abstract and more dynamic and consequently easier for students to understand initially than the more abstract notion of functions as sets of ordered pairs. Vinner (1983) argued that "because of the well known difficulties that students have with the Dirichlet-Bourbaki definition it is better to avoid it in all courses preceding analysis, topology and algebra at the university level" (p. 305). Thorpe (1989) reinforced this perspective by pointing out that the emphasis of the 1950s and 1960s on the logical definition of function was one of our "great errors" of that era (p. 13). He recommended that functions be referred to as rules, or machines, but "certainly not as sets of ordered pairs!" (p. 13). Similarly, the *Standards* (NCTM, 1989) recommend that

> [To] establish a strong conceptual foundation before the formal notation and language of functions are presented, students in grades 9-12 should continue the informal investigation of functions that they started in grades 5-8. Later, concepts such as domain and range can be formulated and the $f(x)$ notation

can be introduced, but care should be taken to treat these as *natural extensions to the initial informal experiences* (italics added). (p. 154)

Although it is difficult to predict just what form the emphasis on functions will take in the future, it seems clear that functions will continue to play a prominent role in school textbooks. Presently, practical and graphical contexts seem to be more in vogue than logical and structural contexts. In a recent text, Demana and Waits (1990) placed a heavy emphasis on graphing functions as they defined *function* in the following way.

> A *function* is a relation with the property: If (a, b) and (a, c) belong to the relation, then $b = c$. The set of all first entries of the ordered pairs is called the *domain of the function*, and the set of all second entries is called the *range of the function*. (p. 18)

Although the authors use set notation to define functions and relations, their approach, in fact, relies much more heavily on graphical representations than on the value of a set-oriented approach. Much of the graphing employs the use of graphing calculators. This suggests that technology may be yet another factor to consider in developing an appropriate perspective for treating functions in secondary school mathematics.

The Function Concept and Teachers' Thinking

Several themes involving mathematical and pedagogical discussions of functions and the representations of functions in school textbooks have emerged and re-emerged during this century. Reports such as those of the National Committee on Mathematical Requirements (Hedrick, 1922), the CEEB (1959), and the NCTM (1989) have helped determine and shape these themes, as have recent technological developments involving the personal computer and graphing calculator. Themes prominent during the first half of the century included thinking about function as a unifying mathematical concept and as a way to describe real-world phenomena. These themes were consistent with early definitions of function that centered on the ideas of dependence and correspondence between variables. The extent to which these themes impacted on the teaching of functions in school mathematics is not completely clear, however.

During the 1950s and 1960s, the theme of function as a unifying concept continued but the emphasis shifted toward thinking about function in the context of a precise mathematical structure. Clearly this theme had impact on school mathematics, at least in terms of the presentations of functions in school textbooks. The impact that it had on teachers' thinking and on the actual teaching and learning of functions is not clear. These basic themes continue today as new emphases are placed on the way in which students' conceptions of functions develop and on the various interpretations teachers and students have of functions represented as graphs, tables, rules, or sets of ordered pairs.

Much of the recent literature (NCTM, 1989; Sfard, 1989; Thorpe, 1989; Vinner, 1983; Vinner & Dreyfus, 1989) stressed that the less abstract notion of functions as rules allows students to gain a strong conceptual background in functional thinking before progressing to the more abstract and general notion of functions as sets of ordered pairs. That is, the historical development of functions, first as dependence relations describing real-world phenomena, then as algebraic expressions, then as arbitrary correspondences, and finally as sets of ordered pairs, may be the most appropriate pedagogical development as well. Yet this approach to the teaching of functions seems inconsistent with the set-theoretic approach used in many of today's textbooks. Whether the issue of consistency is one that resides in teachers' thinking about the teaching of functions is, at this point, purely speculative.

Another recent development that potentially impacts on the teaching of functions is the ever-changing growth of technological revolution. Robert Davis's (1967) analysis of the impact of technology on mathematics education in *The Changing Curriculum* may have been off the mark in terms of its timeline, but not in its intent, as the impending revolution that was once a fantasy is now at the doorstep of reality. Calculators and computers provide teachers and students with new tools to assist them in understanding relationships between quantities through the use of graphical, tabular, and algebraic representations. Consequently, technology provides a vehicle for students to develop a deeper understanding of functions by permitting a focus on representation without tedious calculations (Barrett & Groebel, 1990; Fey, 1989). Of particular interest is the emphasis on the graphical representations of functions. If teachers believe that graphical representations are fundamental to the teaching of functions, then technology may be viewed as an indispensable tool. On the other hand, if teachers consider graphical representations as interesting but not necessarily a central consideration, then the use of technology may be viewed as a secondary consideration.

Similarly, many questions can be posed regarding teachers' thinking about functions. Do teachers tend to think of functions as graphs, rules, algebraic expressions, correspondences, or sets of ordered pairs, and how do these conceptions tend to influence their teaching of functions? In what way do teachers think of functions as a unifying concept within mathematics, or as a connection between mathematics and the real world? What types of representations come to mind when they think about functions? In what way do they see technology as being relevant for teaching functions? These are but a few of the questions we see emerging in a research agenda for teachers' thinking about functions. In the next section we consider research that can serve as a potential basis for studying these and other questions.

RESEARCH THEMES FOR TEACHERS' THINKING ABOUT FUNCTIONS

Ms. Maxwell: What do we mean, class, by linear function? How would we define it, Jake?

Jake: I don't know, I forgot.

Ms. Maxwell: Elsie?

Elsie: Well, it has something to do with a straight line.

Ms. Maxwell: That's true. But we need more. Todd?

Todd: Things like $f(x) = 2x + 3$ and $f(x) = 4x - 10$. These are linear functions, aren't they?

Ms. Maxwell: Yes. That's good. Now let's see if we can graph some linear functions. Look at this problem. (Cooney, 1980, pp. 458–459)

Had the observer of the above classroom interchange been able to probe Ms. Maxwell's thinking about functions, questions such as the following could have been explored:

Was she aware of the difference between the cognitive demand placed on students in producing a definition versus an example?
What cues, if any, did she consider when she accepted an example instead of a definition?
Was she aware of a range of representations that might have been used to assess students' understanding of functions?
To what extent was her mathematical understanding of functions an influencing factor in her decision?
To what extent were classroom circumstances an influencing factor in her decision?

Responses to these and other questions can provide insight as to possible linkages between teachers' thinking about functions and their teaching of functions. We explore the implications of such linkages in this section. We begin by considering research related to teachers' thinking.

Research on Teachers' Thinking

Clark and Peterson (1986) provided an extensive review of the literature on teachers' thinking. However, the implications of that review for teachers' thinking on functions specifically is not obvious, primarily because of two factors. First, most of the research involved experienced elementary teachers—teachers for whom functions would, in all likelihood, receive scant attention. Second, most of the studies did not involve mathematics, or at least not significant mathematics. Nevertheless, research on teachers' thinking can be used at least as a heuristic and a guide for generating a reasonable research agenda on teachers' thinking about functions. It is from this perspective that we proceed.

As indicated by Clark and Peterson (1986), one means of gathering information about teachers' thinking is to examine the nature of their planning, for "research on teacher planning provides a direct view of the cognitive activities of teachers as professionals" (p. 267). An interesting case in point is their discussion of a study by Zahorik (p. 263). According to Clark and Peterson, "Zahorik found that the kind of decision mentioned by the greatest number of teachers concerned pupil activities (81%). The decision most frequently made first was content (51%), followed by learning objectives (28%). Zahorik concluded that teachers' planning decisions do not always follow linearly from a specification of objectives and that, in fact, objectives are not a particularly important planning decision in terms of quantity of use" (p. 263). A question worth considering is whether and how content decisions can be distinguished from decisions about selecting activities. For example, if a teacher wants students to differentiate relations that are functions from those that are not, is the teacher attending to (a) mathematical distinctions based on a set-theoretic perspective, (b) the importance of graphical representations as a means of making differentiations, or (c) generating an activity that particularly lends itself to classroom interaction or small-group work? These kinds of considerations challenge us to define what we mean by content decisions versus decisions about activities. These types of decisions are quite different from ones involving the allocation of time for teaching quadratic or exponential functions.

A secondary teacher recently indicated to us that her lesson planning consisted primarily of *doing* mathematics so that she wouldn't "fall flat on her face in front of the class." We think this perspective on planning is not atypical. While it may be that planning is not a major contributor to developing teachers' repertoire of instructional alternatives (Clark and Peterson, 1986), it may nevertheless be essential for mathematics teachers so that they feel comfortable with the mathematics they will be teaching. The student teachers we have observed spend considerable time working out assigned problems to be "ready mathematically," rather than on developing instructional strategies for teaching mathematics or to accommodate individual students. As suggested by Clark and Peterson (1986), experienced teachers seem to have a "sixth sense" about how a lesson should go, rather than an explicit plan for teaching the lesson. Consequently, it remains to be seen what kinds of decisions teachers make when planning lessons on a specific topic such as functions and what factors contribute to those decisions.

Much of the research cited by Clark and Peterson on teachers' interactive decisions indicated that, in the main, the teachers' decisions had a "business as usual" orientation. One teacher reported that alternative approaches would be considered only in the context of an unanticipated event. In general, their model relied on teachers' judgments about student behavior as a basis for making decisions.

The fact that teachers rely heavily on student behavior begs the question as to what aspects of student behavior are being considered. The teacher might focus on the students' cognitive behavior, for example, on whether a student was demonstrating an understanding of function (consider the dialogue with Ms. Maxwell), on whether the student was demonstrating a particularly unique solution to a problem involving functions, or on whether the student was misbehaving. The determination of what the teacher is focusing on is critical and would reveal much about how he or she defines classroom events. We suggest that any model of teachers' thinking that is based on teachers reacting to student behavior should be adapted to accommodate the range of possible foci and the circumstances under which teachers do not "conduct business as usual." That is, the determination and study of what teachers bring with them into the classroom when teaching functions and the extent to which their decisions are governed by a knowledge of functions (including the various representations presented earlier), by more general concerns related to the teaching of mathematics, or by their perceptions of the roles of schools more generally should be central to any research agenda on teachers' thinking about functions.

A central question is the extent to which the function concept per se is a primary factor that influences instructional decisions. It is one thing to decide that the chapter on quadratic functions should receive more (or less) attention than the chapter on exponential and logarithmic functions; it is quite another to make decisions on the extent to which a set-theoretic approach, a graphical representation, or functional thinking à la practical applications should be emphasized.

In the next section we consider teachers' knowledge and beliefs and their possible implications for teachers' thinking about functions and the teaching of functions.

Research on Teachers' Knowledge and Beliefs

It is our assumption that teachers' thoughts, decisions, and actions—that is, the conscious and deliberate part of teaching—are predicated on what teachers know and believe about their teaching, about mathematics, and about the students they teach. Although knowledge and beliefs are sometimes difficult to differentiate, they nevertheless constitute an important factor that influences how and what mathematics is taught in the classroom. In some cases, beliefs may be dependent on the existence or, perhaps, the absence of knowledge. Teachers who have a limited knowledge of functions, for example, may believe that the teaching of functions should emphasize symbolic manipulation, such as finding $f(2)$ when $f(x) = 2x^2 + 3x - 4$. That is, their teaching of functions is instrumental because their own understanding is instrumental. Although limited mathematical knowledge is not logically connected to narrowly conceived beliefs about either mathematics or the teaching of mathematics, the connection, however construed, is worthy of study.

Fennema and Franke (1992) proposed a model for analyzing mathematics teachers' knowledge. The model emphasizes the importance of considering the dynamic, interactive, and contextual nature of teachers' knowledge. The model outlines four components of teachers' knowledge: (a) knowledge of mathematics, (b) knowledge of pedagogy, (c) knowledge of students' cognitions, and (d) teachers' beliefs. The model provides foci for considering different aspects of teacher cognition that influence the teaching of mathematics, particularly when a specific content area such as functions is being considered.

After reviewing research on the relationship between teacher knowledge and practice, Fennema and Franke (1992) concluded "that when a teacher has a conceptual understanding of mathematics, it influences his/her classroom instruction in a positive way" (p. 14). They also cite research showing that effective teachers know how to put mathematics into a framework understandable to the learners, based both on their knowledge of pedagogy and on their knowledge of how students learn mathematics. The authors place particular emphasis on the importance of teachers developing knowledge about how students learn particular topics, but point out that there is not much information available on how students learn most topics in the school curriculum.

It would be interesting to determine which of the four facets of the Fennema and Franke model most strongly influences teachers' decisions about the teaching of functions. To what extent can teachers envision the appropriateness of various representations of function and their applicability to different instructional settings? To what extent can teachers enhance the treatment presented by a textbook in order to accommodate particular objectives? To what extent is the teacher's knowledge about functions and the way to teach functions deep enough to permit fundamental shifts in a textbook presentation—particularly in the face of the diversity in students' needs? A constructivist perspective, for example, necessitates considerable teacher flexibility in translating among various functional representations if significant mathematical learning is to result. To what extent do teachers have this flexibility? Alternately, a teacher may be so consumed with spur-of-the-moment decisions regarding organizational or managerial matters that the potential contribution of mathematical knowledge about a specific topic is minimized. If teachers' knowledge is to be seriously considered as a potentially significant factor affecting instruction, then the context in which that knowledge is held should also be considered. Much of this context is affected by what the teacher believes about that knowledge. Green (1971) identified three dimensions of teachers' beliefs that warrant consideration. The first involves the notion that beliefs are quasilogical in nature. While beliefs are not precise entities, they nevertheless have certain logical connections although the logic may not be strictly Aristotelian in nature. A teacher may, for example, believe that the graphical representations of functions are quite important and, consequently, that graphing calculators are an essential tool in the teaching of mathematics. A slight shift from this position would be to believe that "If graphical representations are used, then the graphing calculator should be used." This shift touches on Green's second dimension, which suggests that beliefs have a spatial orientation in addition to a logical one, since they are either central or peripheral to the believer. This second dimension points to the fact that while beliefs may be logically connected, the commitment of the believer is a separate issue. Thus, in the second case, the teacher may not believe that, given other representations, graphical representations of functions are very important; hence, the teacher may have no real commitment to use them. The third dimension that Green addresses is that beliefs are held in clusters, although one cluster of beliefs may exist in isolation or even in conflict with another. Thus, for example, a teacher may believe that the graphical representations of functions are important and may, in fact, emphasize the graphing of functions; on the other hand, the same teacher may never feel a need to help students see the connections between graphs and their algebraic representations—for example, the connection between the number and nature of roots of quadratic equations and the graphs of the corresponding functions.

An area of research on teachers' thinking about functions that deserves attention involves the interplay between the knowledge and the beliefs that teachers have about functions. It may be that a teacher believes that functions are a unifying concept within mathematics, that functions are important because of their connections to other disciplines, or that functions are a means of representing real-world phenomena, yet lacks the knowledge that allows the teacher to appreciate or communicate these connections. Owens (1987) found, for example, that preservice teachers generally had little idea of how mathematics connected to their lives beyond their academic preparation to become mathematics teachers. In general, mathematics teachers struggle to provide students with significant applications involving school mathematics. Consequently, such teachers who proclaim the belief that applications are of primary importance in the teaching of mathematics may be exhibiting nonevidential beliefs (Green, 1971), that is, beliefs based more on faith or authority than supported by reason.

Another consideration in thinking about teachers' beliefs is their orientation toward mathematics more generally. Perry (1970) developed a scheme to describe various stages of intellectual development. Oversimplified, his scheme consists of three main categories, which he describes as those of the dualist, the multiplist, and the relativist. Perry thinks of a dualist as one who sees authority as omniscient. A dualistic orientation toward the teaching of mathematics suggests that the teacher sees mathematics as a subject in which single answers are the dominant objective of mathematical thought and the determination of the "correctness" of that answer lies with an authority — the teacher or the textbook. There would be little reason for a dualist to present alternative representations of functions if the first representation "worked." Perry defines a multiplist as one who sees legitimate uncertainty in authority, allowing for multiple interpretations of events. Thus, the multiplist would "permit" multiple representations or multiple means of solving problems; the "authority" sees some truth in all of the representations but begs the question as to whether one is better suited to a given context than another. In contrast, a relativist sees authority as dependent upon the context and, hence, is prone to evaluate different representations and make judgments of the appropriateness and effectiveness of each based on the context in which the representation or solution strategy is used. While these different representations are oversimplified, they nevertheless serve the purpose of capturing the different perspectives that teachers have about mathematics.

Research in mathematics education involving Perry's scheme does not always paint a rosy picture of the nature of mathematics as it is communicated in the classroom. McGalliard (1983) and Kesler (1985) described the teaching of geometry and algebra, respectively, as being primarily from a perspective that reflects Perry's dualistic/multiplistic categories. Their focus was on the nature of the mathematics communicated to students in the classroom. Although teachers may hold relativistic conceptions of mathematics, yet communicate dualistic/multiplistic ones because of contextual circumstances, the point is rather moot. For whatever reason, the mathematics that "happens" in the classroom is narrow in scope and orientation if we are to believe that many of our research findings in fact represent typical teachers and classrooms.

Both Green's model and Perry's scheme provide an orientation toward conceptualizing the way in which the teacher defines classroom events. In particular, we feel that both have relevance for conceptualizing research on teachers' thinking about functions. Green's model can provide an orientation to what the teacher values, that is, what precepts about mathematics in general and functions in particular seem central to the teacher's orientation toward teaching mathematics. Perry's scheme provides a perspective from which we can consider how teachers conceive of mathematics as a body of knowledge.

In the first section of this chapter, we identified various types of representations and orientations toward functions that evolved from a somewhat historical perspective. These can also serve a useful purpose in conceptualizing a teacher's orientation and in determining whether the teacher sees logical connections among such representations. It would be interesting to learn whether teachers consider graphical representations as central or peripheral to their concept of function. If peripheral, we would assume that graphical representations will not be a central theme and will be given only cursory attention. Consider, for example, the problem of solving the inequality $x^3 - x^2 - 6x > 0$. It is our observation that some teachers teach this algebraically, considering various cases involving the factors of the left member of the inequality. In contrast, other teachers encourage students to approach the problem using a graphical representation by considering when the graph of the function $f(x) = x^3 - x^2 - 6x$ is above the x axis.

The problem lies not so much in having different orientations but rather in whether the orientations are isolated one from the other, or whether one is so dominant that it excludes others. Ideally, we would like teachers to adhere to the central belief that all representations are useful depending on the context. It is this sense of flexibility that encourages students to make connections among concepts such as that between the number of real roots of a quadratic equation, the value of the discriminant, and the position of the graph on a coordinate system. When these connections are not explicitly part of the instructional program, for whatever reason, then students must make the connections themselves—an unlikely event—or run the risk of experiencing mathematics as a fragmented body of knowledge.

MAPPING THE TERRAIN OF A RESEARCH AGENDA

There are at least two components that potentially have impact on teachers' thinking about functions: their knowledge and their beliefs. Little research has been conducted on how teachers conceptualize functions, despite all of the rhetoric of the past century regarding the importance of functions. What orientation regarding functions do they bring with them to the classroom? Are they capable of shifting from one orientation to another given the nature of the problem or task? How does the function concept develop among college students (preservice teachers) or how is it extended among experienced teachers? If we are to understand teachers' thinking about functions, either in planning for instruction or in the interactive phase of teaching, our understanding should involve the knowledge base that teachers have about the content. If it is the case that the teaching of functions involves only a limited range of representations, it is incumbent on the researcher to understand whether the "narrowness" is based on a limited knowledge of functions, on strong beliefs about functions or their teaching, or on the circumstances that exist in the classroom at the time of observation.

Although it might be argued that teachers' orientation toward functions is a part of their knowledge, we think it best to separate knowledge and beliefs at least to the extent that one can get a "feel" for the teachers' general orientation toward mathematics and functions. We see Green's analysis (1971) and Perry's scheme (1970) as relevant to this aspect of teachers' thinking about functions. Research on teachers' thinking should attempt to identify what teachers hold as central to their view of mathematics, its teaching, schooling in general, and within this context, what orientation toward functions is dominant in their teaching. But it is more than a matter of determining what is central; it has to do with determining the way knowledge is held and where, in the teacher's view, authority resides for understanding a solution or a mathematical procedure. The NCTM (1991) *Professional Standards for Teaching Mathematics* is explicit about this—that the study of mathematics should empower students to draw and validate their own conclusions based on mathematical reasoning—that is, that the authority of mathematical "truths" resides in the mathematics itself, not with a person or textbook.

Perhaps the most significant question has to do with the centrality of making decisions on what teachers know and believe about functions versus other perhaps more pragmatic concerns about how to conduct the business of the classroom. The fact that teachers in the Clark and Peterson study (1986) made "decisions" based on "business as usual" suggests that they may have been focusing on the smoothness of the lesson rather than on the students' understanding of the content per se. We suspect that this is frequently the case in the mathematics classroom. If the lesson is progressing smoothly, with a minimum of occlusions, then the teacher tends not to intervene with questions that could disrupt the routine. From this perspective, it makes little sense for the teacher to interject multiple interpretations of functions unless they serve the end of solving some problem; that is, there may be little value in coming to grips with multiple representations for the sake of learning itself.

Jones (1990) identified three dimensions that potentially characterize teachers' beliefs: the centrality of mathematics, the importance of social goals, and reflectivity. These dimensions seem quite relevant to the present discussion. To what extent is teachers' thinking about functions based on mathematical considerations? Is the mathematics central to their thinking and planning for the teaching of functions? Alternatively, the teacher may be concentrating on social goals and on finding ways of maintaining a positive and constructive learning environment—even at the risk of sacrificing significant mathematics. Finally, Jones discussed what he terms the reflectivity of teachers, that is, their willingness to engage in reflection about their teaching. This would seem to be a basic ingredient for the kind of flexibility in thinking about functions that reformers in mathematics education have been calling for.

CONCLUSION

Research on teachers' thinking and, concomitantly, research on teachers' beliefs are basically in their infancy. It may be helpful in developing an understanding of teachers' thinking and beliefs to examine them in the context of a specific mathematical domain. But there are risks involved as well. Those risks have to do with whether the research is sensitive not only to learning about what teachers know, think, and believe about functions, but also to an appreciation of the context in which that knowing, thinking, and believing occurs. Given the myriad of factors teachers have to face in today's classrooms, in what way and in what context does thinking about functions contribute to their teaching? The issue has to do with the strength of their beliefs about functions, the knowledge they can bring to bear in teaching functions, the vision they hold for mathematics teaching and learning, and, most certainly, with how they see the teaching and learning of mathematics fitting into the notion of schooling itself.

Research on teachers' thinking that neglects the importance of these contexts runs the risk of studying trees but having no basic understanding of the forest in which those trees grow. Research on teachers' thinking should be highly contextual if the findings are to be believable and instructive both for subsequent research and for practice. As was shown in the first section of this chapter, proclamations about the teaching of functions are not always predictable in terms of how they will influence practice because of the contextuality of teaching. Researchers should take note of this contextuality and should attend to all of its ramifications with respect to conducting research on teachers' thinking about functions.

ACKNOWLEDGMENT

We wish to thank Joao Pedro da Ponte from the University of Lisbon for his suggestions regarding the section that traces the historical development of the function concept.

REFERENCES

Ball, D. L. (1990). Prospective elementary and secondary teachers' understanding of division. *Journal for Research in Mathematics Education, 21*, 132-144.

Ball, D. L. (1991). Research on teaching mathematics: Making subject matter knowledge part of the equation. In J. E. Brophy (Ed.), *Advances in research on teaching: Teachers' subject matter knowledge and classroom instruction* (Vol. II, pp. 1-48). Greenwich, CT: JAI.

Barrett, G., & Groebel, J. (1990). The impact of graphing calculators on the teaching and learning of mathematics. In T. Cooney (Ed.), *Teaching and learning mathematics in the 1990s* (pp. 205-211). Reston, VA: National Council of Teachers of Mathematics.

Betz, W. (1931). *Algebra for today: Second course*. Boston: Ginn.

Breslich, E. R. (1928). *Developing functional thinking in secondary school mathematics*. In National Council of Teachers of Mathematics (Ed.), *The third yearbook* (pp. 42-56). New York: J. J. Little and Ives.

Brown, S. I., Cooney, T. J., & Jones, D. (1990). Mathematics teacher education. In W. R. Houston, M. Haberman, & J. Sikula (Eds.), *Handbook of research on teacher education* (pp. 639-656). New York: Macmillan.

Buck, R. C. (1970). Functions. In E. G. Begle (Ed.), *Mathematics education: The sixty-ninth yearbook of the National Society for the Study of Education* (pp. 236-259). Chicago: NSSE.

Butler, C. H., & Wren, F. L. (1941). *The teaching of secondary mathematics*. New York: McGraw-Hill.

Carpenter, T. P., Fennema, E., Peterson, P. L., & Carey, D. A. (1988). Teachers' pedagogical content knowledge of students' problem solving in elementary arithmetic. *Journal for Research in Mathematics Education, 19*, 385-401.

Clark, C. M., & Peterson, P. L. (1986). Teachers' thought processes. In M. C. Wittrock (Ed.), *Handbook of research on teaching* (3rd ed., pp. 255-296). New York: Macmillan.

College Entrance Examination Board. (1959). *Report of the Commission on Mathematics, program for college preparatory mathematics*. New York: Author.

Cooney, T. J. (1980). Research on teaching and teacher education. In R. J. Shumway (Ed.), *Research in mathematics education* (pp. 433-474). Reston VA: National Council of Teachers of Mathematics.

Cooney, T. J., Grouws, D. A., & Jones, D. (1988). An agenda for research on teaching mathematics. In D. A. Grouws, T. J. Cooney, & D. Jones (Eds.), *Perspectives on research on effective mathematics teaching* (pp. 253-261). Reston, VA: National Council of Teachers of Mathematics.

Davis, R. B. (1967). *The changing curriculum*. Washington, DC: Association for Supervision and Curriculum Development.

Demana, F., & Waits, B. K. (1990). *Precalculus mathematics: A graphing approach*. Reading, MA: Addison-Wesley.

Dolciani, M. P., Berman, S. L., & Freilich, J. (1962). *Modern algebra: Structure and method*. Boston: Houghton Mifflin.

Dolciani, M. P., Berman, S. L., & Wooton, W. (1963). *Modern algebra and trigonometry: Structure and method. Book two*. Boston: Houghton Mifflin.

Dolciani, M. P., Wooton, W., Beckenbach, E. F., & Sharron, S. (1983). *Algebra 2 and trigonometry.* Boston: Houghton Mifflin.

Fennema, E., & Franke, M. (1992). Teachers' knowledge and its impact. In D. A. Grouws (Ed.), *Handbook of research on mathematics teaching and learning* (pp. 147–164). New York: Macmillan.

Fey, J. T. (1989). School algebra for the year 2000. In S. Wagner & C. Kieran (Eds.), *Research issues in the learning and teaching of algebra* (pp. 199–213). Reston, VA: National Council of Teachers of Mathematics; Hillsdale, NJ: Lawrence Erlbaum Associates.

Foerster, P. A. (1984a). *Algebra I.* Menlo Park, CA: Addison-Wesley.

Foerster, P. A. (1984b). *Algebra and trigonometry.* Menlo Park, CA: Addison-Wesley.

Green, T. F. (1971). *The activities of teaching.* New York: McGraw-Hill.

Hamley, H. R. (1934). *Functional and relational thinking in mathematics.* New York: Bureau of Publications, Teachers College, Columbia University.

Hawkes, H. E., Luby, W. A., & Touton, P. B. (1909). *First course in algebra.* Boston: Ginn.

Hedrick, E. R. (1922). Functionality in the mathematical instruction in schools and colleges. *Mathematics Teacher, 15,* 191–207.

Henderson, K. B., Pingry, R. E., & Klinger, D. L. (1968a). *Modern algebra: Structure and function. Book I.* St. Louis, MO: McGraw-Hill.

Henderson, K. B., Pingry, R. E., & Klinger, D. L. (1968b). *Modern algebra: Structure and function. Book II.* St. Louis, MO: McGraw-Hill.

Jacobs, H. R. (1979). *Elementary algebra.* San Francisco: W. H. Freeman and Company.

Jones, D. (1990). *A study of the belief systems of two beginning middle school mathematics teachers.* Unpublished doctoral dissertation, University of Georgia.

Kennedy, J., & Ragan, E. (1969). Function. In National Council of Teachers of Mathematics (Ed.), *31st yearbook: Historical topics for the mathematics classroom* (pp. 312–313). Washington, DC: National Council of Teachers of Mathematics.

Kesler, R. (1985). *Teachers' instructional behavior related to their conceptions of teaching and mathematics and their level of dogmatism: Four case studies.* Unpublished doctoral dissertation, University of Georgia.

Kleiner, I. (1989). Evolution of the function concept: A brief survey. *College Mathematics Journal, 20*(4), 282–300.

Kline, M. (1972). *Mathematical thought from ancient to modern times.* New York: Oxford University Press.

Lovell, K. (1971). Some aspects of the growth of the concept of function. In M. F. Roskopf & L. P. Steffe (Eds.), *Piagetian cognitive-development research in mathematics education* (pp. 12–33). Washington, DC: National Council of Teachers of Mathematics.

Malik, M. A. (1980). Historical and pedagogical aspects of the definition of function. *International Journal of Mathematics Education in Science and Technology, 11,* 489–492.

Mallory, V. S., & Fehr, H. F. (1940). *Senior mathematics for high schools.* Chicago: Benj. H. Sanborn.

Marsh, W. R. (1905). *Elementary algebra.* New York: Charles Scribner's Sons.

May, K. O., & Van Engen, H. (1959). Relations and functions. In National Council of Teachers of Mathematics (Ed.), *The 24th Yearbook: The growth of mathematical ideas grades K-12* (pp. 65–110). Washington, DC: National Council of Teachers of Mathematics.

McGalliard, W. A. (1983). *Selected factors in the conceptual systems of geometry teachers: Four case studies.* Unpublished doctoral dissertation, University of Georgia.

National Council of Teachers of Mathematics. (1989). *Curriculum and evaluation standards for school mathematics.* Reston, VA: Author.

National Council of Teachers of Mathematics. (1991). *Professional standards for teaching mathematics.* Reston, VA: Author.

Osborne, A. R., & Crosswhite, F. J. (1970). Forces and issues related to curriculum and instruction, 7–12. In National Council of Teachers of Mathematics (Ed.), *A history of mathematics education in the United States and Canada* (pp. 153-235). Washington, DC: National Council of Teachers of Mathematics.

Owens, J. E. (1987). *A study of four preservice mathematics teachers' constructs of mathematics and mathematics teaching*. Unpublished doctoral dissertation, University of Georgia.

Perry, W. G. (1970). *Forms of intellectual and ethical development in the college years*. New York: Holt, Rinehart and Winston.

Rosskopf, M. F. (1970). Mathematics education: Historical perspectives in the teaching of secondary school mathematics. In National Council of Teachers of Mathematics (Ed.), *Thirty-third yearbook of the National Council of Teachers of Mathematics* (pp. 3-29). Washington, DC: National Council of Teachers of Mathematics.

School Mathematics Study Group. (1965). *Elementary functions*. New Haven, CT: Yale University Press.

Schorling, R. (1936). *The teaching of mathematics*. Ann Arbor, MI: Ann Arbor Press.

Schorling, R., Smith, R. R., & Clark, J. R. (1949). *Algebra: First course*. New York: World Book Company.

Sfard, A. (1989, July). *Transition from operational to structural conception: The notion of function revisited*. Paper presented at the 13th Annual International Conference for the Psychology of Mathematics Education, Paris.

Smith, S. A., Charles, R. I., Dossey, J. A., Keedy, M. L., & Bittinger, M. L. (1990). *Algebra and trigonometry*. Menlo Park, CA: Addison-Wesley.

Thompson, A. G. (1992). Teachers' beliefs and conceptions: A synthesis of the research. In D. Grouws (Ed.), *Handbook of research on mathematics teaching and learning* (pp. 127-146). New York: Macmillan.

Thorpe, J. A. (1989). Algebra: What should we teach and how should we teach it? In S. Wagner & C. Kieran (Eds.), *Research issues in the learning and teaching of algebra* (pp. 11-24). Reston, VA: National Council of Teachers of Mathematics.

Townsend, E. J. (1915). *Functions of a complex variable*. New York: Henry Holt.

Vinner, S. (1983). Concept definition, concept image and the notion of function. *International Journal of Mathematics Education in Science and Technology, 14*, 293-305.

Vinner, S., & Dreyfus, T. (1989). Images and definitions for the concept of functions. *Journal for Research in Mathematics Education, 20*, 356-366.

Welchons, A. M., & Krickenberger, W. R. (1949a). *Algebra: Book one*. Boston: Ginn.

Welchons, A. M., & Krickenberger, W. R. (1949b). *Algebra: Book two*. Boston: Ginn.

Wells, W., & Hart, W. W. (1929). *Modern second course in algebra*. Boston: D. C. Heath.

Willoughby, S. S. (1967). *Contemporary teaching of secondary school mathematics*. New York: John Wiley & Sons, Inc.

Young, J. W. A. (1927). *The teaching of mathematics*. New York: Longmans, Green.

Youschkevitch, A. P. (1976). The concept of function up to the middle of the 19th century. *Archive for History of Exact Sciences, 16*, 37-85.

IV TEACHER KNOWLEDGE

7 Integrating Research on Teachers' Knowledge of Functions and Their Graphs

F. Alexander Norman
University of Texas at San Antonio

To this point, there has been little research examining teachers' knowledge and beliefs about functions and graphs. This chapter draws parallels from research on students' learning of algebra suggesting that, to some extent, teachers' knowledge and beliefs are not very different from those of postsecondary students. However, we conclude that, although some inferences can be made from research on student learning, much more research is needed in the area of teachers' understanding of functions and graphs. A curriculum research and development model, based on the principles of cognitively guided instruction, is described as a paradigm for research on teachers' knowledge and beliefs.

The principal objectives of this chapter are (a) to review and interpret research on student learning and teacher thinking with a view toward suggesting directions for research in the context of functions and graphs, and (b) to describe a model that effectively integrates aspects of research on learning and teaching. The chapter is divided into two parts. The first part describes some of the research on teachers' knowledge of and beliefs about mathematics, research on students' learning of algebraic concepts (particularly those related to functions and graphs), and implications for research on teachers' cognitions. In the second part, we examine both a model for integrating research in learning and teaching and the implications of this model for teacher educators.

At this juncture one might be inclined to ask two questions: What benefit is there in integrating the research on learning and teaching? Don't we already know what secondary mathematics teachers know and think about algebra — surely, having a college degree in mathematics, they are expert at it?

Indeed, regarding the first question, there are those (see, e.g., Gage, 1964) who argue that theories of instruction are not usefully informed by theories of learning. Furthermore, it might be argued that both research in learning and research in teaching have proceeded quite well on their own during the last decade. Without a universal research "paradigm" in the Kuhnian sense for either area, it can be viewed as a rather risky venture to attempt an integration of research theories that have not yet reached the level of scientific maturity of other disciplines. On the other hand, Shulman (1986), Romberg and Carpenter (1986), and Fennema, Carpenter, and Lamon (1988) argued effectively for the opposing view. For a fuller discussion of the eclectic programs operating in the research arena in mathematics education, see Cooney and Wilson (Chapter 6 in this volume).

The answer to the second question is a bit easier — an equivocal no. We think we know what a typical algebra teacher knows. But there is certainly little empirical evidence in the research literature to validate this. Further, mathematics educators have not at this point determined what it is that we want, or think we want, teachers to know, and, more importantly perhaps, how we want them to know it.

In the third edition of the *Handbook of Research on Teaching*, Shulman (1986) cited two major difficulties engendered by the direction of current research programs investigating teacher cognitions. One is the narrow scope of investigations of teaching situations in which teacher thinking is a critical component. This first difficulty can be seen quite clearly in the minimal attention given thus far to examinations of mathematics teachers' knowledge. The problem is even more critical regarding teachers' knowledge of algebra, in particular functions and graphs, as research in this area is virtually nonexistent.

The second problem is "the growing distance between the study of teacher cognition and those increasingly vigorous investigations of cognitive processes in pupils" (p. 24). A similar observation is made by Romberg and Carpenter (1986) in their chapter in the same *Handbook*. They depict research in students' learning and research in teaching as representing two well-developed but disparate strands of inquiry. Fortunately, since the publication of the last *Handbook*, there has been a discernible move toward integrating the research on the learning and teaching of mathematics generally and, as described later, in algebra in particular.

RESEARCH ON TEACHER COGNITION AND STUDENT LEARNING

Much of the research in teaching has been of the sort commonly known as process/product research. This research relates a specified, observable teacher behavior (such as amount of praise giving) with an identifiable pupil outcome (such as score on a standardized test). The teacher behavior is the process; the pupil outcome is the product. This approach, grounded in the behaviorist tradition, suffers from a number of deficiencies (see Berliner, 1979) inherited from the behaviorist view of learning, including the inability to handle the complex cognitive interactions that occur in the classroom environment and, more importantly for our purposes, the virtual exclusion from consideration of teachers' and students' thought processes.

Following the early proponents of cognitive psychology (e.g., Bruner, Goodnow, & Austin, 1956), educational theorists and researchers like Shavelson (1973) and Shulman and Elstein (1975) began to rethink the role of cognitive processes in teaching and learning. Thereafter, research on teachers' thinking grew rapidly. Extensive general reviews of this research can be found in Clark and Peterson (1986) and Shavelson and Stern (1981). Cooney and Wilson (this volume, Chapter 6) provide a description of the relevance of this research to mathematics education.

According to Clark and Peterson, most of the research on teachers' thought processes is relatively recent. In fact, in the second edition of the *Handbook of Research on Teaching* (Travers, 1973), there was neither a chapter on nor a reference to teachers' thought processes. However, in spite of a growing number of studies, there does not appear to be a consistent body of research on teacher thinking. Clark and Peterson go on to identify several deficiencies in the research literature, including the fact that most of the research in teacher cognition involves experienced elementary or middle school teachers. Clearly such deficiencies impose some fairly restrictive limitations of their own upon inferences that can be drawn regarding algebra teachers' thinking. Most obviously, teachers of algebra generally are not elementary or middle school teachers and thus usually have considerably different educational and academic histories. Moreover, most of the studies involved experienced teachers, which means that we know little about how teachers' thinking evolves over time.

Despite these problems, the research substantiates the importance to instruction of teachers' thought processes. The growing emphasis in teacher education programs on encouraging students to be reflective about their teaching — for example, through keeping diaries or by building metacognitive skills — is one such indication of the increasing focus on teacher thinking.

Teachers' Understandings of Functions and Graphs

Before examining the domain of algebra, a few general comments are needed to provide a perspective for examining teachers' understanding of functions and graphs. A convenient way of categorizing teachers' knowledge is to identify it as practical knowledge, pedagogical knowledge, or content knowledge. Our main concern in this chapter is with teachers' pedagogical and content knowledge. (A recent review of research investigating teachers' practical knowledge can be found in Feiman-Nemser & Floden, 1986.)

Similarly, we may wish to describe the quality of the knowledge held by teachers. For this, we use the distinction between instrumental understanding and relational understanding (Skemp, 1978). For example, an instrumental understanding of the processes required to solve a particular problem might entail a simple, algorithmic application of rules or formulas. The student may know how to solve the problem, but has no clear concept of why the particular application works. In contrast, relational understanding incorporates a deeper understanding of the mechanisms underlying the solution process, as well as an understanding of relationships among the relevant concepts involved.

In general, the research literature has provided a good foundation for the description of teachers' cognitive behaviors. However, according to Shulman (1986), aspects of research in teacher cognition that have not provided much useful information lie in the area of "teachers' cognitive understanding of subject matter content and the relationships between such understanding and the instruction teachers provide for students" (p. 25).

This comment rings especially true relative to the research on the content and pedagogical knowledge of algebra teachers. In contrast to the body of research on students' understanding and learning of algebra, research focusing on teachers is scarce. A recent ERIC search conducted in preparation for this chapter revealed only a handful of papers that specifically address teachers' knowledge—that is, their subject matter or pedagogical knowledge—of functions and graphs. Fortunately, though, the rich literature on students' understanding suggests potential parallels for teachers' understanding, whereas research on teachers' cognition in other areas of mathematics provides useful clues for making inferences about the state of teachers' algebraic knowledge.

For example, we note that over the last decade there have been a number of investigations in a variety of mathematical contexts of the knowledge held by teachers. These studies include areas as diverse as the van Hiele levels of geometric thought (Mayberry, 1983), rational number concepts (Post, Harel, Behr, & Lesh, 1988), problem solving (Brown, 1986), division in different mathematical contexts (Ball, 1990), unitizing strategies in algebraic contexts (Norman, 1986, 1987), strategies for solving proportion problems (Fisher, 1988), and others. Several studies have more direct implications for teachers of algebra, including those by Begle (1972) and Eisenberg (1977), which found no correlation between teachers' knowledge of algebra and student performance. A more recent study revealed, in self-reports of participating secondary mathematics teachers, areas that teachers found most difficult to teach. Not surprisingly, courses at the Algebra II level and above—all courses that require extensive understanding of functions and graphs—were found to give teachers the most trouble (Goldin & Ellis, 1983). In a study that focused on mathematical knowledge of functions, Norman (1992) reported a wide variability among teachers' conceptualizations of function and their degree of understanding of those concepts. Many teachers seemed to maintain a primarily instrumental understanding of function.

However, it should be noted that in most recent studies (e.g., Dubinsky, Hawks, & Nichols, 1989; Even, 1988), the mathematics "teachers" have been undergraduate preservice or novice teachers rather than those with considerable, or even a moderate amount of, experience teaching mathematics. This gap in the research literature presents us with further evidence of the fragmentary nature of our knowledge about what teachers know and how they think about algebra.

Although one might be inclined to despair at this state of affairs, there is much that can be reasonably inferred from these studies and from research on students' understanding of algebra. For example, in a study of teachers' knowledge of rational number concepts (Post et al., 1988), the investigators were surprised by the extent to which many teachers were uncertain in applying routine arithmetic procedures, as well as by their rather sparse conceptualizations of rational number. Could similar incomplete understandings be expected of algebra teachers? In other words, might some algebra teachers be operating at a level of instrumental rather than relational understanding? In fact, all of the studies mentioned in the previous paragraph suggested that some teachers do not have the rich concepts, in a variety of mathematical domains, that one would expect from those assumed to have a reasonably high degree of mathematical expertise. Given that there is some evidence to indicate that this is also the case for teachers of algebra, it is not a great leap to presume that some algebra teachers have similarly impoverished conceptualizations of the notions of function and graphs.

However, to temporize on this issue, we note from personal experience and anecdotal accounts that there are many teachers who have a deep and fundamentally sound knowledge of algebra—who possess the relational understanding that we would expect our teachers to have. It seems, then, that questions of extreme importance to researchers should include those related to the evolution of teachers' knowledge over the course of their first few years of teaching. The educational maxim that "one never *really* understands a subject until one teaches it" is a hypothesis that warrants serious and critical investigation. An important point to note here is that the extent of our knowledge of teachers' understanding of functions and graphs, in terms of both content knowledge and pedagogical knowledge, is insufficient to justify any broad assessment of that understanding. Thus, is it is critical that we begin a comprehensive investigation of questions related to teachers' knowledge of algebra.

There is a second strand of research that may provide clues as to teachers' understanding of functions and graphs. This is the substantial research related to students' learning and understanding of algebra. Unquestionably there are distinct differences in the populations of students and teachers, but there are also similarities that allow us to make some reasonable inferences about the depth and nature of teachers' understanding of functions and graphs.[1] The research on algebra learning is extensive, and although we intend to restrict ourselves here primarily to that research involving graphing and functions, it is useful to discuss briefly the general focus of recent work in this area. (A more complete discussion can be found in Dugdale, Chapter 5 in this volume, and in several reviews mentioned later.)

An important theme among recent researchers in algebra learning (and other areas as well) has centered around a constructivist view of learning. This perspective is an outgrowth of the Gestalt paradigm (see Wertheimer, 1959) and goes hand in hand with the cognitive science approach to learning that has rapidly taken hold in the past three decades. A fundamental assumption of cognitive science is that "mental structures and cognitive processes . . . are extremely rich and complex—but that such structures can be understood, and understanding them will yield significant insights into . . . thinking and learning" (Schoenfeld, 1987). The basic assumption underlying the constructivist perspective is that students construct their mental structures (i.e., understandings) via their own idiosyncratic mental processes. Thus, the role of the researcher is to try to grasp the nature of the processes learners use to build their knowledge and the nature of the knowledge that learners possess.

This constructivist perspective finds expression in a variety of research studies in algebra learning. Among these, a major research focus has been the investigation of perceptual and structural aspects of algebra and how learners process algebraic language. Many of the difficulties students have with algebra are related to the meaning of the symbols used in algebraic expressions, their recognition and use of structure, and the transitions from one mathematical context to another. Recent reviews of this research can be found in Kieran (1989) and Herscovics (1989). Research investigating "cognitive obstacles" to the smooth transition from one mathematical context to another (typically of an arithmetic to algebraic nature) typifies studies in this area. Such research has implications for the understanding of

---

[1] These inferences may well be valid only for novice teachers.

function, one of the primary conceptualizations of which is formulaic and necessarily entails the presentation and interpretation of symbolic algebraic expressions.

A Question of Structure. Do teachers evince any of the difficulties with algebraic structure that students do? It is unlikely that college students would make the concatenation error of interpreting the algebraic monomial $3x$ as "thirty-something." But supposedly experienced calculus students have misinterpreted the expressions $x(t)$, as in a typical parametric system of equations, as $x \times t$; $\cos(x) \cdot \cos(y)$ as $\cos^2(xy)$; and x as a variable (Norman, 1990). Although these errors are not based simply on a misinterpretation of symbolism, they are nonetheless indicative of the difficulties related to algebraic structure.

Another difficulty reported widely (e.g., Wenger, 1987) with which we are all too familiar relates to algebraic errors of the form

$$F(a * b) = F(a) * F(b)$$

in which $*$ is a binary operation and F a function that is not linear (additive) with respect to $*$. By the time our secondary preservice mathematics teachers graduate, they rarely write $(x + y)^n = x^n + y^n$ because we have taught them well never to write that (unless, of course, $xy = 0$ or the domain is Z_2 or . . .). Yet this does not inhibit some students from making similar errors in other equally inappropriate contexts—for example, $\ln(x + y) = \ln(x) + \ln(y)$ is often observed among calculus students, and far too often among more advanced mathematics students, we sometimes see expressions like

$$e^{\int P(x)dx \, + \, C} = e^{\int P(x)dx} + e^C$$

How can this be? Well, most mathematics majors have experienced a number of instances that do admit such linearity—linear transformations, homomorphisms, differential and integral operators, additive functions. Even earlier in high school, the students probably saw such linearity in the definition of functional operations, for example, $(p + q)(x) = p(x) + q(x)$. In fact, to some students these may all appear to be no more than a generalization of the distributivity laws of real number operations that have been so keenly emphasized since grade school. We have identified a number of similar sorts of structural misinterpretations among experienced mathematics students. For example, the *perceptual* differences between the expressions *(f • g)(x)* and *(f ∘ g)(x)* are quite subtle, while the *conceptual* differences are profound. This is confounded even further, if one chooses to represent functional multiplication (or composition) as a simple juxtaposition, *fg*. One might argue that students simply need to associate a given mathematical context with certain conventional algebraic structures. In fact, it is likely this is what most students do. Unfortunately, that is often all that these students do, for they have never developed a fuller and richer understanding of the mathematics that underlies the structure.

Clearly, the problem here lies in an interpretation of structure. Different mathematical contexts require different interpretations for structurally isomorphic symbolic expressions. Students, and teachers, who do not fully understand the mathematical contexts are then at great risk of making errors due to structural misinterpretations.

Let us presume then, from the two lines of research just mentioned (i.e., research on teachers in mathematical domains other than functions and graphs, and research on postsecondary students' knowledge of functions and graphs), that, initially at least, teachers' knowledge of functions and their graphs is probably not too radically different from what research tells us about college-level students' knowledge of the same. Of course, an obvious and important research implication is the need to investigate teachers' understanding and to confirm or reject this hypothesis.

Students' Understanding of Functions and Graphs. Now let us look at what we know about students' understanding of functions and graphs, keeping in mind our hypothesis that teachers' understanding may not be very different. Some of the recent and "classic" research on students' understanding (and misunderstanding) of the function concept (e.g., Dreyfus & Eisenberg, 1983; Graham & Ferrini-Mundy, 1989; Orton, 1970; Thomas, 1975) focused on postsecondary students, who are not far removed, if at all, in mathematical experience from novice high-school mathematics teachers — some of these students, no doubt, become mathematics teachers. These studies indicate that, although some students have a reasonably rich understanding of the function concept, many students have interpretations that are fairly limited in scope and do not exhibit a fully relational understanding of functions. After all, the notion of function is a complex one and includes a number of related subconcepts such as set correspondences, dependence relationships, domain, and range, as well as a number of representational schemes — Cartesian and other graphical devices, numerical tables, schematic diagrams, algebraic formulas or other rule statements, and so forth. Moreover, there are a number of different conceptualizations of functions that vary in appropriateness, depending upon the particular mathematical domain.

The notion of function has been under active investigation by researchers for many years and its centrality in mathematics has been long recognized by educators. In fact, the first NCTM Yearbook written by a single author (Hamley, 1934) had as its title *Relational and Functional Thinking in*

Mathematics. This volume included the history and meaning of the term "function," as well as a history of functional thought among students and the psychological bases for understanding function. More recently, Malik (1980) and Markovitz (1986) have given accounts of the historical development of the notion of function and implications for the teaching of the concept. This development parallels the development of many mathematical concepts in its growth from more concrete to more abstract conceptualizations. For example, among the earliest conceptualizations was the almost kinesthetic notion of a functional relationship as a descriptor of how one change affects, and effects, others. The most modern formulation of a function as a subset of the Cartesian product of sets is much more abstract.

What, then, are students' concepts of functions? One way to answer this question would be simply to ask students to give their definitions of function. However, Vinner and others have pointed out distinctions between the *definition* that one has for a concept and the *image* one has for that same concept (Tall & Vinner, 1981; Vinner, 1983; Vinner & Hershkovitz, 1980). For example, we might define a function as "a set F of ordered pairs for which $(a,b),(a,c) \in F => b = c$," yet maintain the image of a function as "a Cartesian graph for which every vertical line intersects the graph in no more than one point." In this case, both the concept definition and the concept image are mathematically correct but qualitatively different. Vinner and Dreyfus (1989) found that among some students (and teachers!) there were differences between their concept definitions and concept images, and in many of these cases the concept images—the operational mental constructs—were not correct. This is not surprising, however, for even mathematicians may suffer similarly: for example, when contemplating continuous, nowhere differentiable functions—a concept that certainly does not fit well with most of our images of a continuous function.

Most evidence suggests that students prefer to view a function as a rule of correspondence, generally associated with an algebraic formula that is operational over the entire domain of the function (see, e.g., Markovitz, Eylon, & Bruckheimer, 1983; Marnyanskii, 1975). This problem persists even among college calculus students. For example, Graham and Ferrini-Mundy (1989) have found that graphical representations without an associated formula were not classified as functions and that piecewise continuous functions (even those for which formulas were given over the intervals of continuity) caused considerable difficulty for students.

Some other findings are particularly relevant when considering teachers' concepts of functions. Markovits (1986) found that students had some difficulty in constructing graphs of functions that satisfied a set of given constraints and even more difficulty in providing algebraic descriptions (i.e., formulas) corresponding with the constraints. These results validate earlier studies (Wagner, Rachlin, & Jensen, 1984) suggesting that students may not have developed reversibility in the task of relating the algebraic expression of a function and the graphical representation. Most move from formula to graph but not vice versa. Along the same lines, Dreyfus and Eisenberg (1983) found that college students would often exhibit a fixation on linearity in their production of examples of functions, sometimes at the expense of the requirement that the example be a function. These are critical notions for teachers of algebra, who must have a high degree of facility in moving among different representations of functions and who must also be thoughtful and sometimes clever in manufacturing examples of functions for their students.

In addition, there are many questions related to graphing that are intimately tied to the understanding of functions. The literature on graphing is extensive, so we discuss only that which is most pertinent to this chapter—namely, that concerning graph interpretation and the link between graphical and algebraic situations. We have already noted that students sometimes have difficulty in making the appropriate link between graphical and algebraic situations. Just as there is a direct link between the graph of a function and the function's symbolic representation, there is an equally important link between the actual production of a graph, the explicit variables, and the implicit relationship between them. As suggested by Janvier (1984), "the concept of variable cannot be developed without a graphical means for apprehending variation" (p. 57).

Graphical interpretation includes not only the identification of local properties (extrema, rates of change, points of discontinuity) but also an association of global features with the situation that the graph portrays. The complexities that arise in the interpretation of graphs have been shown to have a profound effect on students' ability to interpret graphs (Janvier, 1978; Kerslake, 1977, 1981; Konshak & Monk, 1976). It has been shown, for example, that situational distractors can interfere significantly with proper interpretation of graphs (Bell & Janvier, 1981; Janvier, 1978). Reading and interpreting graphs is affected by the scaling of axes. Some students appear to have difficulty in such interpretations if the axes are not uniformly scaled (Shaw, Padilla, & McKenzie, 1983).

Graphs can provide a powerful analytical tool for studying complex relationships as well as a useful means of communicating otherwise difficult-to-describe information. As a means of enhancing students' understanding of functions and supporting the development of other mathematical ideas, one of the primary roles of the mathematics teacher is to provide students with activities that involve nontrivial ways of reading, interpreting, and constructing graphs. Teachers who share with students some of the difficulties cited previously may not be able to provide students with the crucial activities needed to help them develop a rich understanding of functions and their graphs.

Multiple Representations. Understanding the relationships among the variety of representations for functions is an important component in the fully developed concept of functions. Although work by Kaput (1989) provided a theoretical foundation for linked representation systems, there are examples of software in use today that take advantage of the computer's ability to provide representational links for students. *Green Globs* (Dugdale, 1983) is one of the "old" but still quite popular packages; newer software such as *The Algebraic Proposer* (Schwartz, 1987, 1988) and *Word Problem Assistant* (Thompson, 1989) may also prove valuable. Dreyfus and Eisenberg (1987) found that students using the *Green Globs* software developed an enhanced understanding of the algebraic-graphical function tie. Using the graphing capabilities of microcomputers, particularly in the context of calculus, David Tall developed software that provides students with environments that have led to their increased discussion and conjecturing about functions. He also had considerable success working with students in developing their geometric intuition about functions (see, e.g., 1986, 1987). Although this research has been very promising, we must be alert to potential dangers that can arise through inappropriate interpretations of computer-generated data. For example, Goldenberg (1987, 1989) pointed out some of the illusory effects that arise with certain dynamic parametrizations of computer-generated families of graphs.

In part because of the emergence and widespread availability of computing technology, numerical, graphical, and symbolic representations can be linked in ways that were previously impossible. This has profound implications for mathematics teachers. First, in the context of this new technology, many teachers are relatively inexperienced and do not yet have the knowledge to effectively use the power that computing can bring to the learning of mathematics. Second, many teachers have not had formative computing experiences as learners, thus putting them at a considerable disadvantage if they are ever to consider using the technology to teach. Third, some teachers may suffer, as do students, from such a deficient concept of function that the linking of different representations is made a very difficult task.

Taken together, the research available on teachers' knowledge of mathematics and the research on students' understanding of functions and graphs suggest that careful attention be given to the examination of teachers' cognitions in these areas. The centrality of the function concept in school mathematics—in all mathematics—demands that teachers themselves have deep knowledge of functions, that teachers be cognizant of their students' understanding and potential pathologies in this area, and that teachers have the ability to bring their own knowledge to bear on the problem of facilitating their students' learning.

Teachers' Beliefs About Functions and Graphs

In the last decade there has been a growing emphasis among researchers on the mathematical belief systems of students and teachers. There have been studies examining teachers' beliefs in a variety of mathematical contexts — beliefs about division and multiplication (Graeber, Tirosh, & Glover, 1986), problem solving (Cooney, 1985), responsibility for student success or failure (Pratt, 1985), small-group instruction (Good, Grouws, & Mason, 1990), and so on — as well as studies that have looked at mathematical attitudes and anxieties (Schoeps, Royster, Prichard, & Norman, 1990), the effects of methods courses on teachers' views of teaching mathematics (Owens & Henderson, 1985), and the role of formal mathematics study on beliefs (Rector & Ferrini-Mundy, 1986). There have also been studies that have examined teachers' overall views of mathematics and mathematics teachers (e.g., Thompson, 1982, 1984). There is an equally rich research literature on students' attitudes and views.

However, by comparison, the research on teachers' beliefs about functions and graphs makes the meager research on teachers' knowledge appear positively mountainous. We are aware of no research that specifically focuses on the area of functions and graphs. At least one study of college students' and preservice teachers' attitudes toward mathematics addressed the perceived level of importance of functions and graphs (Schoeps et al., 1990). Preliminary findings indicate, not surprisingly, that most students view interpretation of graphs as important, but knowledge of functions as less so.

The current situation is that, in fact, we know very little about mathematics teachers' beliefs about functions and graphs. As we have suggested several times in postarithmetic mathematics, the primacy of the function concept demands that what teachers know and believe about functions be more fully investigated and articulated.

Research Questions

An examination of related research on student learning and teachers' cognitions and beliefs suggests parallel themes concerning teachers' knowledge and beliefs about functions and graphs that can be carefully researched and documented. This is a fertile ground for mathematics educators and researchers. Indeed, we conclude that both the dearth of research on teachers' knowledge and beliefs about functions and graphs and the critical importance of these concepts to the understanding of mathematics require the immediate attention of researchers.

There are many substantial questions to be addressed:

Are the cognitive learning processes of teachers and the specific knowledge required of them different from those of students?

If so, what are the implications for teacher education programs?

How important a concept do teachers perceive function to be?

In view of the different emphases suggested by the National Council of Teachers of Mathematics' (NCTM) *Curriculum and Evaluation Standards* (1989), how might teachers' beliefs about functions and graphs affect their acceptance of a mathematics curriculum with a different orientation?

What are teachers' views on the importance of introducing the notion of functions via multiple representations?

How do teachers perceive the role of textbook definitions of function in the classroom?

Do their concept definitions match their concept images?[2]

Do teachers exhibit a relational understanding of functions and graphs? If so, what specific mathematical and educational experiences might have led to this deeper understanding?

How do teachers view their own knowledge of functions and graphs?

How do teachers view their students' knowledge of functions and graphs?

How does the instructional process influence teachers' knowledge and beliefs about functions? How do teachers' cognitions and beliefs evolve? What factors influence this evolution? And in what ways?

Although answers to these questions can only come from careful investigations, we hope, in the next section, to show how what we do know can be synthesized in a way that provides direction to research programs, to instructional practice, and to curriculum development.

NEW DIRECTIONS

We noted earlier that the research on teaching and the research on learning seemed to have followed two distinct and disparate paths with very little articulation of the two. We also noted that since the publication of the latest *Handbook of Research on Teaching*, on which the previous point was based, there has been a discernible move in the direction of integrating the two areas of research. In March 1987, the NCTM, through its Research Agenda Project, sponsored a working conference focusing on the learning

---

[2]At least one study (Vinner & Dreyfus, 1989) indicates that this is not necessarily so, though the subjects here were middle-school teachers who were not mathematics majors.

and teaching of algebra. The conference, hosted by the University of Georgia, brought together an international group of mathematicians, mathematics educators, cognitive psychologists, curriculum specialists, practitioners, and educational technologists for the purpose of (a) discussing potentially productive lines of research in the learning and teaching of algebra and (b) developing a research agenda for the 1990s. At the same time the University of Missouri hosted a similar conference, also a part of the Research Agenda Project, focusing on effective mathematics teaching. Proceedings from these conferences, *Effective Mathematics Teaching* (Grouws, Cooney, & Jones, 1989) and *Research Issues in the Learning and Teaching of Algebra* (Wagner & Kieran, 1989), both attest to the recognition by researchers of the importance of joint research ventures among specialists in learning, teaching, and curriculum development, and point to areas where research is needed and where results of past research in learning and teaching might productively be applied. One outgrowth of the conferences was a recognition of the need for establishing research networks and for developing more coherent lines of communication and cooperation among researchers and practitioners (Sowder, 1989).

In January 1989, the Wisconsin Center for Education Research (WCER) sponsored the Hawaii Algebra Conference, a week-long working conference whose purpose was to convene researchers in algebra learning and researchers in mathematics teaching for the purpose of brainstorming—about mechanisms for scholarly collaboration and new research paradigms—and sharing expertise. One result of that conference has been the establishment of an "invisible college" of researchers who share common interests in algebra learning and teaching, thus facilitating communication. A secondary, but no less important, benefit of the success of that conference is the continuing financial support provided by the WCER for collaborative projects involving researchers in learning and teaching. These activities exemplify the current move toward greater collaboration among researchers and a burgeoning emphasis on integrating perspectives in research and teaching.

A Holistic Perspective on Learning and Teaching

It is tautological that the ultimate goal of the educational process is student learning. Learning in mathematics, and more generally, is an extremely complex and delicate process, which affects and is affected by the actions and thoughts of learners, teachers, researchers, cognitive scientists, and mathematicians (among others). In the past there has too often been not only a lack of communication among those involved in trying to understand the learning process, but worse, a lack of recognition that there might even be something valuable to be gained from interacting. Fortunately, this has begun to change.

In operating within a fundamentally constructivist perspective on learning and utilizing the advances made by a maturing cognitive science, mathematics educators and researchers are in a perfect position to develop research and curricular programs that effectively deal with the problems of learning and teaching. We know, for example, that the research in learning has given us a deeper understanding of the way students think and how they process the information that goes into concept building. The learning of complex concepts requires complex processes that, when identified and interpreted, have curricular or instructional ramifications, which in turn have implications for learning.

The role that teachers play in this process is not, nor should it be, simply a technical one. Teaching, like learning, involves a myriad of complex cognitive processes. Teachers are continuously being called upon to analyze intricate situations and make decisions. Shulman (1984) has described the intricacies of decision making in teaching as more complex than that of physicians. We have already mentioned the critical part that thinking plays in teachers' instructional planning and classroom interactions.

Cognitively Guided Instruction. Earlier we mentioned studies in which no direct correlation could be found between the level of teachers' content knowledge and the level of performance by their students on mathematics tests. However, a teacher's content knowledge is only one facet of the understanding teachers must have in order to teach effectively. Recent studies (Carpenter, Fennema, Peterson, Chiang, & Loef, 1989; Peterson, Carpenter, & Fennema, 1988) showed that teachers' knowledge of student thinking in certain mathematical situations does correlate with student achievement. Moreover, in these studies, which introduce a new paradigm for curriculum development, the researchers did not prescribe teaching behaviors. Rather, teachers were simply provided with research-based information on students' learning of mathematics. From there teachers worked in their own idiosyncratic ways to construct a mathematics curriculum that reflected their own enhanced understanding of student cognition. The principal tenets of this paradigm, denoted cognitively guided instruction (CGI), are described in the final sections of this chapter. (For a more complete description, see Carpenter & Fennema, 1988.)

It is my opinion, from the perspective of a mathematician and mathematics educator, that mathematics is not learned from teaching but from doing. I will not deny that students can be taught to recite the multiplication table, perform complicated long division algorithms, or, upon the presentation of a particular graph, identify it as a function. However, these things have little to do with mathematics and certainly do not alone provide evidence that any mathematics has been learned. Mathematical concepts at any level, whether they involve small integer products or tensor products, require active mental engagement. The definition of a particular concept, for instance, may be readily absorbed by a student, but the concept image must be constructed. The definition might be taught, but it is the image that is learned.

Cognitively guided instruction reflects this perspective on learning. Rather than training teachers to behave in certain ways, teachers are provided the opportunity to understand what students are thinking and doing—the goal being the facilitation of students' mental involvement in mathematics. It should be noted that an important aspect of CGI is attention to affective factors (e.g., students need to believe they can understand mathematics) as well as cognitive aspects. Fennema, Carpenter, and Peterson (1989) delineated instructional components of CGI that include the importance of making instructional decisions based on knowledge of each student's cognitions; the active mental engagement of students in mathematical activity; a stress on relationships among concepts, problem solving, and skills; assessment of students' thoughts and knowledge; and encouraging students to monitor their thinking and to become responsible for their own learning.

Although CGI does not subscribe to a single mechanism for accomplishing these tasks—teachers are expected to develop their instruction in ways in which they are comfortable and within the constraints of their classrooms—there are some important considerations related to teachers that must be taken into account. Two of these considerations reflect the two main strands discussed earlier in this chapter, teachers' knowledge and teachers' beliefs.

In order for teachers to operate at a level sufficient to implement CGI in any classroom, they must possess adequate knowledge of the mathematical content as well as knowledge of students' learning. Futhermore, teachers will need the skills to make rapid assessments of student cognition and to make appropriate instructional decisions based on this knowledge. There are research questions that need to be asked, and answered, concerning effective methods for facilitating teachers' acquisition of this crucial knowledge.

In addition to content and pedagogical knowledge, teachers must recognize and share the fundamental tenets of the constructivist perspective and have faith in the outcomes of cognitive science research in order to appreciate fully, as well as work entirely within, the domain of CGI. Fennema et al. (1989) described continua of opposing beliefs (such as the belief that students construct their own knowledge versus the belief that students receive knowledge) that indicate the degree of resonance that teachers had with CGI philosophy. Teachers having beliefs congruent with a CGI perspective conducted their classrooms in ways more consistent with CGI, reflected greater knowledge of the content domain and of their students' thinking, and produced students who were better able to solve problems than those in classes in which the teacher was less inclined to use CGI.

Figure 7.1 represents the essence of the CGI model for curriculum development. Notice that there appears no place in this model for what we normally think of as researchers, curriculum designers, instructional materials developers, or textbook authors. In fact, they are here. These roles are simply embedded among the many roles that the teacher plays. Teachers who are to work successfully within this model must become researchers in the sense that they need to be astute observers of children's actions and thoughts, as well as investigators of the possible reasons for these thoughts and actions. Teachers are empowered by their knowledge of their students' learning to design the curriculum that best fits the cognitive and affective needs of the students. Teachers may not take on the role of textbook author in the usual sense, but teachers must become the recognized source of knowledge in the classroom. Most current evidence indicates that textbooks rather than teachers are viewed as the ultimate arbiter of knowledge in the classroom—by students and teachers alike!

Although the CGI research reported so far has shown very promising results, some temporizing comments need to be made. For instance, the CGI model has up to this point been tested only in elementary mathematics classes. Can the model be extended to higher level grades or generalized to other domains? Teachers' knowledge and beliefs, even after sensitization to the CGI program, are not likely to remain static (in fact, the model indicates an interaction between knowledge and beliefs), but rather will be shaped by the outcomes of their decisions. Given that one's knowledge and beliefs determine decisions that are made, should we not assess the ways in which the outcomes of the decision-making process impinge upon and possibly

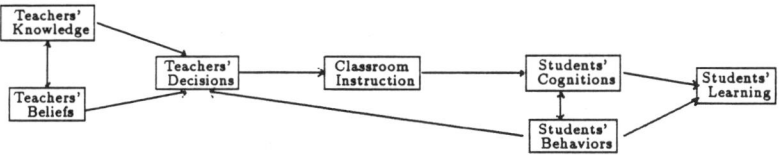

FIG. 7.1. Cognitively guided instruction (CGI) model for curriculum development. *Note.* From "Research in Cognitively Guided Instruction" by T. P. Carpenter and E. Fennema, 1988, in E. Fennema, T. P. Carpenter, and S. J. Lamon (Eds.), Integrating Research on Teaching and Learning Mathematics (p. 9), Madison: Wisconsin Center for Education Research. Copyright 1988 by the National Center for Research in Mathematical Sciences Education (NCRMSE). Reprinted by permission.

alter one's knowledge and beliefs? In the CGI model, experts in children's learning (i.e., researchers/teacher educators) play a crucial role—more than the simple dissemination of literature—in children's learning. Should not these roles be included as part of the CGI model—and consequently be subject to analysis and evaluation?

The current model of CGI describes the interrelated factors that influence teachers' decisions and students' learning, providing a feedback loop between teacher and learner that is mediated by instruction, students' cognition, and their behavior. This model certainly could be localized at different levels, including that of university teacher preparation. However, we would like to suggest a modification of the CGI model that reflects a more nearly complete articulation among those having a substantial influence on the instruction and learning processes. We do not claim completeness because many important factors, such as family influences, administrative and social policy, community beliefs about education, and so forth, are not included. This model is depicted schematically in Fig. 7.2.

There are actually two identical but slightly modified components of CGI embedded in this model. These relate interactions between teacher educator and teacher, and between teacher and pupil. While these can be seen locally to be similar to the CGI model described earlier, differentiating the two is of fundamental importance. This model also reflects the role of the researcher and indicates certain directional influences among learners, researchers, teachers, and teacher educators. There are *instructional* effects directed from teacher educator to teacher to pupil. Here, instruction is the

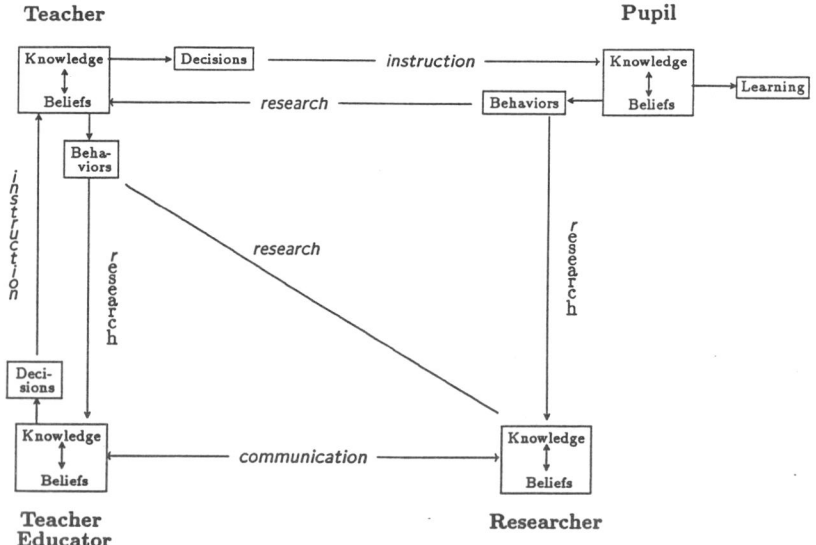

FIG. 7.2. Extended CGI model for curriculum development.

result of teachers' decision making, which in turn impinges upon the students' (including teachers as learners) cognitions, behaviors, and learning. There are also paths representing *research* effects. Here the term research is used in the broadest sense and includes informal research (such as teachers' observations and reflections on student thinking and behavior), as well as the more classical lines of research that one expects from professional researchers. It should be understood that this is a collapsible model in the sense that, for example, a researcher and teacher educator may be one and the same individual. However, there are times when a researcher provides no direct instructional effects; rather, these are mediated through communication with a teacher educator or teacher.

Functions, Graphs, and Cognitively Guided Instruction

Working under the assumption that CGI provides an effective model of instruction, that is, one in which significant learning takes place, we would like to consider some of the implications of this model for researchers, teacher educators, and teacher education programs. To illustrate these implications, we discuss them in the context of functions and their graphs, the focus of this chapter.

First, since the understanding of students' mathematical thinking is so critical to cognitively guided instruction, the continuing analysis and description of students' cognition regarding functions and graphs has direct implications for researchers. Furthermore, in view of the current dearth of information about algebra teachers' mathematical knowledge, one objective of future research should be to develop a complete and coherent picture of teachers' knowledge. There are a number of mechanisms for doing this; one avenue that might prove fruitful is the generation of detailed teacher knowledge profiles. This theory-based assessment of teachers' content knowledge involves a range of content questions, pedagogical explanations, and extensive interviewing (Post et al., 1988). The development for algebra teachers of knowledge profile assessments, possibly focusing on critical algebraic concepts (such as variable, equation, function, graph), would be quite helpful to both researchers and teacher educators.

Second, mathematics teachers at both secondary and postsecondary levels will need to take a more research-oriented perspective in their observations of students' cognitive and mathematical behavior and develop means for integrating this knowledge in a meaningful way into their instruction. The concept of teacher as researcher has been developing over the past few years. An example of this focus is the Research Interpretation Project, which is producing a collection of papers authored jointly by researchers and teachers. The teachers provide their interpretations from the perspective of the practitioner within the context of mathematics classrooms. As teachers become increasingly involved in the research process, both informally and formally, researchers in teaching and teacher education need to examine more critically this new role of teacher as researcher.

This role might be played out in the context of functions and graphs by having teachers examine their own and their colleagues' understanding of these concepts or, perhaps, through a critical examination of how teachers attempt to teach their concepts to their students. For example, a teacher might identify the focus of lessons dealing with functions and the particular attributes of functions that the teacher chooses to emphasize.

A related consideration is the application of CGI at the level of teacher preparation. There are at least two important issues here. One is the role that teacher educators could or should play in influencing teachers' belief systems about mathematics, about learning, and about teaching. Just as it was shown that teachers' comparative agreement with the precepts of CGI influenced their implementation of instruction, it is likely that the beliefs of teacher educators about these precepts will affect the ways in which they sensitize their students (i.e., prospective teachers) to the principles of CGI. Related to this is the role of teacher educators in "educating" their colleagues who teach mathematics in ways that are not consonant with accepted, or even reasonable, principles of learning and teaching. This is an issue worthy of careful thought. A second issue involves the role of teacher as learner. Teacher educators, in order to implement CGI, will need to be knowledgeable, not only about the thinking and learning of children, which has been extensively researched, but also about the cognitive processes of postsecondary students. This, by the way, suggests a number of research questions that need to be addressed: In what way are the cognitive processes of adult learners similar to or different from those of pre-adult learners? What teacher behaviors are most effective for enhancing adult learning? What teacher cognitions and teacher beliefs are relevant? As pointed out previously, in the context of mathematics generally and functions and graphs in particular, little is known about teachers' knowledge and beliefs, and even less about how the knowledge and beliefs of teacher educators might bear on that of teachers. Consequently, teacher educators need to take a more critical view of teachers as learners.

There are implications for teacher education programs as well. The educational experiences of secondary mathematics teachers often include much formal mathematics, some generic education coursework, and very little focus on mathematics education. (For a description of typical programs in mathematics teacher education, see Hooten, 1982.) Of the latter, when a mathematics education component exists, it usually involves the analysis of teaching methodologies and puts little or no focus on the enterprise of learning. The inference that many students draw from these experiences is that good teaching is the product of acquiring a particular constellation of teaching skills. Although this is part of the story, it is certainly not the total picture (see Schoenfeld, 1988). Perhaps now is the time to begin a reevaluation of the mathematical experiences we wish our teachers to have. As we have mentioned before, our teacher education programs may be producing teachers that are good *at* mathematics, but who lack a fundamental understanding of what mathematics is all about.

These ideas certainly apply equally at levels of high specificity. For example, in the mathematics teacher education programs with which we are familiar, certain basic mathematical concepts — such as the concept of function — are assumed to be already well understood by mathematics majors. Thus, little emphasis is placed on deepening and enhancing these concepts. If it is determined that some teachers have a deficient understanding of, say, the function concept, then there are implications both for teaching and curriculum at the university level.

A Study of Teachers' Knowledge of Algebra

Researchers at the University of North Carolina at Charlotte are currently developing a long-term study that deals with some of the issues raised here and is an outgrowth of efforts to further integrate research in learning and teaching. The ultimate goal of the study is to determine whether the CGI model can provide an effective integration of teaching and learning in secondary and postsecondary mathematics classrooms. The details of the study are described elsewhere (Norman & Prichard, 1990), but a brief description of the objectives of the study is appropriate. The research component has two primary goals:

1. To bring current research in algebra learning at all levels into the classroom through the conduct of teaching experiments.
2. To further research in the algebraic knowledge of postsecondary students and college graduates, with a special focus on preservice and inservice secondary mathematics teachers.

There are several foci for this research. The first involves an investigation of algebra teachers' understanding of fundamental algebraic concepts — the most central of which is the concept of function — and an examination of the ways that their content knowledge is utilized in the classroom. A second focus is the utilization of what is learned about teachers' cognitions to guide teacher educators in designing and implementing instruction for teachers that will enhance their understanding of algebraic concepts and the teaching of them. A third focus is on providing teachers access to research in students' learning of algebra and exposure to the principles guiding CGI.

Cognitive and instructional outcomes will then be analyzed. The fourth component involves actively widening the circle of researchers interested in further investigation of the model through similar or novel experimentation.

It is hoped that this study and others will begin to (a) delineate more clearly the algebraic knowledge held by teachers, (b) provide a better understanding of the instructional process, and (c) reinforce the move toward a holistic approach to mathematics education.

Closing Remarks

In the conclusion of their review of research in mathematics education, Romberg and Carpenter (1986) suggested seven areas to which researchers and scholars might turn their attention. These comments are quite relevant to the discussion at hand and can be rephrased in terms of teacher knowledge and beliefs.

1. The scope of research on teachers' learning must be expanded. There is very little in the mathematics research literature documenting either what teachers know or the nature of their knowledge. Teachers' thinking, mathematical or otherwise, is critical to the instructional process. The more we know about teachers' specific knowledge and cognitive processes, the better positioned we will be to understand how teachers deal with the complex activity of teaching. In particular, obtaining information about mathematics teachers' knowledge of functions and their graphs and how this knowledge is acquired is essential if we are to improve the mathematical education of our teachers.

2. Teaching research should consider how teachers' learning proceeds. For the most part, teaching research in the past has not adequately taken into account the role of student learning. It has completely ignored the processes by which preservice and inservice teachers learn. Although we might speculate, quite reasonably, that teachers' understanding does not differ dramatically from that of students, we still need a more coherent picture of what teachers know and how they process information in the unique environment that is teaching.

3. Models bridging the learning-teaching gap need to be constructed. CGI offers a first attempt at such a model—but this is likely not the only one. Others need to be developed, and they all need to be carefully researched. In addition, we need more research that synthesizes what is known about learning and teaching. This synthesis can provide a theoretical base on which such models can be built.

4. Mathematical content should be an important focus in such models. Mathematical thinking forms a complex web of concepts, processes, and

understandings. A missing link in our understanding of mathematical thinking arises in the lack of current research on teachers' knowledge of and beliefs about mathematics. The breadth and depth of mathematical understanding that teachers have determines the richness of their mathematical communication with students. Mathematics educators need to seriously consider what mathematics is, what it means, and why doing mathematics is so important. These are profound questions, and models of learning and teaching that do not address them run the risk of taking a too narrow and simplistic view of a very complex domain.

5. The role of computers and technology must be considered. The emergence and proliferation of computing technology have profound implications for students and teachers alike. In one sense, teachers are still little more than students themselves, as are most of us, when it comes to understanding the technology itself, much less its pedagogical implications. The new technologies have great power to affect not only methods for teaching mathematics, but the mathematical content to be taught, who will have access to that knowledge, and who will teach it. In fact, computers have changed the face of mathematics itself. Teachers and others of us involved in mathematics education must build on a vision of the mathematics of the future and be prepared to realize that vision as competently as we can.

6. New assessment tools must be developed. There are three issues here. First, teachers will have to consider alternative methods of assessment if their instruction is orchestrated from a perspective heavily based on attention to student learning. Second, teacher educators must also consider methods of effectively assessing teachers' understanding of mathematics and knowledge of student learning processes—this latter knowledge is not purely academic and cannot be acquired simply from a reading of the literature. It comes hand-in-hand with the experience of working with students and making careful analyses of students' cognitions and behaviors. This suggests something other than paper-and-pencil, or monitor-and-keyboard, assessments of abstract knowledge of the research. Third, assessment of teachers' professional development must take into account fundamentally different perspectives on teaching. Assessment restricted to a checklist of appropriate teacher behaviors cannot capture or fairly assess the rich understanding that a teacher may be facilitating for children.

7. We need to establish research programs. We have begun to address this issue. The "invisible college" of researchers in the learning and teaching of algebra that has evolved out of the Hawaii Algebra Conference referred to earlier is one example. The present volume is the result of that conference, in which researchers from very different perspectives were able to come together and share their experiences and ideas. Such scholarly collaboration is a step in the right direction.

Finally, we would like to suggest an additional challenge that is of fundamental and unquestionable importance and that, in the light of the focus of the 1989 conference, is an issue that deserves special concentration.

8. We need to undertake a reevaluation of the mathematics curriculum at all levels. First, if secondary school mathematics is to change in the ways advocated by NCTM's *Standards*, then the algebraic knowledge required of students will no longer be algorithmic, but rather will be of a deeper conceptual nature entailing more diverse interpretation. An obvious consequence of this is that, mathematically, teachers will need to be deeper, more creative thinkers, and better able to cope with a conceptual rather than computational approach to mathematics. Second, although the secondary curriculum has not changed radically in its focus, the nature of the mathematics taught in standard college classes is on the verge of a potential revolutionary change. For example, in the calculus (e.g., see Douglas, 1986), streamlining the content and increasing the use of graphing calculators and computers may eliminate some of the problems students have in constructing graphs, but at the same time these changes will require that students develop a greater ability to interpret graphs. Thus, some of the algebra skills that students can perform with only an instrumental understanding may become less important, whereas skills that require a deeper level of conceptualization move to the forefront.

Teachers' knowledge of functions and graphs is a small, albeit very important, part of the spectrum of mathematical knowledge relevant to secondary mathematics teachers. We do not know much about the nature of this knowledge nor about the beliefs that teachers hold. Although this is disappointing in some respects, there is also the sense that this is an area ripe for investigation. This is an exciting challenge! We have reasonably well developed paradigms for research in learning and teaching and now are constructing ways of integrating this research. There is much work to be done, but there is also the promise that the strands of understanding being developed might be woven together to provide a rich cognitive tapestry.

REFERENCES

Ball, D. L. (1990). Prospective elementary and secondary teachers' understanding of division. *Journal for Research in Mathematics Education, 21*(2), 132–144.

Begle, E. G. (1972). Teacher knowledge and student achievement in algebra. *School Mathematics Study Group Reports: Number 9*. Stanford, CA: Stanford University.

Bell, A., & Janvier, C. (1981). The interpretation of graphs representing situations. *For the Learning of Mathematics, 2*(1), 34–42.

Berliner, D. C. (1979). Tempus educare. In P. L. Peterson & H. J. Walberg (Eds.), *Research on teaching* (pp. 120–135). Berkeley, CA: McCutchan.

Brown, C. A. (1986). A study of the socialization to teaching of a beginning secondary mathematics teacher (Doctoral dissertation, University of Georgia, 1985). *Dissertation Abstracts International, 46/09,* 2605-A.

Bruner, J. S., Goodnow, J. J., & Austin, G. A. (1956). *A study of thinking.* New York: John Wiley & Sons.

Carpenter, T. P., & Fennema, E. (1988). Research and cognitively guided instruction. In E. Fennema, T. P. Carpenter, & S. J. Lamon (Eds.), *Integrating research on teaching and learning mathematics* (pp. 2–19). Madison: Wisconsin Center for Education Research.

Carpenter, T. P., Fennema, E., Peterson, P. L., Chiang, C.-P., & Loef, M. (1989). Using knowledge of children's mathematics thinking in classroom teaching: An experimental study. *American Educational Research Journal, 26*(4), 499–532.

Clark, C. M., & Peterson, P. L. (1986). Teachers' thought processes. In M. C. Wittrock (Ed.), *Third handbook of research on teaching* (pp. 255–296). New York: Macmillan.

Cooney, T. J. (1985). A beginning teacher's view of problem solving. *Journal for Research in Mathematics Education, 16*(5), 324–336.

Douglas, R. G. (Ed.). (1986). *Toward a lean and lively calculus.* (MAA Notes: Number 6). Washington, DC: Mathematical Association of America.

Dreyfus, T., & Eisenberg, T. (1983). *The function concept in college students: Linearity, smoothness, and periodicity.* Unpublished manuscript.

Dreyfus, T., & Eisenberg, T. (1987). On the deep structure of functions. In J. C. Bergeron, N. Herscovics, & C. Kieran (Eds.), *Proceedings of the Eleventh International Conference on the Psychology of Mathematics Education* (Vol. I, pp. 190–196). Montreal, Quebec: University of Montreal.

Dubinsky, E., Hawks, J., & Nichols, D. (1989, July). Development of the process conception of function in pre-service teachers in a discrete mathematics course. *Proceedings of the 13th Annual Meeting of the International Group for Psychology in Mathematics Education.* Paris.

Dugdale, S. (1983). *Green globs: A microcomputer application for graphing* [Computer program]. Sunnydale, NY: Sunburst Communications.

Eisenberg, T. A. (1977). Begle revisited: Teacher knowledge and student achievement in algebra. *Journal for Research in Mathematics Education, 8*(3), 216–222.

Even, R. (1989). Pre-service teachers' conceptions of the relationships between functions and equation. In A. Borbas (Ed.), *Proceedings of the 12th International Conference for the Psychology of Mathematics Education* (pp. 304–311). Veszprem, Hungary.

Feiman-Nemser, S., & Floden, R. E. (1986). The cultures of teaching. In M. C. Wittrock (Ed.), *Third handbook of research on teaching* (pp. 505–526). New York: Macmillan.

Fennema, E., Carpenter, T. P., & Lamon, S. J. (Eds.). (1988). *Integrating research on teaching and learning mathematics.* Madison: Wisconsin Center for Education Research.

Fennema, E., Carpenter, T., & Peterson, P. (1989). Teachers' decision making and cognitively guided instruction: A new paradigm for curriculum development. In N. F. Ellerton & M. K. Clements (Eds.), *School mathematics: The challenge of change* (pp. 174–187). Geelong, Victoria, Australia: Deakin University Press.

Fisher, L. C. (1988). Strategies used by secondary school mathematics teachers to solve proportion problems. *Journal for Research in Mathematics Education, 19*(2), 157–168.

Gage, N. L. (1964). Theories of teaching. In E. R. Hilgard (Ed.), *Theories of learning and instruction: Sixty-third yearbook of the National Society for the Study of Education: Part 1.* (pp. 268–285). Chicago: University of Chicago Press.

Goldenberg, E. P. (1987). Believing is seeing: How preconceptions influence the perceptions of graphs. In J. C. Bergeron, N. Herscovics, & C. Kieran (Eds.), *Proceedings of the Eleventh International Conference for the Psychology of Mathematics Education* (Vol. I, pp. 197–203). Montreal, Quebec: University of Montreal.

Goldenberg, E. P. (1989). Mathematics, metaphors, and human factors: Mathematical, technological, and pedagogical challenges in the educational use of graphical representations of functions. *Journal of Mathematical Behavior, 7*(2), 135–173.

Goldin, G. A., & Ellis, J. R. (1983). *Performance difficulties reported by first-year public school science and mathematics teachers in Illinois.* Dekalb: Northern Illinois University. (ERIC Document Reproduction Service No. ED 237 319)

Good, T. L., Grouws, D. A., & Mason, D. A. (1990). Teachers' beliefs about small-group instruction in elementary school mathematics. *Journal for Research in Mathematics Education, 21*(1), 2–15.

Graeber, A. O., Tirosh, D., & Glover, R. (1986). Preservice teachers' beliefs and performance on measurement and partitive division problems. In G. Lappan & R. Even (Eds.), *Proceedings of the Eighth Annual Meeting of PME-N* (pp. 262–267). East Lansing: Michigan State University.

Graham, K. G., & Ferrini-Mundy, J. (1989, March). *An exploration of student understanding of central concepts in calculus.* Paper presented at the Annual Meeting of the American Educational Research Association, San Francisco.

Grouws, D. A., Cooney, T. J., & Jones, D. (Eds.). (1989). *Effective mathematics teaching.* Reston, VA: National Council of Teachers of Mathematics; Hillsdale, NJ: Lawrence Erlbaum Associates.

Hamley, H. R. (1934). *Relational and functional thinking in mathematics: Yearbook Number 9.* Washington, DC: National Council of Teachers of Mathematics.

Herscovics, N. (1989). Cognitive obstacles in the learning of algebra. In S. Wagner & C. Kieran (Eds.), *Research issues in the learning and teaching of algebra* (pp. 60–86). Reston, VA: National Council of Teachers of Mathematics; Hillsdale, NJ: Lawrence Erlbaum Associates.

Hooten, J. R. (Ed.). (1982). Some contemporary problems in mathematics education [Special issue]. *Journal of Research and Development in Education, 15*(4), 64–89.

Janvier, C. (1978). *The interpretation of complex Cartesian graphs: Studies and teaching experiments.* Unpublished doctoral dissertation, University of Nottingham, UK.

Janvier, C. (1984). Constructing the notion of variable using history and bottles. In J. M. Moser (Ed.), *Proceedings of the Sixth Annual Meeting of PME-NA* (pp. 57–63). Madison: University of Wisconsin.

Kaput, J. J. (1989). Linking representations in the symbol systems of algebra. In S. Wagner & C. Kieran (Eds.), *Research issues in the learning and teaching of algebra* (pp. 167–194). Reston, VA: National Council of Teachers of Mathematics; Hillsdale, NJ: Lawrence Erlbaum Associates.

Kerslake, D. (1977). The understanding of graphs. *Mathematics in School, 6*(2), 22–25.

Kerslake, D. (1981). Graphs. In K. M. Hart (Ed.), *Childrens's understanding of mathematics: 11–16* (pp. 120–136). London: John Murray.

Kieran, C. (1989). The early learning of algebra: A structural perspective. In S. Wagner & C. Kieran (Eds.), *Research issues in the learning and teaching of algebra* (pp. 33–56). Reston, VA: National Council of Teachers of Mathematics; Hillsdale, NJ: Lawrence Erlbaum Associates.

Konshak, A. L., & Monk, G. S. (1976). *Mathematics assessment team: Progress report.* Unpublished manuscript, University of Washington at Seattle.

Malik, M. A. (1980). Historical and pedagogical aspects of the definition of function. *International Journal of Mathematical Education in Science and Technology, 11*, 489–492.

Markovits, Z. (1986). Functions today and yesterday. *For the Learning of Mathematics, 6*(2), 18–24, 28.

Markovits, Z., Eylon, B. S., & Bruckheimer, M. (1983). Functions—linearity unconstrained. In R. Herskowitz (Ed.), *Proceedings of the Seventh International Conference for the Psychology of Mathematics Education* (pp. 271–277). Rehovot, Israel: Weizmann Institute of Science.

Marnyanskii, I. A. (1975). Psychological characteristics of pupil's assimilation of the concept of function. In J. W. Wilson (Ed.), *Soviet studies in the psychology of learning and teaching mathematics: Volume XII-Problems of instruction* (pp. 163–172). Chicago: University of Chicago, School Mathematics Study Group and Survey of Recent East European Mathematical Literature.

Mayberry, J. W. (1983). The van Hiele levels of geometric thought in undergraduate preservice teachers. *Journal for Research in Mathematics Education, 14*(1), 58–69.

National Council of Teachers of Mathematics. (1989). *Curriculum and evaluation standards for school mathematics.* Reston, VA: Author.

Norman, F. A. (1986). Students' unitizing of variable complexes in algebraic and graphical contexts. In G. Lappan & R. Even (Eds.), *Proceedings of the Eighth Annual Meeting of PME-NA* (pp. 102–107). East Lansing: Michigan State University.

Norman, F. A. (1992). Teachers' mathematical knowledge of the concept of function. In G. Harel & E. Dubinsky (Eds.), *The concept of function: Aspects of epistemology and pedagogy,* MAA Notes, Vol. 25 (pp. 215–232). Mathematical Association of America.

Norman, F. A. (1987). An examination of students' uses of unitizing strategies in algebraic and graphical contexts (Doctoral dissertation, University of Georgia, 1987). *Dissertation Abstracts International, 48*/03, 587-A.

Norman, F. A. (1990, April). *Aspects of learning in the calculus: A brief look at attitudes and other recent research.* Paper presented at the SIG-RME research presession of the 68th Annual Meeting of the National Council of Teachers of Mathematics. Salt Lake City, Utah.

Norman, F. A., & Prichard, M. K. (1990). *Investigations in the learning and teaching of algebra.* Unpublished manuscript, Department of Mathematics, University of North Carolina at Charlotte.

Orton, A. (1970). *An investigation into growth of the concept of function.* Unpublished manuscript, University of Leeds, Great Britain.

Owens, J., & Henderson, E. (1985). Effects of secondary mathematics methods course on preservice teachers' views of mathematics. In S. K. Damarin & M. Shelton (Eds.), *Proceedings of the Seventh Annual Meeting of PME-NA* (pp. 224–229). Columbus: Ohio State University.

Peterson, P. L., Carpenter, T. P., & Fennema, E. (1988). Teachers' knowledge of students' knowledge and cognitions in mathematics problem solving: Correlational and case analyses. *Journal of Educational Psychology, 81*(4), 558–569.

Post, T. R., Harel, G., Behr, M. J., & Lesh, R. (1988). Intermediate teachers' knowledge of rational number concepts. In E. Fennema, T. P. Carpenter, & S. J. Lamon (Eds.), *Integrating research on teaching and learning mathematics* (pp. 194–219). Madison: Wisconsin Center for Education Research.

Pratt, D. L. (1985). Responsibility for student success/failure and observed verbal behavior among secondary science and mathematics teachers. *Journal for Research in Mathematics Education, 22*(9), 807–816.

Rector, J., & Ferrini-Mundy, J. (1986). Formal mathematics study and teachers' beliefs and conceptions: Interactions and influences. In G. Lappan & R. Even (Eds.), *Proceedings of the Eighth Annual Meeting of PME-NA* (pp. 256–261). East Lansing: Michigan State University.

Romberg, T. A., & Carpenter, T. P. (1986). Research on teaching and learning mathematics: Two disciplines of scientific inquiry. In M. C. Wittrock (Ed.), *Third handbook of research on teaching* (pp. 850–873). New York: Macmillan.

Schoenfeld, A. H. (1987). Cognitive science and mathematics education: An overview. In A. H. Schoenfeld (Ed.), *Cognitive science and mathematics education* (pp. 1–31). Hillsdale, NJ: Lawrence Erlbaum Associates.

Schoenfeld, A. H. (1988). When good teaching leads to bad results: The disasters of "well taught" mathematics courses. *Educational Psychologist* (Learning from instruction: The study of students' thinking during instruction in mathematics. Special issue), *23*(2), 145-166.

Schoeps, N. B., Royster, D. C., Prichard, M. K., & Norman, F. A. (1990). *Attitudes of college mathematics students*. Unpublished manuscript, Department of Mathematics, University of North Carolina at Charlotte.

Schwartz, J. L. (1987). The representation of function in The Algebraic Proposer. In J. C. Bergeron, N. Herscovics, & C. Kieran (Eds.), *Proceedings of the Eleventh International Conference for the Psychology of Mathematics Education* (Vol. I, pp. 235-240). Montreal, Quebec: University of Montreal.

Schwartz, J. L. (1988). *The algebraic proposer* [Computer program]. Hanover, NH: True BASIC, Inc.

Shavelson, R. J. (1973). What is *the* basic teaching skill. *Journal of Teacher Education, 24*(2), 144-151.

Shavelson, R. J., & Stern, P. (1981). Research on teachers' pedagogical thoughts, judgments, decisions, and behavior. *Review of Educational Research, 51*, 455-498.

Shaw, E. L., Padilla, M. J., & McKenzie, D. L. (1983). *An examination of the graphing abilities of students in grades seven through twelve*. Paper presented at the annual meeting of the National Association for Research in Science Teaching, Dallas.

Shulman, L. S. (1984). It's harder to teach in class than to be a physician. *School of Education News, Autumn*, 3. Berkeley: University of California.

Shulman, L. S. (1986). Paradigms and research programs in the study of teaching: A contemporary perspective. In M. C. Wittrock (Ed.), *Third handbook of research on teaching* (pp. 3-36). New York: Macmillan.

Shulman, L. S., & Elstein, A. S. (1975). Studies of problem solving, judgment, and decision making: Implications for educational research. In F. N. Kerlinger (Ed.), *Review of research in education* (Vol. 3, pp. 3-42). Itasca, IL: F. E. Peacock.

Skemp, R. (1978). Relational and instrumental understanding. *Arithmetic Teacher, 26*(3), 9-15.

Sowder, J. (1989). *Setting a research agenda*. Reston, VA: National Council of Teachers of Mathematics; Hillsdale, NJ: Lawrence Erlbaum Associates.

Tall, D. (1986). *Building and testing a cognitive approach to the calculus using interactive computer graphics*. Unpublished doctoral dissertation, University of Warwick, Coventry, UK.

Tall, D. (1987). Constructing the concept image of a tangent. In J. C. Bergeron, N. Herscovics, & C. Kieran (Eds.), *Proceedings of the Eleventh International Conference for the Psychology of Mathematics Education* (Vol. III, pp. 69-75). Montreal: University of Montreal.

Tall, D., & Vinner, S. (1981). Concept images and concept definition in mathematics with particular reference to limits and continuity. *Educational Studies in Mathematics, 12*, 151-169.

Thomas, H. L. (1975). The concept of function. In M. Rosskopf (Ed.), *Children's mathematical concepts: Six Piagetian studies* (pp. 145-172). New York: Teachers College Press.

Thompson, A. G. (1982). Teachers' conceptions of mathematics and mathematics teaching: Three case studies (Doctoral dissertation, University of Georgia, 1982). *Dissertation Abstracts International, 48*/07, 2267-A.

Thompson, A. G. (1984). The relationship of teachers' conceptions of mathematics and mathematics teaching to instructional practice. *Educational Studies in Mathematics, 15*(2), 105-127.

Thompson, P. (1989). *Word problem assistant* [Computer program]. Normal: Illinois State University, Department of Mathematical Sciences.

Travers, R. M. W. (Ed.). (1973). *Second handbook of research on teaching*. Chicago: Rand McNally.

Vinner, S. (1983). Concept definition, concept image, and the notion of function. *International Journal of Mathematical Education in Science and Technology, 14*, 293–305.

Vinner, S., & Dreyfus, T. (1989). Images and definitions for the concept of function. *Journal for Research in Mathematics Education, 20*(4), 356–366.

Vinner, S., & Hershkovitz, R. (1980). Concept images and common cognitive paths in the development of some simple geometrical concepts. In R. Karplus (Ed.), *Proceedings of the Fourth International Conference for the Psychology of Mathematics Education* (pp. 177–184). Berkeley: University of California, Lawrence Hall of Science.

Wagner, S., & Kieran, C. (Eds.). (1989). *Research issues in the learning and teaching of algebra*. Reston, VA: National Council of Teachers of Mathematics; Hillsdale, NJ: Lawrence Erlbaum Associates.

Wagner, S., Rachlin, S. L., & Jensen, R. J. (1984). *Algebra learning project: Final report*. Athens: University of Georgia, Department of Mathematics Education.

Wenger, R. H. (1987). Cognitive science and algebra learning. In A. H. Schoenfeld (Ed.), *Cognitive science and mathematics education* (pp. 217–252). Hillsdale, NJ: Lawrence Erlbaum Associates.

Wertheimer, M. (1959). *Productive thinking*. New York: Harper.

V CLASSROOM INSTRUCTION

8 Functions, Graphing, and Technology: Integrating Research on Learning and Instruction

Carolyn Kieran
*Department of Mathematics & Computer Science,
University of Quebec at Montreal*

This chapter is divided into two parts. The first part provides a conceptual backdrop for discussing research on the teaching and learning of functions and their representations. It includes the following: a comparison between set-theoretic and dependency notions of functions; an overview of the historical development of the concept of function; a detailed description of a "process-object" model that explains past student learning of functions; and an examination of the skills students need in order to translate not only between various representations of functions and the underlying situation but also between one representation and another. The second part of the chapter looks at recent projects involving functions and graphing in technology-supported environments. These projects have been classified according to the extent to which graphs are related to algebraic representations: the use of graphs before the teaching of algebra or in activities that do not require a knowledge of algebra; the use of graphs in first-year algebra courses; and the use of graphs with students who have already completed at least one course in algebra. The chapter concludes with some questions for further research on graphing and functions.

The presence of computer- and calculator-assisted graphing tools in our schools is a strong impetus to making graphs and their functions a centerpiece of mathematical instruction. The capability of computers to dynamically display simultaneous changes in graphical, algebraic, and tabular representations suggests a mathematically rich environment for learning about functions. However, the lessons learned from past curricular reforms compel us to make pedagogical decisions in the light of a range of mathematical, historical, psychological, and technological considerations.

These considerations form the first part of this chapter. Thus, the first part of the chapter is rather lengthy, focusing not only on the underlying mathematical content and how it developed historically, but also on what we know of students' difficulties in coming to learn it. The second part of the chapter illustrates how technology is currently being used in a range of mathematics classrooms from pre-algebra to calculus as a means of helping students better understand the role and value of graphical representations.

FUNCTIONS AND THEIR REPRESENTATIONS

Two major themes emerge from an examination of the background literature on functions and their representations. The first centers on the tension between relational-dependency and set-theoretic notions of functions—also known as dynamic-static, procedural-structural, or process-object interpretations. The second focuses on the various skills involved in both representing functions and translating from one representation to another. However, there has been very little explicit intersection between these two themes. Up to now, except for the descriptions by Kaput (1989) and others of function representations that are either action oriented or display oriented, few attempts have been made to characterize the various representation–translation skills required by students in terms of, say, their ability to conceive of functions either as processes or as objects. Because of the separation of these two themes in the past literature on functions, much of the material in this chapter relates directly to either one theme or the other; nevertheless, I occasionally attempt to make explicit some connections between the two.

Set-Theoretic Versus Dependency Notions of Functions

When Freudenthal (1973, 1982) characterized functions, he emphasized the notion of dependency:

> The world is a realm of change, describing the world is describing change, and to do this one creates variable objects—physical, social, mental, and finally mathematical ones. . . . Our world is not a calcified relational system but as I called a realm of change, a realm of variable objects depending on each other. Functions is a special kind of dependences, that is, between variables which are distinguished as dependent and independent—an old fashioned looking terminology, which, however, stresses the phenomenologically important element: the directedness from something that varies freely to something that varies under constraint. [When] mathematized: the function

from A to B is an act that assigns to each element of A an element of B. Functions are all around in mathematics and its applications, albeit labelled in various ways—mapping, transformation, permutation, operation, process, functional, operator, sequence, morphism, functor, automaton, machine—which are used according to needs and opportunities. . . . Functions can be considered as special relations. Relation from A to B is any subset of the Cartesian product $A \times B$. Such a relation f is called a function from A to B if for every $a \in A$ there is exactly one $b \in B$ such that $\lceil a, b\rceil \in f$. This definition is logically equivalent with the former; phenomenologically it is not, and didactically phenomenologically not at all. It obscures the essential action of assigning, directed from A to B. . . . One can oppose these two definitions to each other as dynamic versus static. From the fact that all mathematics can be reduced to set theory, one may not conclude that it should be done and even less that it is a didactic necessity. (Freudenthal, 1982, pp. 12-13)

In underscoring the pedagogical accessibility of an approach to functions that is based on the notion of change and dependency, Freudenthal (1982) suggested introducing functions to elementary school children by means of graphs. More is said in a later section on the particular approach favored by Freudenthal. Davis (1982) also emphasized the importance of introducing the concept of function as a dependency relation at a relatively early age (around 10 years):

"Set" does NOT become a fundamental idea until relatively late in the course of a person's mathematical growth. "Function" becomes a key idea far earlier chronologically than "set" does. Any time we want one number to depend upon another—as the postage charge depends upon the weight of a letter, say—we are dealing with the idea of function. Any time we want some decision to depend upon some particular outcome or value or data or situation, we are dealing with the idea of function. Functions are not only from rational numbers to rational numbers. When we assign truth values to statements, or numerical areas to geometric shapes, we are dealing with functions. Whenever we have any sort of mapping, we are dealing with this same fundamental idea. The refinement may differ, but the basic notion is recognizably the same. Without functions we can have no graphs, no formulas—and, really, no binary operations such as addition or subtraction. Mathematically, "function" is a fundamental idea from quite an early point in the individual's study of mathematics. Indeed, it is one of the most important of all. (p. 52)

Unfortunately, as Shuard and Neill (1977) pointed out, the idea of functional dependence has been totally eliminated from the current definition of function. In the process of generalizing the function definition, mathematicians have banished the rule that was the essential idea of the function.

Historical Development

The historical development of the concept of function is a relatively recent phenomenon, dating from the 17th century onward. Prior to that period, there appeared, during the 13th and 14th centuries in the works of philosophers in Paris and Oxford, a graph of the relationship of the velocity of a freely falling body as a function of time; however, there was no algebraic language with which to express this functional correspondence (Kleiner, 1989). Kleiner pointed to four events that were instrumental to the development of the function concept:

1. Extension of the concept of number to embrace real and (to some extent) even complex numbers (Bombelli, Stifel);
2. The creation of a symbolic algebra (Viète, Descartes);
3. The study of motion as a central problem of science (Kepler, Galileo);
4. The wedding of algebra and geometry (Fermat, Descartes). (p. 283)

It should be pointed out that the last development was crucial, but that it relied on the introduction of variables and the expression of the relationship between variables by means of equations. What was still lacking, however, for the final stage of the development of the concept of function was the notion of independent and dependent variables.

It was not until 1755 that Euler, whose own view of functions evolved over a period of several years, produced the following definition:

> If, however, some quantities depend on others in such a way that if the latter are changed the former undergo changes themselves then the former quantities are called functions of the latter quantities. This is a very comprehensive notion and comprises in itself all the modes through which one quantity can be determined by others. If, therefore, x denotes a variable quantity then all the quantities which depend on x in any manner whatever or are determined by it are called its functions. (Rüthing, 1984, pp. 72-73)

This concept was subsequently modified in the 1830s by Dirichlet, who viewed a function as an arbitrary correspondence between real numbers, and was generalized a hundred years later by Bourbaki, who defined function as a relation between two sets.

This brief historical account suggests how the use of mathematics as an instrument to study natural phenomena was first manifested by means of graphs of physical relationships. This kind of mathematical representation predated the use of algebraic language to describe the same events. It is also significant that these first graphs represented change as a function of time. As is seen later in the chapter, the use of time as an independent variable can be an effective pedagogical approach when first introducing the study of qualitative graphs to represent functional change. Other predecessors of functional representations include the tables of values generated by the Babylonians and the "anticipations" of functional relationships in Euclid's *Elements* (Kleiner, 1989).

The way in which the function concept developed historically has been reflected in school mathematics textbooks. In textbooks from the end of the 19th century until the middle of this century, a function was considered a change, or a variable depending on other variables. Hight (1968) provided us with a typical schoolbook definition from the 1880s: "Two variables may be so related that a change in the value of one produces a change in the value of the other. In this case the second variable is said to be a function of the first" (p. 575). Modifications occurred from about the middle of this century, illustrating the Bourbaki influence. In almost all school curricula the function is now defined as a relation between members of two sets (not necessarily numerical) or members of the same set, such that each member of the domain has only one image. Some textbook definitions do include mention of a rule; however, the notion of dependency is lost.

Psychological Considerations

The way in which mathematical concepts, including the function concept, evolved historically led to the elaboration by Sfard (1991) of a parallel model of mathematical conceptual development. This model can provide us with a global picture of the demands made on learners by various approaches to the teaching of functions. Consequently, it offers the potential to inform pedagogical decision-making in this area.

Process-Object Model. Sfard's (1991) historical-psychological analysis of different mathematical definitions and representations shows that abstract notions such as number and function can be conceived in two fundamentally different ways: structurally (as objects) or operationally (as processes). Note that, even though other researchers (e.g., Dreyfus, 1990; Harel & Kaput, 1990; Thompson, 1985) referred to the process-object duality in mathematics, I am emphasizing Sfard's work here because few others have characterized this duality at the same level of detail as she has. Sfard contrasted the distinctions between the two approaches in the following way:

> There is a deep ontological gap between operational and structural conceptions. . . . Seeing a mathematical entity as an object means being capable of referring to it as if it was a real thing—a static structure, existing somewhere in space and time. It also means being able to recognize the idea "at a glance" and to manipulate it as a whole, without going into details. . . . In contrast,

interpreting a notion as a process implies regarding it as a potential rather than actual entity, which comes into existence upon request in a sequence of actions. Thus, whereas the structural conception is static, instantaneous, and integrative, the operational is dynamic, sequential, and detailed. (p. 4)

Sfard claimed that the operational conception is, for most people, the first step in the acquisition of new mathematical notions. The transition from a "process" conception to an "object" conception is accomplished neither quickly nor without great difficulty. Sfard hypothesized three phases in the evolution of the process-object continuum: interiorization, condensation, and reification. After they are fully developed, both conceptions are said to play important roles in mathematical activity.

For the concept of function, Sfard suggested that a structural conception might be that of a set of ordered pairs, à la Bourbaki, and that an operational conception might include viewing a function as a computational process. Sfard also discussed various representations for functions from a structural/operational perspective:

The computer program seems to correspond to an operational conception rather than to a structural, since it presents the function as a computational process, not as a unified entity. In the graphic representation, on the other hand, the infinitely many components of the function are combined into a smooth line, so they can be grasped simultaneously as an integrated whole; the graph, therefore, encourages a structural approach. The algebraic representation can easily be interpreted both ways: it may be explained operationally, as a concise description of some computation, or structurally, as a static relation between two magnitudes. (p. 6)

Sfard pointed out that the historical evolution of the notion of function in the 17th and 18th centuries was tightly connected to the development of algebraic symbolism. One of the problems with the early definitions was that they leaned heavily on the concept of variable, which was itself very fuzzy. Sfard emphasized that Euler's operational definition, stated earlier in this chapter, was one that even Euler tried to structuralize by relating it to a graphical representation. However, "each time a definition had been proposed which would fit the algebraic-operational intuition, after a while somebody would find an example showing that the new description fell short of the structural-graphic version; and vice versa" (Sfard, 1991, p. 15). As we have seen, Dirichlet rebelled against the operational definition, which led eventually to the purely structural definition of Bourbaki, which made no reference to any kind of computational process. According to Sfard, function was at long last converted into a mathematical object on which new operations could be performed.

The stages during which the concept of function evolved historically from operational to structural led Sfard to create the parallel three-phase model of conceptual development that, as we discuss shortly, appears to have a good deal of empirical support. During the first phase, called *interiorization*, some process is performed on already familiar mathematical objects. For the concept of function, a variety of numbers can be used as input for a function machine that performs an arithmetical calculation. The idea of variable and formula can enter into play.

The second phase, called *condensation*, is one in which the operation or process is squeezed into more manageable units. For example, the learner might refer to the process in terms of an input-output relation, rather than indicate any operations. In the case of a computer procedure, a name might be given to the procedure, a name that refers to the product rather than to the underlying process. According to Sfard (1991):

> Thanks to condensation, combining the process with other processes, making comparisons, and generalizing become much easier. . . . When function is considered, the more capable the person becomes of playing with a mapping as a whole, without actually looking into its specific values, the more advanced in the process of condensation he or she should be regarded. Eventually, the learner can investigate functions, draw their graphs, combine couples of functions (e.g., by composition), even to find the inverse of a given function. (p. 19)

The condensation phase lasts as long as a new entity is conceived only operationally. The importance of this bridging phase cannot be emphasized too strongly in the design of pedagogical appproaches.

The third phase, *reification*, involves the sudden ability to see something familiar in a new light. Whereas interiorization and condensation are lengthy sequences of gradual, quantitative rather than qualitative changes, reification seems to be a leap: A process solidifies into an object, into a static structure. The new entity is detached from the process that produced it. Sfard stated:

> In the case of function, reification may be evidenced by proficiency in solving equations in which "unknowns" are functions (differential and functional equations, equations with parameters), by ability to talk about general properties of different processes performed on functions (such as composition and inversion), and by ultimate recognition that computability is not a necessary characteristic of the sets of ordered pairs which are to be regarded as functions. (1991, p. 20)

Operational/Structural Conceptions in Learners. Empirical support for the Sfard model with respect to the concept of function is derived from several studies, including her own. Sfard (1987) attempted to find out whether sixty 16- and 18-year-olds, who were well acquainted with the notion of function and with its formal structural definition, conceived of functions operationally or structurally. The majority of the students were found to view functions as a process for computing one magnitude by means of another, rather than as a correspondence between two sets. In a second phase of the study involving ninety-six 14- to 17-year-olds, students were asked to translate four simple word problems into equations and also to provide verbal prescriptions for calculating the solutions to similar problems. They succeeded much better with the verbal prescriptions than with the construction of equations. This evidence suggests a predominance of operational conceptions among Sfard's algebra students. (For further evidence of this phenomenon, see Thomas, 1969.) These findings also support the results of a previous study (Soloway, Lochhead, & Clement, 1982) that showed that students can cope with translating a word problem into an equation when that equation is in the form of a short computer program specifying how to compute the value of one variable based on another.

In several countries the concept of function is introduced in ninth-grade mathematics classes through a formal set-theoretic definition—a many-to-one correspondence between elements of a domain and range. However, functions are also introduced in science classes as a relationship between variables. Markovits, Eylon, and Bruckheimer (1983, 1986) investigated the effect of context (mathematics vs. science) on problems in which students were asked to draw the graph of a function by connecting given noncolinear points in a Cartesian plane. Two groups of ninth graders—a high-ability group and a low-ability group—were tested. Results indicated that most students provided a linear response, that is, almost all the graphs were composed of straight-line segments. Furthermore, the context had an effect on the success rate. High-ability students were more successful with the pure mathematics problems than they were with the problems embedded in a scientific context. This trend was reversed for the low-ability group. These findings suggest that, for the high-ability group, the transition to function as object may have been more advanced than it was for the low-ability group for whom function was still being perceived as a process. Their conclusion that many ninth graders "hold a linear prototypical image of functions" (Markovits et al., 1983, p. 276) is of pedagogical interest: With the advent of technology, it is no longer necessary to restrict instruction to the graphing of simple linear functions. (For other research findings on students' notions of functions, see the review by Leinhardt, Zaslavsky, & Stein, 1990.)

Another study related to process-object preferences in the learning of mathematical concepts is that carried out by Dreyfus and Eisenberg (1981). They investigated the intuitive bases for functional concepts among 440 sixth- to ninth-grade students. They asked questions on image, pre-image, growth, extrema, and slope in three representational settings—graph, diagram, and table of ordered pairs—in both concrete and abstract contexts. They found that high-ability students preferred the graphical setting for all questions, whereas low-ability students preferred the tabular setting. Though neither the graphical setting nor the tabular setting specifies directly how to compute one magnitude by means of another, the findings of this study suggest that low-ability students may be able to derive this information more easily from tabular settings than from graphical settings.

That teachers may not be sensitive to the hypothesized need of students to treat functions as processes in the early stages of learning is shown by the results of a study by Dreyfus and Vinner (1982; Vinner & Dreyfus, 1989). Five groups of students were tested: (a) college students taking a low-level course in mathematics, (b) college students taking intermediate-level courses, (c) college students in high-level courses, (d) college mathematics majors, and (e) junior high school mathematics teachers. One of the questions asked was, "What is a function, in your opinion?" Responses to this question indicated that, despite the emphasis on a set-theoretic definition of the function concept taught in their mathematics courses, the great majority of students did not accept it. The first three groups of students overwhelmingly (73%, 70%, and 63%, respectively) produced variations on a *process* definition (a function is a dependence relation, a rule, an operation, a formula, and so on). Even the mathematics majors were about equally split between a process definition and a formal, *object* definition. Only the teachers showed a strong preference (73%) for a formal interpretation.

Much of the existing research evidence (Kieran, 1992) suggests that high-school mathematics teachers have an object conception of function and that this is what they attempt to teach. The findings of the studies cited above indicate that more instructional effort ought to be placed on helping students bridge the gap between the old and new definitions of function, that is, between process and object interpretations. Sfard (1991) offered one caution, however:

> In order to see a function as an object, one must try to manipulate it as a whole: there is no reason to turn process into object unless we have some higher-level processes performed on this simpler process. But here is a vicious circle: on one hand, without an attempt at the higher-level interiorization, the reification will not occur; on the other hand, existence of objects on which the higher-level processes are performed seems indispensable for the interiorization—without such objects the processes must appear quite meaningless. In other words: *the lower-level reification and the higher-level interiorization are prerequisite for each other!* . . . The thesis of the "vicious circle" implies that one ability cannot be fully developed without the other: on one hand, a person

must be quite skillful at performing algorithms in order to attain a good idea of the "objects" involved in these algorithms; on the other hand, to gain full technical mastery, one must already have these objects, since without them the processes would seem meaningless and thus difficult to perform and to remember. (pp. 31–32)

If this is attempted too quickly, the period of "temporary meaninglessness" that must occur while lower-level reification is being orchestrated with higher-level interiorization may develop into a life-long attitude of apprehension toward mathematics. Unfortunately, there would seem to be no recipes that teachers can follow to assure that students safely make the transition from function as process to function as object.

Related Issues. One of the aims of the preceding sections has been to sensitize those who will be (or already are) using technology in the teaching of functions and graphs to consider the value of a process-object perspective. In the past, functions have been taught primarily by means of their algebraic representations; graphs were often used only to the extent that they might enrich or extend students' understanding of functions as represented by either equations or tables of values. But the advent of technology suggests that we might profitably shift the emphasis from the algebraic representations of functions to their graphical ones.

Such a shift suggests that major attention will have to be given to addressing issues related to "seeing" graphs as mathematical objects that can be manipulated, in the same way that algebraic representations of functions evolve to become manipulable objects. An additional concern relates to the cognitive interaction of graphs as objects with algebraic representations that are still at the process level. For example, consider the potential difficulties in, say, exploring the effects on the graph of the function $f(x) = ax^2 + bx + c$ by varying the values of a, b, and c. By changing the parameters, students are no longer working with variables of the function $f(x)$, but are studying some function-valued function $f_1(a,b,c)$ whose variables are a, b, and c, and whose value is the graphical output (Goldenberg, 1988). In other words, they are confronted with the need to operate on the function in such a way that both the algebraic representation and the graphical representation are conceived as objects. We do not yet know the psychological impact of such a pedagogical approach to the teaching of functions.

Representation and Translation

Much of the past work on functions and graphs focused on the various modes of representation for functions. Verstappen (1982) distinguished three categories for recording functional relations using mathematical language: (a) geometric—schemes, diagrams, histograms, graphs, drawings; (b) arithmetical—numbers, tables, ordered pairs; and (c) algebraic—letter symbols, formulas, mappings. Swan (1982) claimed that the most useful of these representations are tables of data, Cartesian graphs, and algebraic expressions. He summarized in Fig. 8.1, which he adapted from Janvier (1978) and Burkhardt (1981), the skills that students need in order to translate not only between each of these representations and the underlying situation but also between one representation and another.

The instruction assumed to underlie the development of skills summarized in the figure from Swan is one based on the notion of function as a dependency relation in a practical situation, rather than the more modern definition of a function as a subset of a Cartesian product. We briefly discuss these skills and related tasks because their description aids us in filling out the contours of the content domain of functions and the demands it makes on learners.

Tabulating, Plotting, and Reading-Off Values. Swan pointed out that tabulating, plotting, and reading-off values are the skills that are commonly emphasized in most mathematics textbooks, particularly those where scant attention is given to practical situations. Students are routinely asked to generate tables of values satisfying algebraic equations in two variables,

From \ To	Situations Pictures or verbal descriptions	Tables of data	Graphs	Algebraic expressions
		MODELING SKILLS		
Situations Pictures or verbal descriptions	—	measuring	sketching	descriptive modeling
	INTERPRETATION SKILLS			
Tables of data	reading	—	plotting	fitting
Graphs	interpreting	reading off	—	curve fitting
Algebraic expressions	formula recognition	tabulating or computing	curve sketching	—

FIG. 8.1. Skills needed in representation and translation. *Note.* Adapted from "The Teaching of Functions and Graphs" by M. Swan, 1982, in G. van Barneveld and H. Krabbendam (Eds.), *Proceedings of the Conference on Functions*, p. 155, Enschede, The Netherlands: National Institute for Curriculum Development. Copyright © 1982 by Swan. Adapted by permission.

plot points on a suitably scaled Cartesian graph, and read the coordinates of points off a graph, sometimes with the aim of solving an equation or system of equations. According to Swan, the consequences of emphasizing exclusively these skills are that students lose sight of the meaning of the task, rarely meet graphs other than those of straight lines, and get little practice at interpreting graphs in terms of realistic situations—skills that are essential in the sciences.

Research evidence supports the claim that facility with these skills does not necessarily go hand-in-hand with a student's ability to interpret the global features of a graph. Kerslake (1977, 1981), for example, presented 13-, 14-, and 15-year-olds with a task in which they were asked to describe the journey represented by the graph of Fig. 8.2. Only 14% of the 13-year-olds and 25% of the 15-year-olds succeeded. Many described the journey as "climbing up a mountain" or "going up, down, and then up again," illustrating students' confusion of the graph with a "picture" of the situation.

Sketching and Interpreting Graphs. Although sketching and interpreting seem basic to the understanding of graphs, very little attention has been paid to their development in mathematics classes in most North American schools. In contrast, programs in science education, as well as a few mathematics education projects abroad such as those at the Shell Centre in Nottingham and the National Institute for Curriculum Development in the Netherlands, have been known to include some work in this area.

Krabbendam (1982) emphasized the importance of learning to sketch

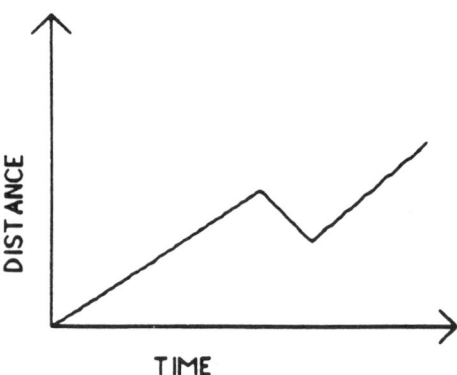

FIG. 8.2. Students were asked to describe the journey represented by this graph. *Note.* Adapted from "Graphs," by D. Kerslake, 1981, in K. M. Hart (Ed.), *Chidren's Understanding of Mathematics: 11–16,* p. 128, London: John Murray. Copyright © 1981 by John Murray Ltd. Adapted by permission.

graphs that are initially non-quantitative. For example, a peace demonstration in Amsterdam in 1981 served as a context for discussing a way of representing the event:

> On Saturday November 21, people from all our nation converged to Amsterdam, Museum Square, where a peace manifestation had been convoked. About 400,000 people—the biggest number ever—responded to the call, arriving in Amsterdam or at the outskirts by train, bus or car. Newspapers afterwards described the gigantic flow and how it had been processed. (Freudenthal, 1982, p. 15)

The students—seventh graders—were asked to draw a graph of the number of people on Museum Square on that day, depending on the hour of the day. The final mathematical representation drawn by the students had no scale on the vertical axis, and only an approximate sketch showing how the crowds increased until about two o'clock and stayed on until four o'clock when it began to rain (see Fig. 8.3).

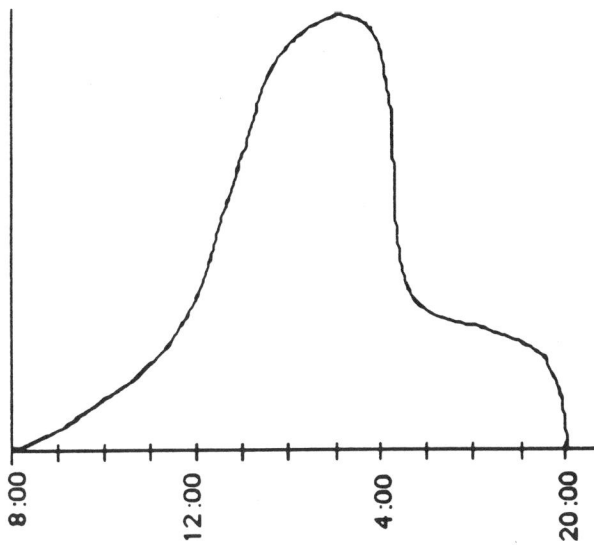

FIG. 8.3. Graph produced by seventh graders to represent how the number of people attending a peace demonstration changed as the day progressed. *Note.* Adapted from "The Non-Quantitative Way of Describing Relations and the Role of Graphs: Some Experiments" by H. Krabbendam, 1982, in G. van Barneveld and H. Krabbendam (Eds.), *Proceedings of the Conference on Functions,* p. 134, Enschede, The Netherlands: National Institute for Curriculum Development. Copyright © 1982 by Krabbendam. Adapted by permission.

Regarding the experience, Freudenthal (1982) made the following remarks:

> As a whole I think this 21 November is a marvelous rich subject, chockfull of good mathematics. But even if restricted to teaching functions it is prolific and illuminating rather than being imposed. A function has to be found out and justified by reasonable guesses, which can be doubted by the one and shall be argued on by the others. Mathematics is much more than meticulous precision of data, methods, and outcomes. Rough solutions of problems and educated estimations of numerical data is a perhaps even more important feature of initial learning than is hairsplitting precision. This fact is an aspect of mathematical attitude, and functions such as [those observed] in the above example is a good help on the way to this important goal. (p. 15)

Other examples of graph interpretation tasks are found in the works of Janvier (1978) and Swan (1982, 1985). For the graph illustrated in Fig. 8.4, Swan asked, "Which sport will produce a graph like this?" The classic example of Janvier's asked students to choose the race track along which the car was traveling when it produced the graph illustrated in Fig. 8.5.

Recent research in developing graph sketching and interpretation skills in the early stages of learning points to the importance of using time as an implicit or explicit independent variable (Krabbendam, 1982) and to the careful choosing of graphs that do not iconically resemble an aspect of the situation being represented (Hart, 1981). Swan (1982) suggested that the microcomputer offers important potential as a teaching aid for the devel-

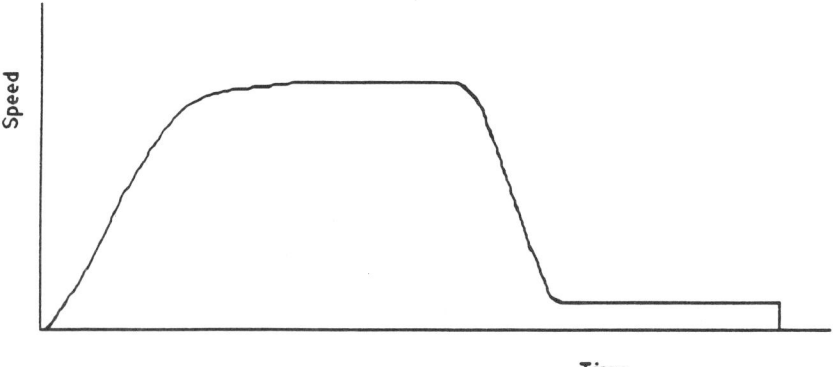

FIG. 8.4. Students were asked which sport would produce this graph. *Note.* Adapted from "The Teaching of Functions and Graphs" by M. Swan, in G. van Barneveld and H. Krabbendam (Eds.), *Proceedings of the Conference on Functions,* p. 157, Enschede, The Netherlands: National Institute for Curriculum Development. Copyright © 1982 by Swan. Adapted by permission.

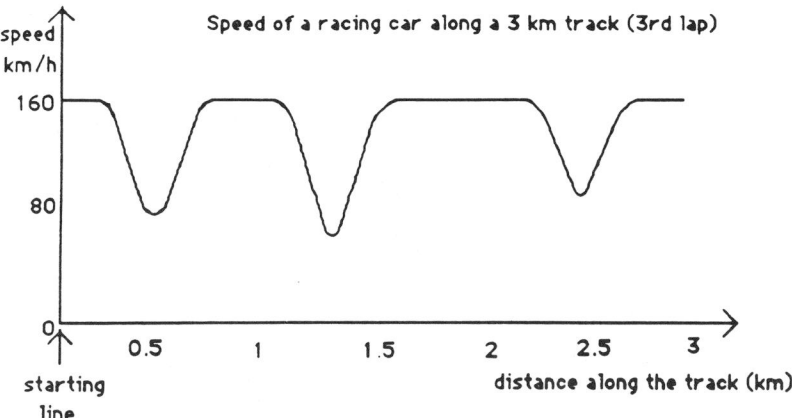

FIG. 8.5. Students were asked to choose the shape of the racetrack along which a car was traveling in order to produce this. Note. Adapted from *The Interpretation of Complex Cartesian Graphs—Studies and Teaching Experiments* by C. Janvier, 1978, unpublished doctoral dissertation, University of Nottingham, England. Adapted by permission.

opment of these two translation skills. Several examples of the use of technology in this area are offered in the second part of the chapter.

Measuring and Fitting. Constructing a table of data from a concrete situation and generating an algebraic representation that fits the tabular data are the two skills referred to here. Results from the Concepts in Secondary Mathematics and Science (CSMS) evaluation, which involved 3,000 13-, 14-, and 15-year-olds (Hart, 1981), indicate that translating a functional relationship between data pairs into algebraic symbols is one of the most difficult of representation tasks for students. An example of a task testing these skills that is not unlike those used in the CSMS project is one involving the pattern of tiles shown in Fig. 8.6.

Students were asked to find the number of white tiles needed to construct "bridges" over 10, 20, and eventually 100 black tiles. They could see a numerical pattern in their tables of data but were unable to translate it into an algebraic expression. A similar finding has also been reported by the National Assessments of Educational Progress (NAEP). One NAEP task, for example, involved completing the table shown in Fig. 8.7 (Carpenter, Corbitt, Kepner, Lindquist, & Reys, 1981). Most of the students with one or with two years of algebra could recognize the pattern—adding 7—from the given numerical values (success rates of 69% and 81% respectively). However, about a quarter of those who succeeded with the numerical part of the table were unable to generate the expression $n + 7$ or the equation $y = x + 7$.

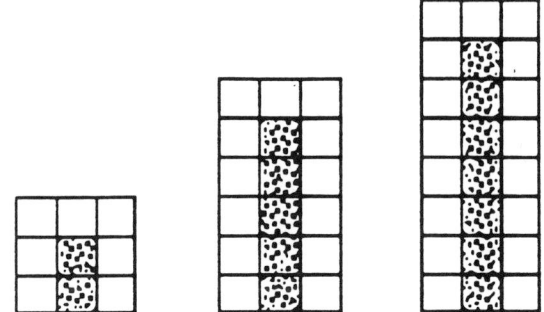

FIG. 8.6. Students were asked to translate the numerical pattern of the relationship between the black and white tiles into an algebraic representation. *Note.* Adapted from "The Teaching of Functions and Graphs" by M. Swan, in G. van Barneveld and H. Krabbendam (Eds.), *Proceedings of the Conference on Functions,* p. 159, Enschede, The Netherlands: National Institute for Curriculum Development. Copyright © 1982 by Swan. Adapted by permission.

X	1	3	4	7	n
Y	8		11	14	

FIG. 8.7. NAEP task on completing the function table and generating an equation. *Note.* Reprinted from *Results from the Second Mathematics Assessment of the National Assessment of Educational Progress* by T. P. Carpenter, M. K. Corbitt, H. S. Kepner, Jr., M. M. Lindquist, and R. E. Reys, 1981, p. 68, Reston, VA: National Council of Teachers of Mathematics. Copyright © 1981 by NCTM. Reprinted by permission.

Curve Sketching and Curve Fitting. Curve sketching and curve fitting involve translating from an algebraic expression to a graph and vice versa. Just as translating from a table to an algebraic expression is a difficult task for students, so are these. Kerslake (1977), whose extensive study of pupils' concepts of graphs formed the graphical component of the CSMS project in Great Britain (Hart, 1981), found that the relation between straight lines and their equations was understood by only 5–30% of students (depending on their age). Another British study, the Assessment of Performance Unit (1980), which tested a random sample of 14,000 16-year-old students in England, Wales, and Northern Ireland, reported that only 22% of its sample responded correctly to the question, "Which one of the following could be the graph of $y = (x - 1)(x + 4)$?" and only 9% to the question, "The graph shown is a representation of the function $f(x)$ where $f(x) = x(a - x)$; what is the numerical value of a?"

8. FUNCTIONS, GRAPHING, AND TECHNOLOGY 205

An activity that researchers at the Shell Centre (Swan, 1982) found effective in developing an awareness of curve fitting is an applied task requiring the generation of an algebraic model to describe a set of observed data. For example, students are presented with the information depicted in Fig. 8.8. They are asked to figure out how the prices of the windows have been worked out. They are encouraged to invent their own hypotheses (what do the prices depend on?), collect their own data (by measuring lengths, areas, and other features), and test these hypotheses (perhaps by drawing a scattergraph of price against the chosen factors).

The computer and the graphing calculator have enormous potential for helping students to develop skills in these two areas by providing opportunities for them to discover the effect of changing parameters and to draw whole families of curves on the same set of axes, as well as to analyze a variety of quantitative situations using curve-fitter programs. Examples are given in the second part of the chapter.

Formula Recognition, Reading Tables, and Descriptive Modeling. Two examples from Swan (1982) are used to illustrate the skills of formula recognition, reading tables, and descriptive modeling. The abilities to recognize formulas and to read tables are both required in the following task: "The largest weight W in kilograms that a plank bridge L metres long, b centimetres broad, and d centimetres thick can support is given by $W = 0.7b/Ld^2$. There has been a misprint in this formula somewhere. What is it?" To answer this question, students must test whether the formula varies according to their intuitions when they substitute numerical values for one of the variables.

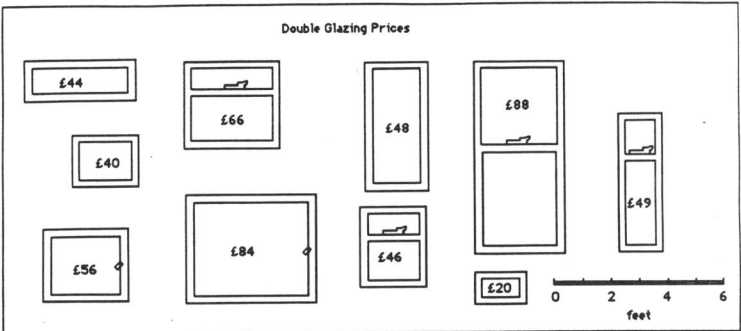

FIG. 8.8. A curve-fitting task requiring students to invent hypotheses, collect data, and test hypotheses. *Note.* Adapted from "The Teaching of Functions and Graphs" by M. Swan, in G. van Barneveld and H. Krabbendam (Eds.), *Proceedings of the Conference on Functions,* p. 162, Enschede, The Netherlands: National Institute for Curriculum Development. Copyright © 1982 by Swan. Adapted by permission.

The skill of descriptive modeling, perhaps the most difficult of all translation skills, is required in the following task: "Two boys decide to earn some holiday money by making and selling chess sets; write down an expression which they can use to tell how many sets need to be sold in order to make the enterprise profitable." Here, students must identify the relevant variables, symbolize them, and then generate the important relationships among them.

Summing Up

In the first section, we noted that there are two different perspectives that underlie mathematical notions of function—the situational one that emphasizes dependency in a relation of change and the formal, set-theoretic one. As Freudenthal and others have pointed out, the former can be considered a dynamic view and the latter, a static one. Instruction tends to focus on the latter approach and to emphasize the role of algebraic representations prior to graphical ones; yet historically, prior to the invention of algebraic language, graphs were used to represent early conceptions of functions. We glimpsed a few of the obstacles involved in translating among the various representations and examined in detail the power of the historical-psychological model of Sfard (1991) to explain the process-object phases of student learning of functions, even when functions are taught using structural, set-theoretic approaches. The application of this model to the initial learning of functions by means of instruction that emphasizes graphical representations permits us to generate a host of questions that can only be answered by careful research.

In the meantime, the arrival of technology in the mathematics classroom has sparked the beginnings of projects involving the use of graphs of functions in ways that were not possible in pretechnology days. The next part of this chapter focuses on recent attempts to enrich the classroom teaching of graphs and functions in ways that not only take advantage of advances in technology but also try to overcome some of the cognitive obstacles that have been uncovered in past research. In the concluding remarks, an effort is made to tie the objectives of these very recent projects to issues of process and object and to raise questions for further research.

TECHNOLOGY IN THE TEACHING OF GRAPHS AND FUNCTIONS

The description of recent projects involving graphs and their functions in technology-supported environments can be loosely divided into three sections: (a) use of graphs before the teaching of algebra or in activities that do not require a knowledge of algebra, (b) use of graphs in first-year algebra courses, and (c) use of graphs with students who have already completed at least one course in algebra. One of the advantages of this kind of categorization is that it can fit quite well with a process-object discussion according to algebraic maturity. It is to be noted that the description of projects that follows is illustrative rather than exhaustive; I attempt to provide as complete an overview as possible, with the caveat that I do not have access to all of the most recent details of ongoing projects.

Graphing Before Algebra

Most graphing activities that do not involve recourse to algebraic representation are those that deal with interpreting and sketching skills — that is, the translation of graphs into pictures or verbal descriptions and vice versa. An example of a program that has been found to assist children in acquiring an understanding of some of these qualitative aspects of graphing is *Eureka*, developed by the Shell Centre in Nottingham, England (Phillips, 1982). On the computer screen, the children are shown a cartoon of a man taking a bath. The lower part of the screen simultaneously illustrates with a graph how the water level varies when a child or the teacher presses one of the following four keys: (a) P — plug in or plug out; (b) T — taps on or taps off; (c) M — man in or man out; and (d) S — man sings or stops singing. Either the picture or the graph can be switched off and a new key sequence pressed. The class has to either sketch a graph corresponding to the situation being enacted or interpret a graph such as the one displayed in Fig. 8.9.

In order for children to make sense of such graphs, it is clear that certain conventions must be learned; for example, a trace moving up from left to right implies an increase along both dimensions and a trace moving down from left to right implies an increase along only the horizontal dimension. These are not trivial learnings for young children, as witnessed by the common tendency, even among older students, to interpret a graph as a picture of an event. Even though the axes are not scaled, interpreting graphs correctly in this qualitative context would also seem to require at least some discussion of scaling conventions.

Another piece of software emphasizing the qualitative interpretation of graphs is *Interpreting Graphs*, a set of two programs (*Relating Graphs to Events* and *Escape*) developed by Dugdale and Kibbey (1986a). The program *Relating Graphs to Events* displays three possible graphs for each event. The graphs are of different colors on the same axes, which have labels but no quantitative markings. The student selects the graph whose shape is most appropriate for the given event. When the student makes an incorrect choice, feedback is provided in the form of information as to what the chosen graph really describes.

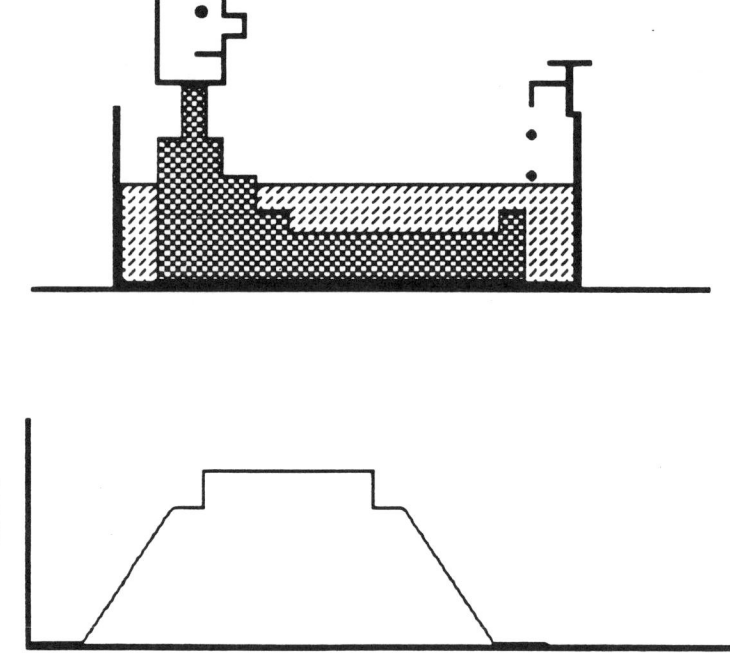

FIG. 8.9. The graph records changes in the water level. *Note.* Reprinted from "An Investigation of the Microcomputer as a Mathematics Teaching Aid" by R. J. Phillips, 1982, *Computers & Education, 6,* pp. 45–50. Copyright © 1982 by Pergamon Press. Reprinted by permission.

In her observations of 10 classes of high school students using this software, Dugdale (1987) noted that *Relating Graphs to Events* introduces ideas to which students are generally not accustomed and also reveals some problems such as students' initial tendency to ignore the labels on the axes. However, Dugdale found that this kind of software is an interesting and appropriate beginning to the learning of graphs and one that teaches students to deal with the ideas presented.

Karplus (1979) made the same point with graphs such as the one illustrated in Fig. 8.10. He emphasized that all that is really important about the functional relationships governing the given phenomenon can be determined by means of such graphs with coordinate axes that do not even have the units marked. He contrasted this teaching approach with the usual one in which students pay primary attention to the accuracy of the numerical values and to locating these carefully on a graph—only to have no concept of the relationship between the variables that are being represented.

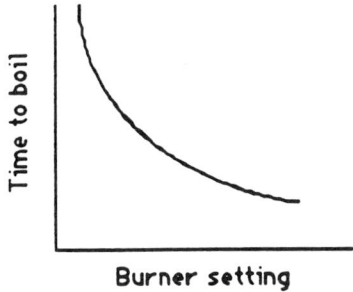

FIG. 8.10. A pot of water is heated to boiling on the kitchen stove. *Note.* Adapted from "Continuous Functions: Students' Viewpoints" by R. Karplus, 1979, *European Journal of Science Education, 1*(3), pp. 397–415. Copyright © 1979 by Taylor & Francis, Ltd. Adapted by permission.

However, for students who have already done some graph construction by plotting points, the adjustments required in learning to interpret graphs qualitatively can be considerable. Preece (1983) researched the graph-interpretation performance of 122 students aged 14 and 15 years with various pieces of software, including the computer simulation program *Pond* (Tranter & Leveridge, 1978), which had been developed to be used by students in investigating changes in populations of pond organisms. With *Pond*, students can, for example, set the initial population and pollution levels and the intensity of fishing. The program then produces a variety of different graphs on the same screen.

One of the tasks that Preece gave to the students included interpreting the graph of Fig. 8.11. She found that the students were not very confident with such tasks. They had had little experience with qualitative interpretation, and the absence of scales on the axes was unnerving for them. Despite the lack of scales, some students insisted on reading the curves in an absolute point-wise manner. Most started by selecting a single curve and gradually progressed to interrelating the curves. Questions involving gradients proved to be a stumbling block for many students. Preece pointed out that more research is needed on multiple-curve displays of interrelated variables and on the effects of individual strategies of graph interpretation. She concluded with the following remarks:

> The educational philosophy underlying many computer simulations such as POND is to provide pupils with an opportunity to devise and test hypotheses—the "what would happen if. . . ." approach. This research suggests that the richness of this type of educational environment cannot be exploited by most pupils . . . and it is very unlikely that they would perform differently . . . unless they received a lot of help from a teacher. (p. 54)

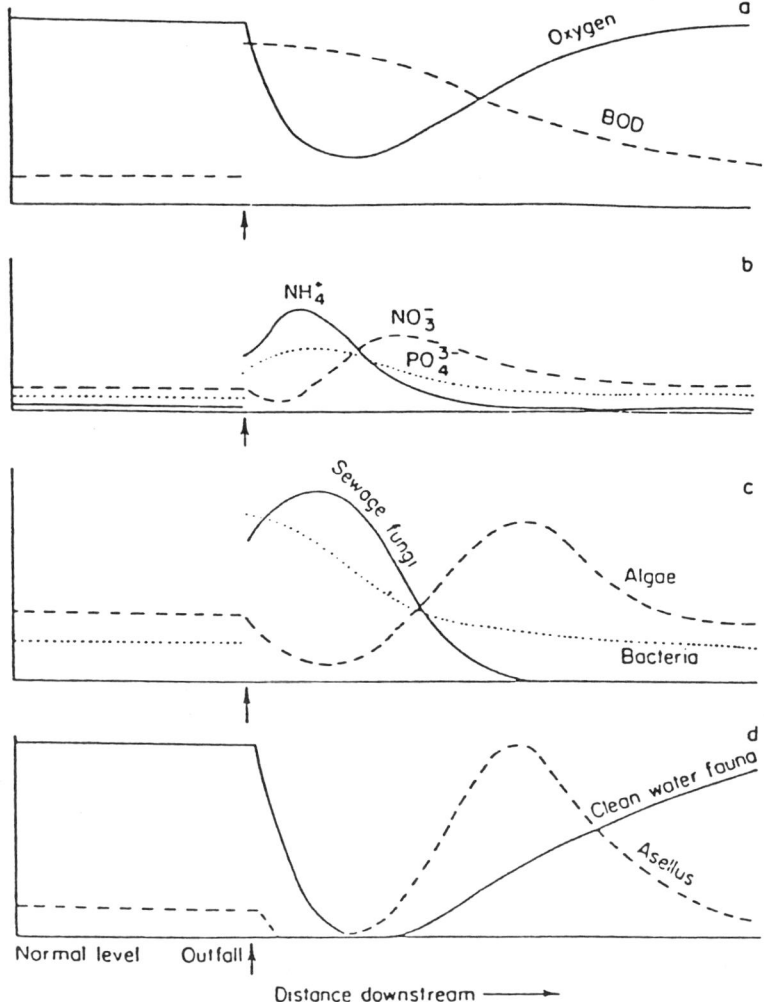

Note:
Asellus is a small animal commonly found in rivers and streams. It is a scavenger on dead plant and animal remains and does not have high oxygen requirements. Algae are green plants and the clean water fauna include fish and other animals.
- Relate the changes in graphs *a* and *b* to the changes in the graphs *c* and *d*. In doing this make clear the cause of each change and the interactions involved. Comment on the complexity of the series of changes illustrated by these data and their causes.

FIG. 8.11. Adaptation of graph-interpretation task given by Preece (1983). *Note.* From "Graphs are not straightforward" by J. Preece, 1983, in T. R. G. Green and S. J. Payne (Eds.), *The Psychology of Computer Use*, p. 43, London: Academic Press. Copyright © 1983 by Longman Group, UK. Adapted by permission.

Graphing in First-Year Algebra Courses

The technology-related projects described in this section involve graphs of functions along with algebraic representations. The students were in first-year algebra courses or had the equivalent in algebra experience.

The "Algebra With Computers" Project. Fey and Heid have been involved in an ongoing effort to reformulate the high-school algebra curriculum by "making the concept of function a central organizing theme for theory, problem solving, and technique in algebra; and developing students' understanding of algebra concepts, and their ability to solve problems requiring algebra, before they master symbol manipulation techniques" (Heid, 1988b, p. 1). In the computer-intensive environments developed by Fey and Heid for eighth and ninth graders, students regularly use a symbol manipulation program, function grapher, table-of-values generator, and curve fitter to analyze a variety of quantitative situations. As Heid reported,

> Prototypical problems would have students examine a given set of values for two related quantities, generate a function rule to describe the relationship, then answer questions about the effect on one variable of an equality condition on the other variable. . . . As students studied the algebraic concepts of variables and functions, they refined their abilities to work within and among various representations for functions. While initial computer explorations introduced students to the use of particular computer tools to analyze problem situations, later explorations engaged them in selecting appropriate computer tools and in using several computer tools in their analysis of a single problem situation. (1988b, p. 2)

Following is an excerpt of a classroom discussion that illustrates one of the ways in which graphs are treated in classroom discussions (Heid, Sheets, Matras, & Menasian, 1988):

The concept of different types of rates of increase came under scrutiny early in the year. One class transcript shows some of the early discussion on steadily increasing and gradually increasing curves. As is typical for this particular teacher, the discussion is fairly informal, with definitions brought up as needed:

The development of the concept of rate of increase starts early in the curriculum. The class is engaged in a discussion of a graph of the average height of females as a function of their age in years.

T: How did we connect the dots?

S: Straight lines.

T: Is that what we did?

S: We did it.

T: What would a straight line going up represent?

S: All of a sudden.

T: Not so.

S: A steady increase.

T: A steady increase [Teacher draws ] It means that something increases all the time.

But here [Teacher draws 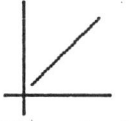] our first part of the curve doesn't have much of an increase. But as we go up, the curve is increasing, but increasing how?

S: Faster.

T: Here the curve is increasing the same way. [Teacher draws

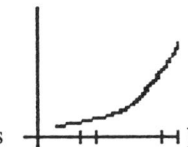]

If we take the same amount of space here and here

[Teacher draws ]

we see that the rate of increase is different. Say we took a 1-hour interval here and here and look at a portion of the graph here. We get the same increase.

[Teacher draws]
The easiest way to find the increase is to subtract the numbers like we did yesterday. For something like age and height, the increase was gradual. If there were any change in the rate at which you grow, they'll be in — starts with "g" . . .

SS: Gradual.

T: So we couldn't draw a curve like this:

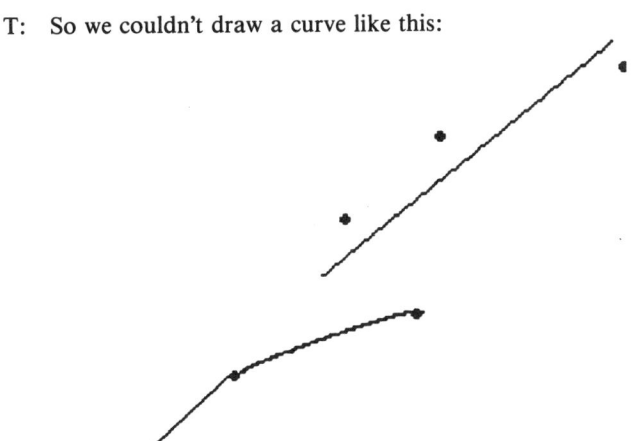

We need
[Excerpt from experimental class—9/23]

Midway through the second semester, this teacher still took advantage of opportunities to discuss rate of change. The discussions at this time directed student attention to the connection between the formula and the rate of increase:

Students had worked in the computer lab answering questions about the following situation:

Suppose that the Orioles' Eddie Murray hits a high pop-up straight above home plate. If the bat meets the ball 1.5 meters above the ground and sends it up at a velocity of 30 meters per second, then the height in meters of the ball t seconds later will be closely predicted by:

$$h(t) = -4.9t^2 + 30t + 1.5$$

[Teacher draws the curve on the board]

T: What does the graph tell you?

S: The further it goes, the faster it goes down.

T: How can you tell it's going down faster on a graph?

S: It's steeper.

T: That's right. The gravity factor [4.9] is next to something that's squared and the numbers get real big when you square them. They go down fast too.

[Excerpt from experimental class—3/27] (pp. 30-31)

The success of the Fey-Heid approach is noteworthy. End-of-year interviews with both project and control students showed that the project students outperformed their counterparts on such mathematical modeling goals as constructing, interpreting, and linking representations. They also surpassed the conventional classes in improvement of problem-solving abilities and did as well on a department final examination—a test of ability with traditional algebraic manipulations. Other results reported by Heid (1989) included the finding that initially the eighth graders experienced some difficulty in working successfully with the graphing tools, but by the end of the year all of them were able to make sense of and skillfully use the function-graphing tools.

The successor to the "Algebra With Computers" project is the "Computer-Intensive Algebra" study. This second project has the same pedagogical aims as the first one; the major differences between the two concern the kinds of research questions being investigated. One of the many issues being looked at is the seeming preference of some students for certain tools rather than others, despite their awareness that the tools they are choosing are not the most efficient ones for the task at hand.

The Resolver. *The Resolver* is a piece of software that is part of the Algebra series of the Education Development Center (EDC)—a series that was designed and programmed by Schwartz and Yerushalmy. Yerushalmy (1988) described *The Resolver* as "mainly an algebraic notepad . . . that can work in two modes: a free mode in which the user can enter any expression and then input transformed expressions; a target mode . . . which allows for target expressions to be written as representing a family of expressions (for example: $Ax + B$) or as the exact expected term (for example: $7x + 5$)" (p. 14). The characteristic of this software that is important for the present discussion is the capacity to provide for each transformation a parallel display of three graphs: (a) one graph of the original expression, (b) one graph of the current transformed expression, and (c) one graph of the difference between the two expressions. Because any legitimate transformation of an algebraic expression does not affect the graph of the expression, the difference graph is intended to provide both qualitative and quantitative information about the correctness of each step. Although *The Resolver* also includes tools to act on the graphs, such as scaling and zooming operations, the main goal of the particular research project described here was to use the graphs as feedback and not as a mathematical object on which to act.

Yerushalmy (1988) observed a group of seventh, eighth, and ninth graders using this software. The single seventh grader of the group had already learned how to transform and simplify expressions; the eighth graders had already worked with functions and graphs in a computer environment using *The Analyzer* (another EDC program); and the ninth graders, although considered weak students, had already learned to simplify and solve basic equations, but had not studied functions and graphs. Yerushalmy found that for the students who were familiar with graphs, the qualitative feedback proved to be interesting and of value. For those students exposed to graphs for the first time, graphical feedback proved to be unconvincing.

Because Yerushalmy's goal was to learn the effects of graphical feedback, she concluded that learning about graphs and functions seemed a vital prerequisite. She listed a set of topics that she hypothesized were required in order to be able to use graphical feedback effectively:

1. Defining rules in three representational modes: by graphical representation on coordinate system, by words, and by algebraic symbols.
2. Acquiring concepts related to graphs and coordinate systems, such as: axis, ordered pairs, and table of values.
3. Shifting from discrete points to the functions and their graphs.
4. Classifying graphs and functions by different criteria.
5. Understanding the role of various parameters of functions.
6. Geometrically transforming functions and graphs and observing parallel changes in the symbolic representation. (p. 33)

Her suggested curriculum does not follow the traditional chapters on functions and graphs in a first-year algebra course. Yerushalmy pointed out that the leading idea of the above list is "that students will be able to navigate with the coordinate system, that they will recognize shapes of graphs as relating to classes of polynomes and rational functions, and that they will understand geometrical modifications such as translation and stretch" (p. 34). She noted that computational methods for determining functions is missing from the list and emphasizes further that "even those students who have learned the traditional course on functions and graphs will probably lack the deep visual knowledge required but they might catch up while training with the graph feedback" (p. 34). Yerushalmy's experience with these students prompted her to remark, first, that she would continue to experiment with the use of graphs among students "who have already learned basic algebra" and, second, that the following is still an open question: "What could be the value of teaching graphs before teaching algebraic transformations?" (p. 34).

Triple Representation Model. The principal aim of the *Triple Representation Model* (TRM) microworld, which is an integral part of the TRM Curriculum developed by Schwarz (1988), is to facilitate transfer from one functional representation to another (the three representations implemented in TRM are the algebraic, graphic, and tabular). The TRM curriculum includes most of the topics taught in introductory units on functions. However, it is designed to avoid some of the obstacles usually encountered in traditional first-year algebra courses that include material on functions, such as students' mechanical rather than meaningful responses, their exclusive attachment to linearity (Markovits et al., 1986), difficulties in transferring information between representations, lack of a dynamic conception of functions, inability to see a function as an object, and the effects of compartmentalization (Vinner & Dreyfus, 1989). The structure of the TRM curriculum is listed next (Schwarz, Dreyfus, & Bruckheimer, 1990). The parts in italics are the main departures from conventional curricula.

1. Intuitive understanding of the concept of function.
   Operative definition of a function; basic concepts—pre-image, image, etc.; tabular and arrow representation; maximum, minimum; increase, decrease; *collection of experimental data; finding of sequential rules by guessing;* "real world" questions.
2. Graphical representation of a function.
   Transfer between verbal, graphical, and tabular representation; interpretation of graphs; *construction of graphs corresponding to a collection of data; limitations of the graphical representation; accuracy.*
3. Algebraic representation of a function.
   Emphasis on finite domain functions (discrete); infinite domain functions; transfer between tabular and algebraic representation; *inductive guessing of algebraic rules.*
4. Transfer between all three representations.
   The absolute value function, 1/x, greatest integer function, x^n, $\sqrt{x}$, polynomials.
5. Problem solving encouraging transfer between algebraic, tabular, and graphical representation.
   Solution of equations and of systems of equations; maximum/minimum problems; geometric transformations on graphs (translations, symmetries); links with algebraic transformations; *open ended search questions.* (p. 251)

Two examples of student activities with TRM are the following (Schwarz et al., 1990).

1. Given a (hidden) function f, find all its zeros and solve the equation $f(x) = 2$. In this example, the algebraic representation was locked, and so was the Draw operation (an operation by which students can magnify, stretch, or shrink a graph of a function defined algebraically). One of the options available was to use the Find Image operation (the tabular mode that enables students to build a table for a function defined in the algebraic representation).
2. Given the function $f: R \rightarrow R$, $f(x) = x^3 - x$; choose limits for x and y for which the graph of the function f will look like the given graph. (The use of the Draw operation allows students to treat

graphs as concrete manipulable objects through a sequence of zoomings and stretches.)

The curriculum was tested in a ninth-grade class where computers were present in a ratio of one per pair of students. One aspect of student learning that Schwarz and his colleagues reported in detail concerns the issue of student attachment to linearity. They compared their findings with those reported by Markovits et al. (1986) and Karplus (1979). Students were given a questionnaire with four items adapted from the Markovits and Karplus questionnaires. It should be noted that the course that Markovits' students followed was an attempt to take into account some of the difficulties and misconceptions about functions that Karplus had noticed. The TRM curriculum took into account Markovits' results. The results of the analysis with respect to students' attachment to linearity would seem to indicate that curriculum development is moving in the right direction.

It was found that 69% of the students reached partial or full curved-line reasoning (i.e., that an infinite number of curves can be drawn to pass through two given points), as opposed to 18% in the Karplus study and 17% in the Markovits study. According to Schwarz and his colleagues, the main distinction between partial and full curved-line reasoning is between conceptual knowledge about the function concept and an integration of conceptual and procedural knowledge. The highest level of reasoning, which is attained when the student interprets data correctly in a functional framework, is considered by the authors to be essential as background in science courses. Schwarz et al. (1990) attributed their successful results to a curriculum that includes extensive qualitative representation of functions by means of graphs. However, they also pointed out that the teachers were continuously guided during their teaching in didactical and technological aspects of TRM and that "great effort had to be put into pedagogical engineering in order to enable teachers to master composite systems which contain group work techniques, computer sessions, and conventional lessons" (p. 260).

Quadfun. Dreyfus and Halevi (1988) described how they used an open-ended microworld for quadratic equations with classes of Grade 10 students of low to average ability. At the outset, the authors pointed out that one of the advantages of an open-ended software-based curriculum is its flexibility; however, such a computer program cannot possibly guide the students in their learning experience because it does not pose any specific problems. This guidance must come, according to Dreyfus and Halevi, from both a set of worksheets and the teacher. The worksheets, by specifying activities as well as their sequencing, help to achieve certain learning objectives. The role of the teacher in such an environment is twofold: one, to act as a resource, giving hints that are adapted to each student's ability and pace; two, to lead discussions with the entire class, to compare different approaches that have been taken by different students, and to stress the general features of the activities students have just completed. Dreyfus and Halevi emphasized that without these summary discussions at the end, much of the effect of the learning process on the student might well be lost.

One of the main difficulties that students typically experience in constructing their mental image of function (Vinner & Dreyfus, 1989) is the establishment of a connection between the formula defining a function algebraically and its graphical representation. Another difficult topic is the role of the parameters a, b, and c in the expression $f(x) = ax^2 + bx + c$ and the effect of changing these parameters on the graphical representation of the function as a parabola.

With these difficulties in mind, the microworld environment was designed in such a way as to allow the pupil to build parabolas by specifying some of their properties, such as the symmetry axis or points through which the parabola passes. The construction of the parabolas proceeds in the graphical and algebraic representations in parallel. Specifically, a quadratic function may be selected in two ways, either by using the command "Parameters," upon which the computer asks for the values of a, b, and c in the algebraic formula or by asking for "Parabola," upon which the computer offers the following list of geometric properties of a parabola: point, openness, y intercept, vertex, symmetry axis. After entering a value for one of these five choices, the computer offers two more possibilities: "More data" or "Draw." If "Draw" is chosen before a parabola is uniquely specified, a family of parabolas is presented symbolically by means of three randomly chosen members of the family. Together with any parabola, the corresponding quadratic expression appears on the screen. If the parabola is overspecified by giving incompatible data, the computer reacts by announcing "Mission impossible." The computer does not react if more data than necessary are given, as long as the data are not contradictory.

Worksheets have been generated with the aim of fostering an intuitive familiarity with the set of all quadratic functions and an understanding of those combinations of data that determine a unique parabola, in addition to how changes in the geometric data influence the parameters a, b, or c of the quadratic function. An example of a worksheet that was given in the third or fourth class session is provided next. Dreyfus and Halevi (1988) reported that, in a typical class period, students worked on two to four such activity sheets.

You are given a parabola with opening $a = -1$.
1. How many such parabolas are there?

2. Among the parabolas with opening a = −1 choose one which goes through the point (1, 2). How many such parabolas are there?
3. Among the parabolas with opening a = −1 choose one with symmetry axis x = 1. How many such parabolas are there?
4. Among the parabolas with opening a = −1 choose one with vertex K (1, 2). How many such parabolas are there?
5. How many givens do you think determine a unique parabola?
6. Can you explain why in Question 4 only two givens determined a unique parabola? (p. 3)

A teaching experiment was carried out with three classes of Grade 10 students. In each class a slightly different teaching approach was used. In the first class, students used computers in pairs or alone, while the teacher and one of the authors acted as consultants. Every lesson was introduced and summarized by a consultant. At first, some students tried to monopolize the consultants, but this situation changed drastically from the third or fourth session onward.

In the second class, which was instructed by one of the authors, one computer was used as a "teaching assistant" (Fraser, 1986). The class worked as a group, with teacher and students taking turns deciding which tasks the computer should carry out. Often students were asked to try the questions on the worksheets and suggest answers before they would be tried on the computer. Although the use of the computer was less frequent, the teacher could direct it to produce exactly the right effect "at the right time." The role of the teacher as a source of information was reduced to almost zero; the information, when it had to come, came from the computer or from the students.

The third class was taught by the regular classroom teacher who had received the worksheets from the researchers. The students worked in pairs at the computers with the teacher acting as a consultant. After the instruction in all three classes was completed, students were given a questionnaire that, according to Dreyfus and Halevi, was rather difficult. Results showed that some but not all students did reach a satisfactory level on both topics: the dependence of quadratic functions on their parameters and the combinations of data determining a unique parabola. Dreyfus and Halevi pointed out that, in view of the fact that these topics are very difficult and well beyond what students of this level usually learn, these results should be considered an achievement. The interactive manner of the learning process, the conceptual type of work, and the immediate feedback they got from the computer program challenged these students to make an effort and consequently reach a depth of understanding well beyond the usual. (The study did not include a comparison of the three teaching approaches to determine which one was the most effective.)

Some of the misconceptions and/or gaps in student knowledge that surfaced during these classroom sessions include the following:

1. During the third class period of the group that was working with the computer as a "teaching assistant," the teacher happened to ask the students to check whether the point (4,5) was on the parabola given by $y = x^2 - 3x + 1$. The students did not know what to do; the teacher then realized that the students had not established any link between the graphical and the algebraic description.
2. Very often students did not distinguish between the vertex and the symmetry axis.
3. For many students, a parabola has a finite horizontal extension, in spite of the fact that obviously any number may be substituted for x in the algebraic formula.

Subsequent classroom discussions helped sensitize students to some of the more difficult aspects of this topic and showed that such a learning environment had the potential for challenging even relatively weak students to deal in depth with these difficult topics. Dreyfus and Halevi suggested that the realization of this potential depends critically on the way in which the learning environment is integrated interactively with a set of worksheets containing activities and problems for the students and with the guiding activity of the teacher in the classroom. They concluded with the remark that "this teacher-environment-worksheet triple is crucial for the successful implementation of open-ended microworld environments in the classroom" (p. 7).

Graphing Games. Friske (1988) distinguished between function graphers and graphing games. In the Dreyfus and Halevi study just discussed, we saw how a special kind of function grapher can be used in the classroom. One of the gaps in student knowledge that was disclosed in that study was the lack of a connection between graphical and algebraic representations, which became evident when students were asked to tell whether the point (4,5) was on the parabola given by $y = x^2 - 3x + 1$. Friske suggested that one of the advantages of graphing games is the development of point-locating strategies—an awareness that might have been helpful as a supplement to the kinds of learning that were taking place in the *Quadfun* microworld. Perhaps the best known of graphing games is *Green Globs*, developed by Dugdale and Kibbey (1986b).

With this software, beginning algebra students can estimate the coordinates of the globs and determine an appropriate equation—usually linear or quadratic. More advanced students can experiment with graphs of polynomial, logarithmic, and trigonometric functions. Dugdale (1982) described how *Green Globs* has been used by students of widely varying mathematical experience and abilities—from the ninth-grade student with no background in algebra whose early strategy was to hit all the globs with horizontal and vertical lines, to the students who after about 6 weeks of experimentation with polynomials in factored form came up with equations such as the one illustrated in Fig. 8.12.

Dugdale (1982) found that this kind of software is not only powerfully motivating for students but, more importantly, that "such activities focus students' attention on their own substantial ability to use mathematics in interesting and creative ways to achieve what is to them a worthwhile goal" (p. 214).

Input-Output Graphs. In this concluding piece of the first-year algebra work in graphing, I summarize a technique for operating on graphs that is

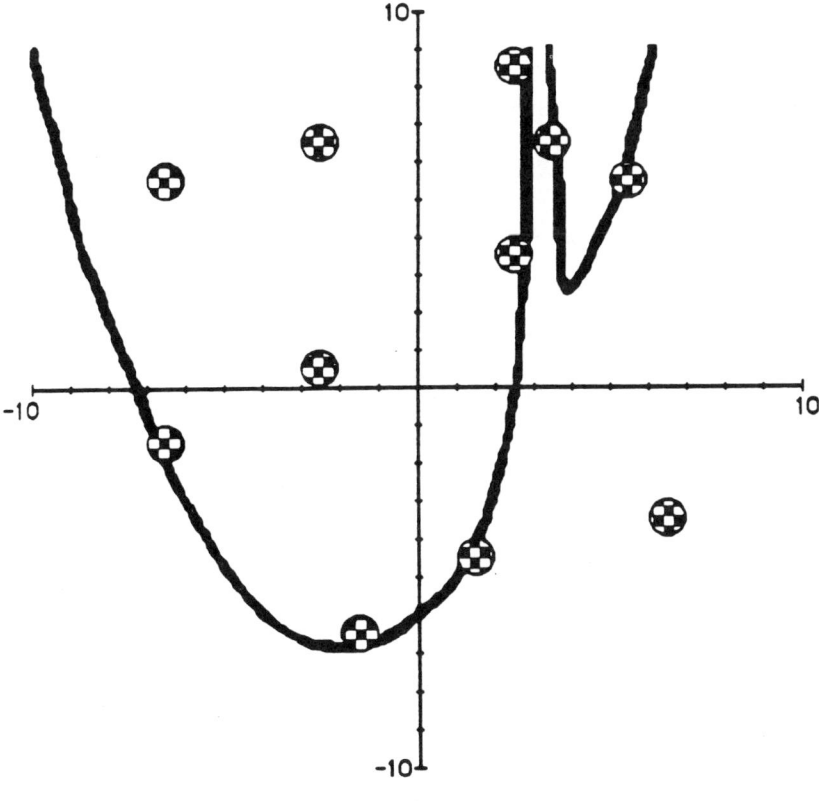

$$y = .25(x+2)^2 - 7 + 1/(x-3)^2$$

FIG. 8.12. In the *Green Globs* game, a student has generated a parabola to hit several globs, and has then added $1/(x - 3)^2$ to make it split upward and hit three more globs near $x = 3$. *Note.* Adapted from "Green Globs: A microcomputer application for graphing of equations" by S. Dugdale, 1982, *Mathematics Teacher*, 75(3), p. 213. Copyright © 1982 by National Council of Teachers of Mathematics. Adapted by permission.

described in detail in Thaeler (1988); it is a technique that could be used equally well with more advanced students. The approach is one in which the graph is operated on as an object; the operation is either an input or output transformation applied to the x variable of the corresponding algebraic representation. The resulting object is a modified graph.

For example, the function $y = -2x^2 + 12x - 16$ is re-expressed in such a way that the input variable x appears only once on the right-hand side (by completing the square): $y = -2(x - 3)^2 + 2$. The next step is to decide which of a set of basic functions is represented by this expression in order to predict the basic shape of the graph. Following this, the step-by-step process of what happens to x to turn it into the output y is described in words. Some of the represented transformations are considered input modifications and some are output modifications (see Fig. 8.13 for the input/output description of this function).

Thaeler pointed out that any modifications occurring before or above the basic function are input modifications. These modifications have an effect on the graph in the x input or left-right direction by, for example, stretching or compressing the graph either toward or away from the y axis, shifting the graph left or right, or flipping the graph left/right. Those modifications that occur after the basic function affect the graph in the y output or up-down direction by, for example, stretching or compressing the graph either toward or away from the x axis, raising or lowering the graph, or flipping the graph up/down.

Continuing with this example, Fig. 8.14 illustrates how to draw the graph of $y = -2(x - 3)^2 + 2$ using the input-output method. One begins with the

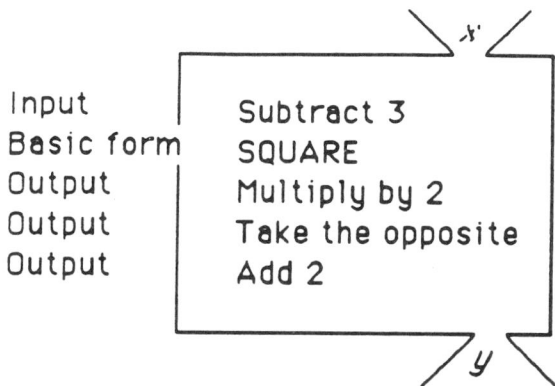

FIG. 8.13. Input-output description of the function $y = -2(x - 3)^2 + 2$. *Note.* Adapted from "Input-Output Modifications to Basic Graphs: A Method of Graphing Functions" by J. S. Thaeler, 1988, in A. F. Coxford (Ed.), *The Ideas of Algebra, K–12,* p. 231, Reston, VA: National Council of Teachers of Mathematics. Copyright © 1988 by NCTM. Adapted by permission.

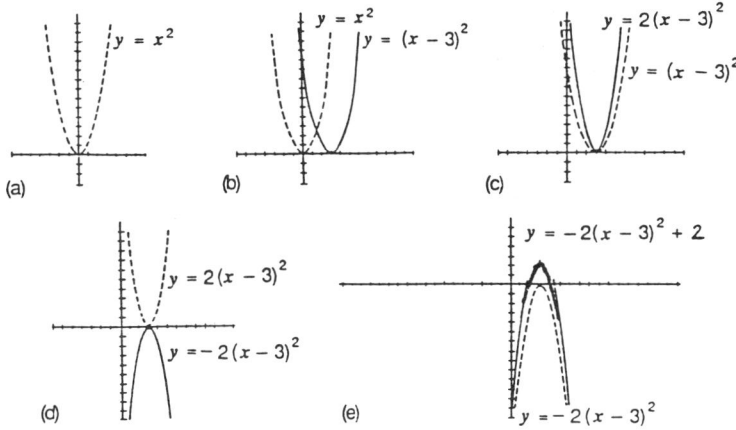

FIG. 8.14. Drawing the graph of $y = -2(x - 3)^2 + 2$ by the input-output method. *Note.* Adapted from "Input-Output Modifications to Basic Graphs: A Method of Graphing Functions" by J. S. Thaeler, 1988, in A. F. Coxford (Ed.), *The Ideas of Algebra, K-12*, p. 238, Reston, VA: National Council of Teachers of Mathematics. Copyright © 1988 by NCTM. Adapted by permission.

standard parabola (Fig. 8.14a); shifts it to the right 3 units—an input modification applied to x (Fig. 8.14b); stretches it vertically by a factor of 2—an output modification that stretches the graph away from the x axis in the y direction (Fig. 8.14c); slips it upside down (Fig. 8.14d); and finally raises it 2 units (Fig. 8.14e).

One of the advantages of this approach to graphing is that students learn transformations that they can apply to any basic graph. They are also able to do the reverse—that is, determine the equation of a function, given its graph. The student seeks to find what shifting, stretching, or compressing is necessary to produce the given graph starting with the basic function. Even though Thaeler (1988) did not describe particular classroom experiences in which this approach was used, I include a description of this method because it provides a clear illustration of how graphs can be operated on as objects and can perhaps come to be better understood by students as objects. Although Thaeler did not suggest what kinds of underlying knowledge about the basic graphs might be needed by students in order for them to make sense of these particular operations, he did maintain that "the input-output approach gives students a valuable tool in understanding and gaining a feel for what the graph of a function is and how each number in the equation affects the final appearance of the graph" (p. 241).

Another advantage to emphasizing those changes that occur to a graph as a result of either input or output modifications is the addressing of the possible perceptual illusions that can occur in graph reading (Goldenberg,

1988). Goldenberg pointed out that algebra experts know, for example, that the graph of a line ($ax + b$) moves up as b is increased. However, because an infinite line presents us with no discrete points to watch, it may also appear to be moving from left to right as the constant term increases (see Fig. 8.15).

Goldenberg emphasized further that changing the scale on one of the axes can suggest to students that a graph has shifted up/down or to the left/right. Similarly, changing the scale uniformly on both axes can also "fool the eye" into thinking that the angle of, say, a parabola has changed. Whether some of these perceptual illusions can be combatted by developing the kind of input-output thinking just summarized (Thaeler, 1988) is an interesting question on which future research study ought to be able to shed some light.

Graphing in Later Courses

Some of the graphing activities that can be engaged in with students who have already completed at least one course in algebra are described in detail in other chapters in this volume. Nevertheless, for the sake of completeness we include brief descriptions of some of this work, as well as discussion of other projects.

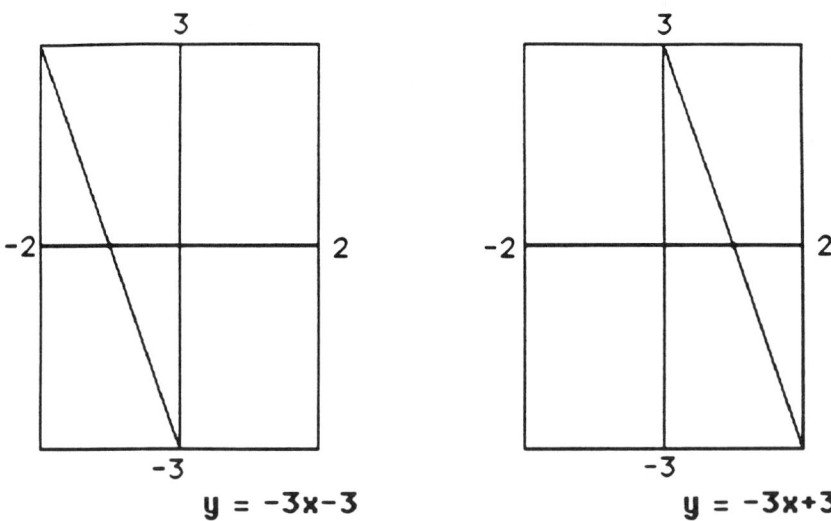

FIG. 8.15. As the constant term increases, the graph may appear to move from left to right. *Note.* Adapted from "Mathematics, Metaphors, and Human Factors: Mathematical, Technical, and Pedagogical Challenges in the Educational Use of Graphical Representation of Functions" by E. P. Goldenberg, 1988, *Journal of Mathematical Behavior, 7,* p. 143. Copyright © 1988 by Ablex Publishing Corporation. Adapted by permission.

Using Computer Graphs to Teach Trigonometric Identities. Dugdale (1989) reported a study comparing two approaches to incorporating graphical representations into a unit on trigonometric identities. The first approach supplemented traditional instruction with related graphing activities. The second used graphical representations as the foundation for trigonometric identities. The teaching activities were embedded into a normal classroom sequence for trigonometry. The 30 students in the class had completed the introductory trigonometric material of a standard Algebra II and Trigonometry textbook and were ready to begin the chapter on trigonometric identities. The class was then split in two. Both groups had regular class sessions, microcomputer laboratory activities guided by worksheets designed by the author, and homework. In both groups, students worked with partners on the microcomputer activities. The major difference between the instructional approaches for the two groups was that the Supplemented Traditional (ST) group recorded graphs that had been plotted by the computer while the Graphical Foundation (GF) group predicted graphs and used the computer to check their predictions. Other aspects of the two approaches are summarized in Fig. 8.16.

A posttest comprising two sections, proving identities and graphical representations, was administered after the teaching experiment. There was no significant difference between the groups as far as proving identities was concerned, but for relating trigonometric functions to their graphical representations, the GF group proved superior — even though these students had been exposed to fewer graphical representations than the ST students. Dugdale suggests that the GF students' experience with graphically predicting the shape of a function accounts for the difference in posttest performance on the nonroutine items of the test, but it is not clear why this group should also have outperformed the other on items involving only recognition of basic graphs and use of routine symbol manipulation. The results of this study illustrate how a well-organized graphical approach that significantly involves learners can produce a rich experience with possible effects that go beyond what could have been predicted.

Graphing Calculators. One of the advantages of graphing calculators over computer-based function graphing software is that every student has access to the tool, both at school and at home. Furthermore, as pointed out by Dick and Musser (1988), a mathematical environment involving graphing calculators "not only unfetters both students and teachers to attack problems and topics that might otherwise prove to be too time-consuming, it also allows students to approach mathematical problems in ways quite different than that provided by the traditional classroom" (p. 1).

One of the recent projects based on the use of graphing calculators involved two secondary mathematics departments in Oregon (Dick &

> **Supplemented Traditional (ST)**
> 1. Trigonometric identities were treated in the traditional fashion, as exercises in symbol manipulation. Graphs were used as an additional representation.
> 2. Symbol manipulation exercises were preceded by a straightforward presentation of the eight fundamental identities and direct instruction covering procedures to be used.
> 3. Computer activity sheets were arranged with a worked-out example preceding each set of exercises. Exercises were routine repetitions of the procedure used in the example.
> 4. Guidance provided on activity sheets focused on what procedures subjects should apply.
> 5. Relationships between graphs (such as the correspondence between the zeros of a function and the asymptotes of its reciprocal function) were presented, but the activities did not require subjects to use these ideas.
>
> **Graphical Foundation (GF)**
> 1. Trigonometric identities were introduced graphically, and the usual symbol manipulations were used to justify the relationships evidenced in the graphing activities.
> 2. Subjects were asked to justify algebraically the equivalence of functions without being instructed how it should be done. Part of the task was to decide what information was applicable and how to use it.
> 3. Computer activity sheets presented non-routine tasks, some of which required analyzing graphic feedback and revising functions to change their graphs.
> 4. Guidance provided on activity sheets focused on what questions subjects should address.
> 5. Subjects were asked to use graphs of functions to predict graphically the shapes of other functions before plotting. For example, from graphs of y = sinx and y = cosx, subjects figured out where y = sinx/cosx would have zeros and asymptotes and predicted the general shape.

FIG. 8.16. Differences between the two instructional treatments of the Dugdale study (1989). *Note.* Reprinted from "Building a Qualitative Perspective Before Formalizing Procedures: Graphical Representations as a Foundation for Trigonometric Identities" by S. Dugdale, 1989, in C. A. Maher, G. Goldin, and R. B. Davis (Eds.), *Proceedings of the Eleventh Annual Meeting of PME-NA*, p. 250, New Brunswick, NJ: Rutgers University. Copyright © 1989 by Dugdale. Reprinted by permission.

Musser, 1988). Six leading teachers used the calculators extensively in second-year algebra, trigonometry, and calculus classes. They found that one of the greatest benefits of graphing calculators is the feasibility of discovery lessons based on finding patterns in the student-generated graphs of related functions. For example, in an introductory class on exponential functions, a teacher was able to have each student graph 20 different exponential functions in less than 15 minutes of class time. The observational data gathered on the qualitative nature of the graphs were analyzed by the students for patterns—an activity that parallels what has been reported to take place in computer-based function-graphing environments.

Another interesting finding of this project relates to student attitudes. Dick and Shaughnessy (1988) reported that male students seemed more

8. FUNCTIONS, GRAPHING, AND TECHNOLOGY

impressed with the utility and power of the calculator, whereas female students found that it makes mathematics more enjoyable. The teachers involved in the project reported that they were taking much more of an exploratory approach to teaching graphics than they ever had before and that the use of the calculator enhanced their teaching of graphing. The teachers' own perception of student attitudes was that the calculator was viewed quite favorably by the vast majority of students; at least two of the teachers reported of an individual student whom they felt changed her plans in favor of enrolling in future mathematics primarily because of the calculator experience.

Other recent projects involving the use of graphing calculators include the ongoing Ohio State University Calculator and Computer Precalculus (C^2PC) Project (Waits & Demana, 1988). Students participate in an interactive lecture-demonstration classroom environment containing a single computer, along with individual graphing calculators. Teachers use a carefully prepared sequence of questions and activities to help students understand or discover important mathematical concepts.

One of the instructional techniques used is to quickly demonstrate with the classroom computer, for example, a set of graphs such as $y = 5 + x^3$, $y = 5 + (x - 2)^3$, $y = 5 + (x + 3)^3$, and so forth. Careful questioning by the teacher then ensues in order to help students explain the significance of the v and h terms in $y = v + (x - h)^3$. After guided activities involving graphing calculators, students are eventually able to sketch the graph of complicated functions without using a graphing utility.

Another example of the use of graphs in the C^2PC project is in algebra problem solving. Harvey and Waits (1989) described how students might proceed for the following algebra problem: "Squares of side length x are removed from a 15-inch by 60-inch piece of cardboard and a box with no top is formed by folding up the resulting tabs; determine x so that the volume of the resulting box is at least 175 inches":

> To solve this problem a student studying from *Precalculus Mathematics: A Graphing Approach* (Demana & Waits, 1988) would first generate the function f (i.e., $f(x) = x(10 - 2x)(25 - 2x)$) that is an algebraic representation of the problem situation. Then the student would draw a complete graph of the function f (i.e., a graph of the function in an optimal viewing rectangle that accurately depicts the "end behavior" of the function). Then, in some order the following would ensue. From the original problem situation he or she would determine that the only relevant values of x are those between 0 and 5 and would redraw the graph so as to show only that portion of the graph of f. On this graph the horizontal line $y = 175$ would be superimposed to determine the points on the graph at which to zoom-in to obtain the approximate values of x for which $f(x) = 175$ and thus to solve the problem. In contrast students enrolled in conventionally taught courses would solve the

inequality $x(10 - 2x)(25 - 2x) > 175$ by forming the equation $x(10 - 2x)(25 - 2x) - 175 = 0$, find the real roots of that equation, and use those roots to develop a sign chart. The C²PC materials also teach many of these conventional techniques, including sign charts, usually along with or after instruction using graphical problem solving procedures. (p. 3)

The classroom examinations that have been administered to C²PC students using graphing calculators have not up to now, according to Harvey and Waits, been able to reveal very much about the knowledge students have about graphical representations or their uses. Harvey and Waits emphasized that, since most of these examinations typically focused on symbolic answers to algebra problems, the effects of their graphical approach to the teaching of precalculus are not yet known. Nevertheless, they have observed that the students

more frequently talk about functions and their zeros instead of equations and their roots, that they are quite comfortable with the ideas of relative and local maxima and minima, that they seem to understand that functions belonging to the same class (i.e., cubic polynomial functions) have similar graphs and vice versa, and that they seem to understand the advantages and shortcomings of using graphical representations while solving problems. (Harvey & Waits, 1989, p. 5)

Function Graphers in Calculus. The conclusion of this section focuses on the use of function graphers in a computer-based, applied calculus curriculum (Heid, 1988a; Sheets & Heid, 1990). The course was designed to acquaint nonmathematics majors, through the use of computer tools, with the importance of calculus concepts in their chosen fields. The curriculum centered on applications and concepts rather than upon a singular development of procedural skills. Because the computer could be used for the execution of routine procedures, concentrated attention was given to developing a broad-based understanding of calculus concepts by using computer tools to generate a variety of representations for each calculus concept. Derivatives, for example, were viewed through their symbolic representation (using a computer symbol manipulation program, *muMath*), through their graphical representation (using a computer function grapher), through their numerical representation (using a table-of-values generator), and through applied settings (using a curve fitter to provide the link between real data and the functions which describe them). Because it was not until the last 3 weeks of class that students were introduced to by-hand techniques for executing calculus procedures, the computer was a necessary tool.

In describing how graphs were used during the first 12 weeks of her experimental classes, Heid (1988a) first pointed out how, in a traditional

8. FUNCTIONS, GRAPHING, AND TECHNOLOGY 229

calculus course, students are seldom asked to draw conclusions on the basis of their graphs or to comment on the relationship between two different graphs. Typical graphing assignments involve the sketching of fairly straightforward equations such as $y = 2x^2 + 5x + 2$ or $f(x) = x^3 + 3x + 1$. On the other hand, in the experimental classes students engaged in the following graphing activities:

> They examined a large variety of computer-generated graphs, reasoned from these graphs, and studied the similarities, differences, and connections between graphs related to each other in formula or in an applied interpretation. For example, the students in the experimental classes studied graphs of linear, quadratic, cubic, quartic, and quintic functions, as well as rational functions of a variety of forms. In addition to analyzing computer-generated graphs, they also constructed their own graphs of the functions using computer-generated information about the zeros and about other values of the derivatives. They analyzed how the properties of the function's graph were reflected in the graphs of its derivative functions. They deduced graphical properties of a function from properties of the inverse function, noting relationships between intercepts, asymptotes, intervals of concavity, and slopes. (p. 8)

In addition, the students also analyzed applications through graphs. For example:

> They interpreted the properties of graphs in applied situations, noting, for example, the meaning of y-intercepts, of differences in y-values, and of intersection points for revenue and cost curves. They used superimposed graphs of revenue and cost functions to locate sales levels with chosen profit values or with maximal profit values. They used graphs to compare optimal sales levels for corresponding revenue and profit functions as well as for families of profit functions associated with cost functions of varying parameters. They devised applied situations consistent with given graphs and generated original questions on the basis of the graphs of given applications. (p. 8)

Heid pointed out that the experimental instruction using computer-generated graphs aimed at encouraging students to reason deeply from and about graphs. By the end of the course, their experience with graphs went far beyond that usually gained by the traditional concentration on the creation of graphs for simple equations. The students themselves felt that the computer aided in their conceptual understanding by refocusing their attention in three ways:

1. The computer relieved them of some of the manipulative aspects of calculus work.

2. It gave them confidence in the results on which they based their reasoning.
3. It helped them focus attention on more global aspects of problem solving.

CONCLUDING REMARKS

The foregoing examples of recent research projects focusing on the use of technology in the teaching of functions and their graphs illustrate how traditional emphases have begun to shift. In the past, graphs were rarely taught with an eye toward viewing their global features. They were used most often as another way of representing a relationship that was initially depicted in algebraic form. Thus, most graphical interpretation activities involved the use of point-wise methods applied to basic functions, such as the linear, quadratic, and trigonometric.

However, in schools that have access to computers and graphing calculators, particularly those in which research projects such as those just described are being conducted, the methods that were traditionally used in the teaching of graphing are being put aside. As seen from the described projects, the newer approaches to graphing focus primarily on three different kinds of activities:

1. Interpreting global features of graphs using unscaled, labeled axes in tasks requiring of students only a minimal knowledge of the conventions used to indicate increasing values on the axes.
2. Employing basic families of graphs to explore the roles of the parameters of the algebraic representations—relying on an implicit awareness that the graphs could be obtained by plotting (x, y) points that satisfy certain equations.
3. Using graphs as a tool for problem solving in applied settings.

We have also witnessed a renewed emphasis on integrating various representations of functions, such as graphical, algebraic, and tabular; this integration is now not only considerably enriched due to activities focusing on the global properties of graphs but also modified because of the capability of computers to simultaneously display changes in these representations.

The objectives of many of the studies presented in this chapter centered on seeing how successfully technology could be used as a tool for enhancing students' notions of graphs and/or for solving applied problems. Most studies achieved their objectives. But what is the overall picture that emerges from an examination of this set of studies? How do we explain, for example, that students who initially learned to interpret graphs qualitatively had an easier time doing so than students who first learned to do some graph construction by plotting points? And from another study, what kinds of prior experience and/or intervention would have helped students better use graphical feedback to monitor the correctness of their algebraic transformations and why? And in those studies in which students were reported to have successfully learned to relate the various representations of functions, was it because they had already acquired a fairly strong base in algebra and/or because they had previously spent considerable time with point-plotting activities?

Not only do these studies raise more questions than they answer (and so they should), but also as a whole they provide only unrelated fragments of a picture. They do not tell a story. Clearly, we are missing something. This elusive ingredient does not appear to be related to the way that the technology was being exploited in these studies; it was used for exploration, not for "show and tell." Furthermore, the technology permitted students not only to avoid the drudgery of excessive calculation and hand-plotting a large set of points but also to manipulate graphs in a way that was simply not possible before. So if it is neither the technology nor the way it is being used that are at fault, what are we missing? I suggest that an element that might have helped to tell more of a story would have been the presence in these studies of a theoretical framework; such an element is essential for explaining the findings of individual studies and relating them to the findings of others.

Many of the studies referred to here included instructional interventions that emphasized the global features of graphs and their corresponding algebraic representations. Because of the nature of these two representations, an appropriate candidate for a theoretical framework for such studies is the process-object model summarized in a previous section. In discussing her model, Sfard (1991) provided us with some examples of process and object descriptions of a few mathematical notions, such as function, symmetry, natural numbers, rational number, and circle. For the concept of function, she looked briefly at three representations—a graph, a computer program, and an algebraic equation—and suggested how each of these representations corresponds respectively to an object approach, a process approach, and both. However, none of these examples has been sufficiently characterized in terms of the process-object duality to serve as the basis of an operational definition for conducting empirical research. Thus, one of the first things that is needed in order to begin to use a process-object model not only for interpreting research findings but also for designing studies is a more detailed elaboration of what constitutes process-type tasks and object-type tasks for each of the various representations of functions—not just algebraic, tabular, and graphic, but also verbal and computer program representations (see Garançon, Kieran, & Boileau, 1990, for an example of an environment involving computer program representations of functional situations).

One approach would be to translate into process-object terms the already existing wide range of function tasks in the research literature, as well as to construct new ones that were not really feasible without the new technologies. It might also be useful to take the various representation and translation skills, as characterized according to Swan (1982) in an earlier section of this chapter, and to redescribe them in terms of process-object demands. (An issue that arises in this regard is whether tasks themselves can demand/suggest either a process approach or an object approach. On this point, Sfard has said that individuals who have only a process interpretation of a particular concept would be unable to succeed at tasks requiring an object conceptualization.)

An aspect of the process-object model as characterized by Sfard, which ought to be kept in mind when attempting to apply this model to the learning of graphical representations, is that "in computational mathematics, the majority of ideas originate in processes rather than in objects" (Sfard, 1991, p. 11) and that a process conception is for most people the first step in the acquisition of new mathematical notions. Nevertheless, Sfard pointed out that "geometric ideas for which the unifying, static graphical representations appear to be more natural than any other, can probably be conceived structurally [as objects] even before full awareness of the alternative procedural [process] descriptions has been achieved" (p. 10). Thus, the following remains an open research question: Does the learning of graphical representations of functions follow the same process-to-object sequence as has been documented in past studies of functions involving primarily set-theoretic and algebraic representations? Does the same phenomenon occur with respect to the learning of graphs; in other words, do students, in the face of an object-oriented teaching approach, initially seek out process-oriented explanations?

The traditional, nontechnology-supported teaching of graphical representations of functions has always emphasized a process approach to the graphical representation along with a process approach for the function's algebraic and tabular representations: For example, "take a linear equation; substitute values for x in the equation and set up a table of x–y values; plot the values on a Cartesian plane; or take any two points, calculate the slope, and draw a graph using the slope and a point; and so on." The technology-supported projects described in this chapter have clearly shown that this route is not the one that has to be followed if we want to encourage students to learn to read the global features of graphs. We have choices now. But what shall we base those choices on?

Clearly, we need to know where we want to go and what we believe is important for our students to learn. Having endorsed in principle that Grade 9–12 students need to be able to "represent and analyze [functional] relationships using tables, verbal rules, equations, and graphs; . . . analyze the effects of parameter changes on the graphs of functions; . . . understand operations on, and the general properties and behavior of, classes of functions; [and so on]" (National Council of Teachers of Mathematics, 1989, p. 154), we next need to design studies that can do more than simply show that technology can help our students to learn such material. Most of the projects described in this chapter were classroom projects; the findings of such studies can at best report only in a sketchy way the nature of the learning that takes place. Thus, we have to think about committing ourselves to very detailed studies of student cognition in this domain — studies that can tell us what works and why, or what does not work and why not. The process-object theoretical framework can, at the very least, help to tie such findings together. Even more than that, this framework can also suggest perspectives not tried before, such as the potential interplay between process and object approaches not only in the initial learning of a domain but also afterward when each approach is more fully developed in the individual. Studies such as the one carried out by Schoenfeld, Smith, and Arcavi (in press), are a good example of what I mean by detailed studies of student cognition. But as those authors point out, "we have just begun to chart the territory."

It is to be noted before concluding that, after the final editing of this chapter and prior to the publication of the volume, some fresh work touching on some of these issues came to my attention. In the latter part of the year 1990, a conference on functions was organized by Harel and Dubinsky (1992), and some of the papers that were presented (Dubinsky & Harel, 1992; Schwartz & Yerushalmy, 1992; Sfard, 1992) included discussions of a process-object nature. Furthermore, a very recent research report on the graphing of functions, written by Moschkovich, Schoenfeld, and Arcavi and appearing in this volume (Chapter 4), described a study that, in both its design and the analysis of its findings, involved the application of the process-object model to various ways of representing functions in a computer environment. Moschkovich, Schoenfeld, and Arcavi emphasized in their conclusions that this theoretical framework "will be a profitable approach for both curriculum development and (student and curriculum) assessment" (p. 97). Their pioneering study, along with the research already carried out by Sfard and others, can serve as examples of the kind of theoretically based work that offer the potential to help research in mathematics education to move forward, especially the more recent research involving technology where the absence of theoretical frameworks has been especially noticeable.

ACKNOWLEDGMENT

I am deeply indebted to Anna Sfard for many of the ideas expressed in the section on process-object.

REFERENCES

Assessment of Performance Unit. (1980). *Mathematical development, secondary survey, report no. 1.* London: Her Majesty's Stationery Office.

Burkhardt, H. (1981). *The real world and mathematics.* Glasgow: Blackie.

Carpenter, T. P., Corbitt, M. K., Kepner, H. S., Jr., Lindquist, M. M., & Reys, R. E. (1981). *Results from the second mathematics assessment of the National Assessment of Educational Progress.* Reston, VA: National Council of Teachers of Mathematics.

Davis, R. B. (1982). Teaching the concept of function: Method and reasons. In G. van Barneveld & H. Krabbendam (Eds.), *Proceedings of the Conference on Functions* (pp. 47-55). Enschede, The Netherlands: National Institute for Curriculum Development.

Demana, F. D., & Waits, B. K. (1988). *Precalculus mathematics: A graphing approach* (preliminary edition). Reading, MA: Addison-Wesley.

Dick, T. P., & Musser, G. L. (1988, July-August). *Symbolic/graphical calculators and their impact on secondary level mathematics.* Paper presented to the theme group on Microcomputers and the Teaching of Mathematics at the Sixth International Congress on Mathematical Education, Budapest, Hungary.

Dick, T. P., & Shaughnessy, J. M. (1988). The influence of symbolic/graphic calculators on the perceptions of students and teachers toward mathematics. In M. J. Behr, C. B. Lacampagne, & M. M. Wheeler (Eds.), *Proceedings of the Tenth Annual Meeting of PME-NA* (pp. 327-333). DeKalb, IL: Northern Illinois University.

Dreyfus, T. (1990). Advanced mathematical thinking. In P. Nesher & J. Kilpatrick (Eds.), *Mathematics and cognition: A research synthesis by the International Group for the Psychology of Mathematics Education* (pp. 113-134). Cambridge, England: Cambridge University Press.

Dreyfus, T., & Eisenberg, T. (1981). Function concepts: Intuitive baseline. In C. Comiti & G. Vergnaud (Eds.), *Proceedings of the Fifth International Conference for the Psychology of Mathematics Education* (pp. 183-188). Grenoble, France: Institut IMAG.

Dreyfus, T., & Halevi, T. (1988, July-August). *Quadfun—A case study of pupil computer interaction.* Paper presented to the theme group on Microcomputers and the Teaching of Mathematics at the Sixth International Congress on Mathematical Education, Budapest, Hungary.

Dreyfus, T., & Vinner, S. (1982). Some aspects of the function concept in college students and junior high school teachers. In A. Vermandel (Ed.), *Proceedings of the Sixth International Conference for the Psychology of Mathematics Education* (pp. 12-17). Antwerp, Belgium: Universitaire Instelling.

Dubinsky, E., & Harel, G. (1992). The nature of the process conception of function. In G. Harel & E. Dubinsky (Eds.), *The concept of function: Aspects of epistemology and pedagogy* (pp. 85-106). Washington, DC: Mathematical Association of America.

Dugdale, S. (1982). Green Globs: A microcomputer application for graphing of equations. *Mathematics Teacher, 75*(3), 208-214.

Dugdale, S. (1987). Pathfinder: A microcomputer experience in interpreting graphs. *Journal of Educational Technology Systems, 15*(3), 259-280.

Dugdale, S. (1989). Building a qualitative perspective before formalizing procedures: Graphical representations as a foundation for trigonometric identities. In C. A. Maher, G. Goldin, & R. B. Davis (Eds.), *Proceedings of the Eleventh Annual Meeting of PME-NA* (pp. 249-255). New Brunswick, NJ: Rutgers University.

Dugdale, S., & Kibbey, D. (1986a). *Interpreting graphs* [Computer program]. Pleasantville, NY: Sunburst Communications.

Dugdale, S., & Kibbey, D. (1986b). *Green globs and graphing equations* [Computer program]. Pleasantville, NY: Sunburst Communications.

Fraser, R. (1986). Program design for the micro as a "teaching assistant." In L. Lewis & B. Feinstein (Eds.), *Proceedings of the International Conference on Courseware Design and Evaluation* (pp. 105-111). Ramat Gan, Israel: Israel Association for Computers in Education.

Freudenthal, H. (1973). *Mathematics as an educational task*. Dordrecht, The Netherlands: D. Reidel.

Freudenthal, H. (1982). Variables and functions. In G. van Barneveld & H. Krabbendam (Eds.), *Proceedings of Conference on Functions* (pp. 7-20). Enschede, The Netherlands: National Institute for Curriculum Development.

Friske, J. S. (1988). Using computer graphing software packages in algebra instruction. In A. F. Coxford (Ed.), *The ideas of algebra, K-12* (1988 Yearbook of the National Council of Teachers of Mathematics, pp. 181-184). Reston, VA: National Council of Teachers of Mathematics.

Garançon, M., Kieran, C., & Boileau, A. (1990). Introducing algebra: A functional approach in a computer environment. In G. Booker, P. Cobb, & T. N. de Mendicuti (Eds.), *Proceedings of the Fourteenth International Conference for the Psychology of Mathematics Education* (Vol. II, pp. 51-58). Mexico: PME Program Committee.

Goldenberg, E. P. (1988). Mathematics, metaphors, and human factors: Mathematical, technical, and pedagogical challenges in the educational use of graphical representation of functions. *Journal of Mathematical Behavior, 7,* 135-173.

Harel, G., & Dubinsky, E. (Eds.). (1992). *The concept of function: Aspects of epistemology and pedagogy*. Washington, DC: Mathematical Association of America.

Harel, G., & Kaput, J. J. (1990). The role of conceptual entities in learning mathematical concepts at the undergraduate level. In G. Booker, P. Cobb, & T. N. de Mendicuti (Eds.), *Proceedings of the Fourteenth International Conference for the Psychology of Mathematics Education* (Vol. I, pp. 53-60). Mexico: PME Program Committee.

Hart, K. M. (1981). *Children's understanding of mathematics: 11-16*. London: John Murray.

Harvey, J. G., & Waits, B. K. (1989, September). *Assessing precalculus student achievement in a course requiring use of a graphic utility: Background and preliminary results*. Paper presented at the symposium on Changes in Student Assessment Occasioned by Function Graphing Tools at the Eleventh Annual Meeting of the North American Chapter of PME, New Brunswick, NJ: Rutgers University.

Heid, M. K. (1988a). Resequencing skills and concepts in applied calculus using the computer as tool. *Journal for Research in Mathematics Education, 19,* 3-25.

Heid, M. K. (1988b, July-August). *The impact of computing on school algebra: Two case studies using graphical, numerical, and symbolic tools*. Paper presented to the theme group on Microcomputers and the Teaching of Mathematics at the Sixth International Congress on Mathematical Education, Budapest, Hungary.

Heid, M. K. (1989, September). *Function graphing tools*. Paper presented at the symposium on Changes in Student Assessment Occasioned by Function Graphing Tools at the Eleventh Annual Meeting of the North American Chapter of PME, New Brunswick, NJ.

Heid, M. K., Sheets, C., Matras, M., & Menasian, J. (1988, April). *Classroom and computer lab interaction in a computer-intensive algebra curriculum*. Paper presented at the annual meeting of the American Educational Research Association, New Orleans, LA.

Hight, D. W. (1968). Functions: Dependent variables to fickle pickers. *Mathematics Teacher, 61,* 575-579.

Janvier, C. (1978). *The interpretation of complex Cartesian graphs—Studies and teaching experiments*. Unpublished doctoral dissertation, University of Nottingham, England.

Kaput, J. J. (1989). Linking representations in the symbol systems of algebra. In S. Wagner & C. Kieran (Eds.), *Research issues in the learning and teaching of algebra* (Vol. 4, Research agenda for mathematics education, pp. 167-194). Reston, VA: National Council of Teachers of Mathematics; Hillsdale, NJ: Lawrence Erlbaum Associates.

Karplus, R. (1979). Continuous functions: Students' viewpoints. *European Journal of Science Education, 1*(3), 397-415.

Kerslake, D. (1977). The understanding of graphs. *Mathematics in Schools, 6*(2), 22-25.

Kerslake, D. (1981). Graphs. In K. M. Hart (Ed.), *Children's understanding of mathematics: 11-16* (pp. 120-136). London: John Murray.

Kieran, C. (1992). The learning and teaching of school algebra. In D. A. Grouws (Ed.), *Handbook of research on mathematics teaching and learning (pp. 390-419)*. New York: Macmillan.

Kleiner, I. (1989). Evolution of the function concept: A brief survey. *College Mathematics Journal, 20*(4), 282-300.

Krabbendam, H. (1982). The non-quantitative way of describing relations and the role of graphs: Some experiments. In G. van Barneveld & H. Krabbendam (Eds.), *Proceedings of the Conference on Functions* (pp. 125-146). Enschede, The Netherlands: National Institute for Curriculum Development.

Leinhardt, G., Zaslavsky, O., & Stein, M. K. (1990). Functions, graphs, and graphing: Tasks, learning, and teaching. *Review of Educational Research, 60,* 1-64.

Markovits, Z., Eylon, B.-S., & Bruckheimer, M. (1983). Functions—Linearity unconstrained. In R. Hershkowitz (Ed.), *Proceedings of the Seventh International Conference for the Psychology of Mathematics Education* (pp. 271-277). Rehovot, Israel: Weizmann Institute of Science.

Markovits, Z., Eylon, B.-S., & Bruckheimer, M. (1986). Functions today and yesterday. *For the Learning of Mathematics, 6*(2), 18-24.

Moschkovich, J., Arcavi, A., & Schoenfeld, A. H. (1992). *What does it mean to understand a domain? A study of multiple perspectives and representations of linear functions, and the connections among them.* Unpublished manuscript, University of California, School of Education, Berkeley.

National Council of Teachers of Mathematics. (1989). *Curriculum and evaluation standards for school mathematics.* Reston, VA: Author.

Phillips, R. J. (1982). An investigation of the microcomputer as a mathematics teaching aid. *Computers & Education, 6,* 45-50.

Preece, J. (1983). Graphs are not straightforward. In T. R. G. Green & S. J. Payne (Eds.), *The psychology of computer use* (pp. 41-56). London: Academic Press.

Rüthing, D. (1984). Some definitions of the concept of function from Joh. Bernouilli to N. Bourbaki. *Mathematical Intelligencer, 6*(4), 72-77.

Schoenfeld, A., Smith, J., & Arcavi, A. (in press). Learning: The microgenetic analysis of one student's understanding of a complex subject matter domain. In R. Glaser (Ed.), *Advances in instructional psychology.* Washington, DC: Mathematical Association of America; Hillsdale, NJ: Lawrence Erlbaum Associates.

Schwartz, J., & Yerushalmy, M. (1992). Getting students to function in and with algebra. In G. Harel & E. Dubinsky (Eds.), *The concept of function: Aspects of epistemology and pedagogy* (pp. 261-289). Washington, DC: Mathematical Association of America.

Schwarz, B. (1988, July-August). *The Triple Representation Model curriculum for the function concept.* Paper presented to the theme group on Microcomputers and the Teaching of Mathematics at the Sixth International Congress on Mathematical Education, Budapest, Hungary.

Schwarz, B., Dreyfus, T., & Bruckheimer, M. (1990). A model of the function concept in a three-fold representation. *Computers & Education, 14,* 249-262.

Sfard, A. (1987). Two conceptions of mathematical notions: Operational and structural. In J. C. Bergeron, N. Herscovics, & C. Kieran (Eds.), *Proceedings of the Eleventh International Conference for the Psychology of Mathematics Education* (Vol. III, pp. 162-169). Montréal, Quebec: Université de Montréal.

Sfard, A. (1991). On the dual nature of mathematical conceptions: Reflections on processes

and objects as different sides of the same coin. *Educational Studies in Mathematics, 22,* 1-36.
Sfard, A. (1992). Operational origins of mathematical objects and the quandary of reification: The case of function. In G. Harel and E. Dubinsky (Eds.), *The concept of function: Aspects of epistemology and pedagogy* (pp. 59-84). Washington, DC: Mathematical Association of America.
Sheets, C., & Heid, M. K. (1990). Integrating computers as tools in mathematics curricula (grades 9-13): Portraits of group interactions. In N. Davidson (Ed.), *Cooperative learning in mathematics: A handbook for teachers* (pp. 265-294). Reading, MA: Addison-Wesley.
Shuard, H., & Neill, H. (1977). *From graphs to calculus.* Glasgow: Blackie.
Soloway, E., Lochhead, J., & Clement, J. (1982). Does computer programming enhance problem solving ability? Some positive evidence on algebra word problems. In R. J. Seidel, R. E. Anderson, & B. Hunter (Eds.), *Computer literacy* (pp. 171-185). New York: Academic Press.
Swan, M. (1982). The teaching of functions and graphs. In G. van Barneveld & H. Krabbendam (Eds.), *Proceedings of the Conference on Functions* (pp. 151-165). Enschede, The Netherlands: National Institute for Curriculum Development.
Swan, M. (Ed.). (1985). *The language of functions and graphs: An examination module for secondary schools.* Nottingham, England: Shell Centre for Mathematical Education; Joint Matriculation Board.
Thaeler, J. S. (1988). Input-output modifications to basic graphs: A method of graphing functions. In A. F. Coxford (Ed.), *The ideas of algebra, K-12* (1988 Yearbook of the National Council of Teachers of Mathematics, pp. 229-241). Reston, VA: National Council of Teachers of Mathematics.
Thomas, H. L. (1969). *An analysis of stages in the attainment of a concept of function.* Unpublished doctoral dissertation, Columbia University, New York.
Thompson, P. W. (1985). Experience, problem solving, and learning mathematics: Considerations in developing mathematics curricula. In E. A. Silver (Ed.), *Teaching and learning mathematical problem solving: Multiple research perspectives* (pp. 189-236). Hillsdale, NJ: Lawrence Erlbaum Associates.
Tranter, J. A., & Leveridge, M. E. (1978). Pond ecology. In M. E. Leveridge (Ed.), *Computers in the biology curriculum.* London: Edward Arnold.
Vergnaud, G. (1987). Conclusion. In C. Janvier (Ed.), *Problems of representation in the teaching and learning of mathematics* (pp. 227-232). Hillsdale, NJ: Lawrence Erlbaum Associates.
Verstappen, P. (1982). Some reflections on the introduction of relations and functions. In G. van Barneveld & H. Krabbendam (Eds.), *Proceedings of the Conference on Functions* (pp. 166-184). Enschede, The Netherlands: National Institute for Curriculum Development.
Vinner, S. (1989). The avoidance of visual considerations in calculus students. *Focus on Learning Problems in Mathematics, 11* (1 & 2), 149-156.
Vinner, S., & Dreyfus, T. (1989). Images and definitions for the concept of function. *Journal for Research in Mathematics Education, 20,* 356-366.
Waits, B. K,, & Demana, F. (1988, July-August). *New models for teaching and learning mathematics through technology.* Paper presented to the theme group on Microcomputers and the Teaching of Mathematics at the Sixth International Congress on Mathematical Education, Budapest, Hungary.
Yerushalmy, M. (1988). *Effects of graphic feedback on the ability to transform algebraic expressions when using computers* (Interim report submitted to the Spencer Fellowship of the National Academy of Education). Haifa, Israel: University of Haifa.

VI CURRICULAR IMPLICATIONS

9 Curricular Implications of Graphical Representations of Functions

Randolph A. Philipp
San Diego State University

William O. Martin and Glen W. Richgels
University of Wisconsin-Madison

The advent of graphing utilities in the form of calculators and computer software allows teachers and students to quickly and accurately construct and manipulate the graphs of functions—a process that was once beyond the scope of school mathematics. This change in accessibility to useful mathematical representations and information has important implications for the curriculum. This chapter addresses these implications in three sections. The first section outlines some of the forces that have shaped the existing curriculum, noting that although there have been calls for increased use of graphs and functions in mathematics, neither has yet played a prominent role. We believe technological developments can change this situation, making important and useful mathematical ideas accessible and interesting to a much broader range of students. The second section utilizes a number of examples to illustrate this potential for change. We contend that a graphical approach makes it easier to work with many existing topics, greatly increases the range of problems which are accessible to students, and reduces the prerequisite of extensively developed manipulative skills, so that students can spend more time modeling and solving problems. The last section discusses some of the issues that must be addressed and the research that must be conducted to support these changes.

Problem 1. Alison has found a beautiful house in a quiet neighborhood, which she has agreed to purchase for $73,000. She has planned to spend around $600 per month for housing, but this figure is flexible. She has $18,000 available for a down payment and can obtain a home loan at an annual rate of 9.875%. Alison is interested in investigating the relationship between the monthly payments she makes, the term, and the total cost of the loan.

A present-value annuity can be used to model this problem. The appropriate formula is:

$$A = R \frac{1-(1+i)^{-n}}{i}$$

where R = amount of each payment
A = present value of annuity (loan amount)
n = number of payments
i = interest rate per period

(A fifth parameter could be introduced: m = number of periods per year.) Two functions are relevant to Alison's loan:

1. Treating the monthly repayment R as a function of the number of repayments n, the formula gives:

$$R(n) = \frac{55000 \frac{0.09875}{12}}{1-\left(1+\frac{0.09875}{12}\right)^{-n}}$$

$$\approx \frac{452.604}{1-1.008229^{-n}}$$

2. The total amount repaid T as a function of the number of repayments, n is given by:

$$T(n) = \frac{55000 \frac{0.09875}{12} n}{1-\left(1+\frac{0.09875}{12}\right)^{-n}}$$

$$\approx \frac{452.604n}{1-1.008229^{-n}}$$

Graphing utilities can quickly and accurately provide information about the functions, as shown in the following figures. Without a graph, particular values would need to be computed by hand or calculator, or perhaps methods from calculus could be used to gain some understanding of the functions. Figures 9.1 through 9.3 are views of the first function, whereas Fig. 9.4 and Fig. 9.5 show the second.

Figure 9.3 provides one scenario for Alison. By paying $518 per month, she can pay off the $55,000 loan in 240 months. Figures 9.1 and 9.4 make

9. IMPLICATIONS OF GRAPHING FUNCTIONS 241

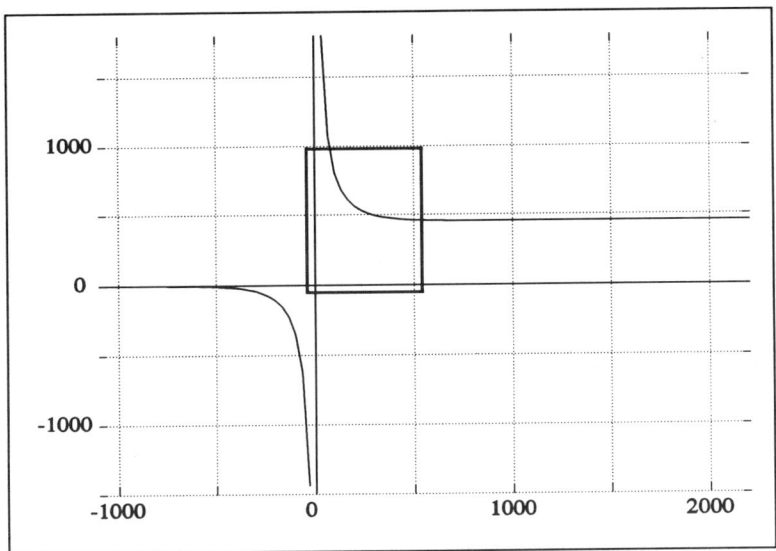

FIG. 9.1. Complete graph of function $R(n)$ relating amount of each monthly repayment to term of loan.

it easy to see that comparatively small increases in monthly repayments result in decreasing the term (number of repayments), which in turn dramatically decreases the total amount repaid. Without the graph it would be necessary to substitute particular values (for example, as illustrated in Figs. 9.2 and 9.5, $n = 120, 240, 360$ for terms of 10, 20, or 30 years respectively) and to make comparisons using a calculator. The graphical representation offers the same information, but can also provide a more complete representation of the relationship between the variables. Analyzing end behavior of the first graph also illustrates another interesting property of the function. There is a minimum monthly payment—the interest incurred on the amount of the loan over one month—that must be made regardless of the number of years one chooses to finance the loan. That is, should one finance the loan for 200 years, 1000 years, or 1,000,000 years, one would still be required to pay $452.60 each month. An interesting class discussion might develop as students try to decide why this seemingly bizarre notion is mathematically reasonable.

We chose this problem to open the chapter not because we felt it is likely to be particularly relevant or interesting for students, but rather because it illustrates several fundamental changes that may occur in the mathematics curriculum as a result of having immediate access to tools for graphing functions. First, a class of functions (exponentials) that frequently arise in mathematical models of our world becomes usable and understandable much earlier, because the dependence on paper and pencil computations is

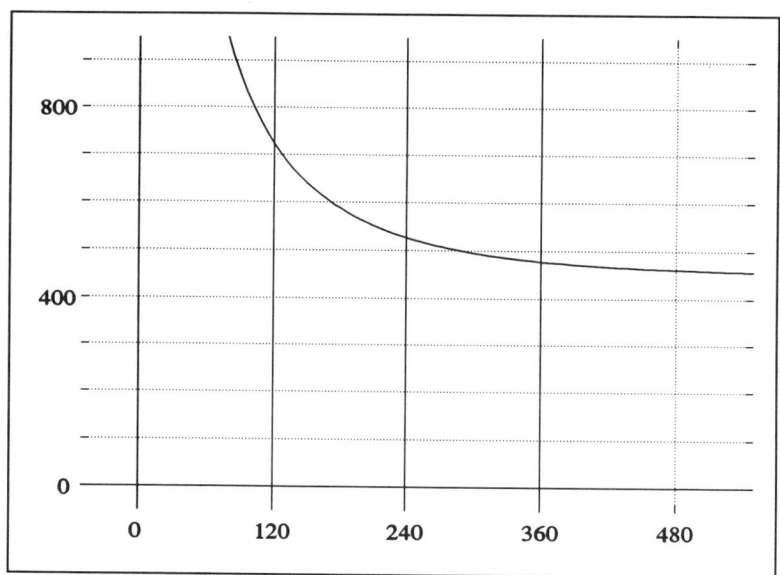

FIG. 9.2. Enlarged view of rectangle shown in Fig. 9.1. The three vertical lines correspond to terms of 10, 20, and 30 years.

greatly reduced. Second, this problem differs from many currently found in the curriculum because it does not have any particular "right" or "best"answer. The solution of the problem involves gaining an understanding of the relationships and interactions of variables in the model, rather than obtaining a particular number or expression. Again, it is the graphical representation that influences the accessibility of the function and the way in which we can use it. One of the authors, while presenting this problem in a graduate seminar in mathematics education, was asked to provide more details, such as alternative investments available, so that an optimum solution could be found. Although it was not clear in this case why more details were requested, this type of question is representative of deeply rooted, implicit ideas about what it means to do mathematics. One such belief is that mathematics involves finding the single best answer to well-defined problems.

The first goal of this chapter is to present the position that the ability to accurately represent the graph of a function quickly and easily offers the potential to significantly alter the mathematics curriculum. As has been done briefly in the preceding example, we intend to describe what some of these changes might look like.

The second goal of this chapter is to address some of the issues that determine whether this potential can be realized. As we examine these issues, we discuss both the forces that might influence this change and the

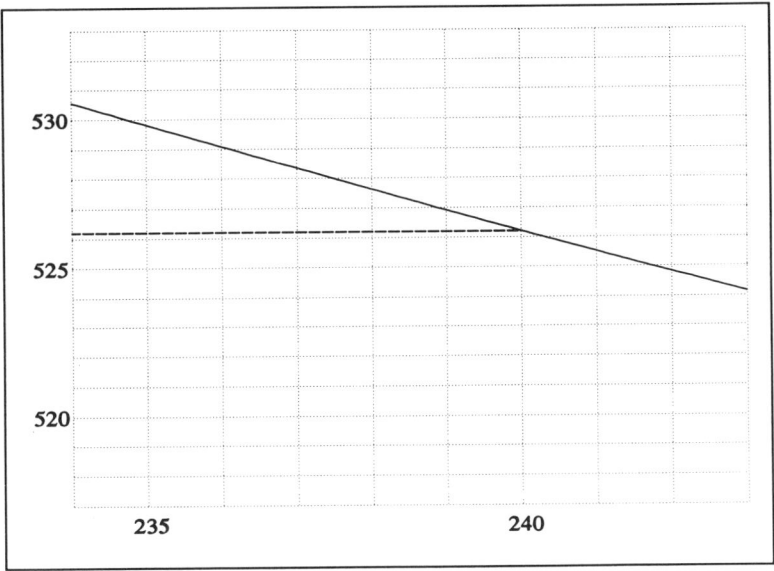

FIG. 9.3. Zoom-in view from Fig. 9.2 indicating monthly repayments required for 20-year term of loan. The horizontal line indicates a monthly payment of just over $526.

research questions that need to be answered. In order to develop a perspective from which to examine these ideas, we present a sketch of the historical evolution of the algebra curriculum, focusing particularly on the role that functions and their graphs have played in school mathematics. Although we believe that graphical representations of functions will play a key role at all grade levels within the curriculum, the focus of this chapter is on the secondary level, especially senior high school.

HISTORICAL PERSPECTIVE

Forces Acting on the Algebra Curriculum

The introduction of algebra into the high-school curriculum came about during the first half of the 19th century when Harvard, Yale, and Princeton added it to their list of required courses for entrance (Jones & Coxford, 1970). The transition from algebra as a college course to algebra as a high-school course was made with relatively few changes (Osborne & Crosswhite, 1970). Factoring, extracting roots, using powers, and other fundamental operations were emphasized, with little or no emphasis placed

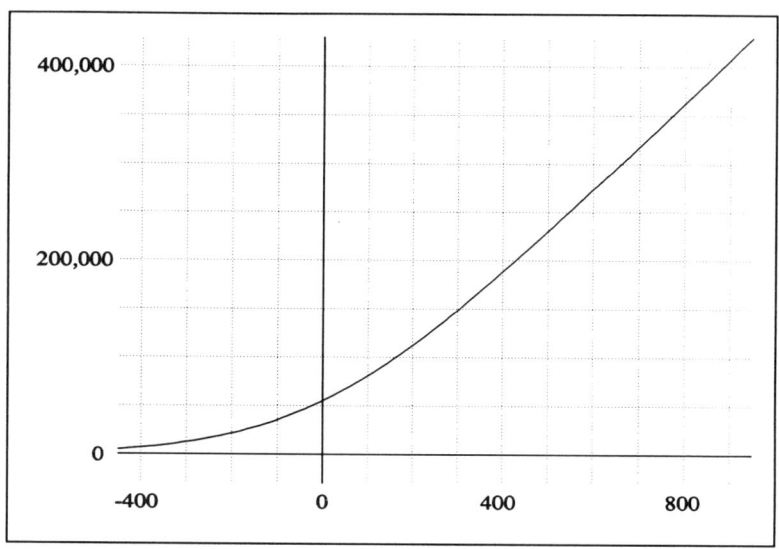

FIG. 9.4. Complete graph of *T(n)* giving total repayments as a function of the term of the loan.

on either graphs or functions. This wholesale adoption of the procedurally oriented college algebra curriculum was supported by mental disciplinarians who believed in the theory of faculty psychology. This theory presumed the existence of a few discrete faculties in the mind, including memory, imagination, observation, will, and reasoning. It was believed by mental disciplinarians that the curriculum should include those topics that best developed such faculties of the mind. Mathematics was high on their list, because memorizing tables would develop the capacity for memory, constructing proofs would develop reasoning, and solving a lot of tedious exercises would develop the will.

The situation had changed little by 1890, when only about 7% of the high-school aged youths attended school. The algebra curriculum still focused on fundamental operations with rational expressions, powers and roots, some factoring, and the solving of literal and numerical equations. Only about 0.1% of text material in the 1890s contained work with graphs, and functions were of little importance (Osborne & Crosswhite, 1970). As the 19th century ended, however, there were many signs that the time had come for an overhaul of this algebra curriculum. A survey of teachers of mathematics in 1890 indicated that they were dissatisfied with the emphasis on manipulative skills in algebra and were asking for a treatment of the subject that would result in deeper meaning and understanding (Osborne & Crosswhite, 1970). National and international organizations, including the

9. IMPLICATIONS OF GRAPHING FUNCTIONS   245

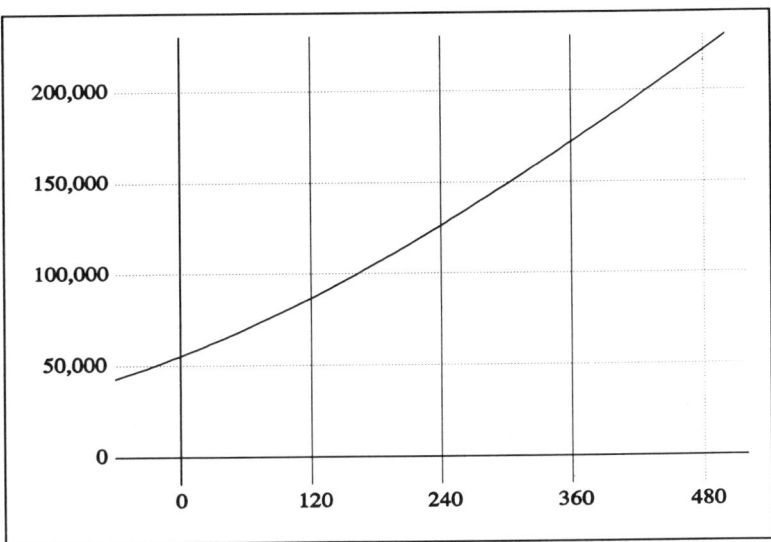

FIG. 9.5. Zoom-in from Fig. 9.4 indicating the problem situation. Vertical lines correspond to terms of 10, 20, and 30 years, while horizontal lines indicate total amounts repaid.

National Council of Education's Committee of Ten on the Secondary School Studies in 1892 and the National Education Association's Committee on College Entrance Requirements in 1899, began to study the mathematics curriculum and to make recommendations for reform, calling for less emphasis on symbol manipulation and more unification of the curriculum (Bidwell & Clason, 1970). Results of studies carried out by James in the 1890s and Thorndike in 1901 cast serious doubt on the theory of faculty psychology, removing one of the obstacles to a concept-oriented curriculum. Meanwhile, in the mathematics community, calls for unifying the curriculum around the function concept were being heard. In an 1893 address to the International Congress of Mathematicians, Professor Felix Klein of Germany called for increased attention to the function concept. A similar point was made in a 1902 presidential address to the American Mathematical Society in which E. H. Moore advocated unifying the mathematics curriculum around the function concept. Although increased attention was given to graphs in secondary schools between 1890 and 1930 (Izzo, 1957), the algebra curriculum of the early 20th century changed very little from the curriculum of the 1890s, in spite of these calls to reform. Sigurdson (1962) concluded that, "By 1930 the idea of unified mathematics and the ideology supporting it had faded from the educational scene" (p. 1997). One explanation for this might be that teachers "who relied almost exclusively upon the textbook as a teaching device" (Osborne & Crosswhite, 1970, p. 160) had few if any textbooks available that implemented these reform ideas (Lennes, 1932).

During the early part of the 20th century the subject-matter specialists exerted considerable influence on the elementary algebra curriculum. In particular, it was the mathematicians who were largely responsible for authoring textbooks (Kahn, 1974). The number of enrolled students who took high-school algebra increased from 45% in 1890 to 57% in 1910 (Breslich, 1933). Whereas in 1890 approximately 6.7% of 14- to 17-year-olds attended high school, by 1910 that percentage had increased to 15.4% (Stanic, 1984).

Between 1910 and 1920, the curriculum theory of social efficiency took the place of mental discipline, a theory that was formally dead (but would be around for a long time in the mind of educators). In describing the theory of social efficiency, Kliebard (1987) wrote,

> Efficiency became more than a byword in the educational world; it became an urgent mission. That mission took the form of enjoining curriculum-makers to devise programs of study that prepared individuals specifically and directly for the role they would play as adult members of the social order. To go beyond what someone had to know in order to perform that role successfully was simply wasteful. Social utility became the supreme criterion against which the value of school studies was measured. (pp. 89–90)

Mathematics, due to its prominent place in the curriculum, was a primary target for reform (Stanic, 1984). It no longer seemed essential for all students to take algebra because they would not all use it in their adult professional life. Thorndike et al. (1923) expressed the opinion that a pupil whose I.Q. is below 110 "will be unable to understand the symbolism, generalizations, and proofs of algebra. He may pass the course, but he will not really have learned algebra" (p. 37). David Eugene Smith, a professor at Teachers College of Columbia University, wrote in 1927 that the time for making a diagnosis of a pupil's abilities and tastes was during junior high school.

> At its close, each pupil should have shown his parents, his teachers, and himself what is his natural bent of mind and what it will probably continue to be. If he has no taste for or ability with respect to mathematics, and does not expect to enter a college or technical school, he should no longer be required to pursue the subject, and similarly with respect to the other great branches. (Smith, 1927, p. 232)

Teachers, too, expressed the belief that not everyone ought to be studying algebra. In a 1926 survey of 416 secondary-school teachers, Counts (cited in

Stanic, 1984) concluded that all high-school teachers, except mathematics teachers, believed that more students ought to be enrolled in their courses. Evidence that this belief was translated into practice can be seen in the fact that the number of enrolled algebra students leveled off at 57% in 1910 and began dropping to 40% in 1922 and 35% in 1928 (Breslich, 1933).

Social efficiency became the dominant curriculum theory between the two world wars, and with it came a change in who wrote the mathematics textbooks, with mathematicians giving way to educators. The social efficiency theory reflected the belief that people should not waste time studying subjects that will not directly help them in their careers. Many recommendations were made for differentiated curriculums. In a study conducted in 1923–1924, George S. Counts reported 18 different curricula in Los Angeles secondary schools and 15 in Newton, Massachusetts (Kliebard, 1987). Many students were discouraged from studying algebra altogether. The lead article in the National Council of Teachers of Mathematics Yearbook on the teaching of algebra reiterated Thorndike's recommendation that only students with I.Q. over 110 take algebra (Jablonower, 1932). Using a mean of 100 and a standard deviation of 16, this meant that under 25% of the population would be recommended for the study of algebra.

Mental discipline dominated curriculum theory of the 19th century, offering support for a procedurally oriented algebra curriculum that made few attempts at developing meaning. The social efficiency model, which dominated between the two world wars, stressed multiple curricula for different segments of the population, depending on their career goals. Although different theories were prominent at different times, none of the theories ever completely disappeared, and the 20th century curriculum could be described as a struggle between these and other forces (Kliebard, 1987). A recent example of mental discipline can be found in *Case Studies in Science Education* by Stake and Easley (1978, cited in Stanic, 1984), who summarized the views of teachers they interviewed as follows:

> Teachers said they could not change the fact that life is going to require youngsters to do a lot of things that do not make sense at the time, and that seem very difficult. Therefore, they argued that there should be a significant body of learning at every grade level which is difficult, which may not make sense at the time, but which has to be learned by every student. (p. 3)

On the other hand, social efficiency is alive and well in the minds of parents, teachers, and students who ask the question, "Why study a topic unless I know I'll use it?"

Beginning in the last century with the college entrance examinations that led to the introduction of high-school algebra, assessment has played a role in the development of the mathematics curriculum. George Counts wrote in

1923, "Undoubtedly, the entrenched position which algebra and geometry hold in the college entrance requirements have much to do with their persistence in the high school curriculum" (quoted in Betz, 1930, p. 108). A second example of the role of assessment is the substantial influence exerted on the curriculum by the College Entrance Examination Board (Barber, 1932). A third example can be found again in the writing of Betz (1930), who claimed that one of the problems encountered in trying to convert an algebra curriculum based on unrelated techniques to a curriculum based on thinking was textbooks and examinations. He wrote, "The individual teacher feels that, since she is held responsible primarily for her examination results, she must follow the beaten path" (p. 113).

Teachers, too, have influenced the algebra curriculum. In fact, the first appearance of graphs in elementary textbooks came about in the 1880s "as a result of classroom experiments and beliefs of individual teachers that graphs could aid in the teaching and learning of mathematics" (Izzo, 1957, p. 1506). Furthermore, it can be inferred that teachers' beliefs about who should learn algebra affected the decline in the percent of students enrolled in algebra between 1915 and 1930.

Graphical Representations of Functions

Current practice indicates that teachers rely heavily on the textbook, regarding it as "the authority on knowledge and the guide to learning" (Romberg & Carpenter, 1986, p. 867). This seems to have been the case in the past as well. Lennes, in 1932, wrote,

> I know of no method of obtaining a correct judgement of the way a subject is taught that is anywhere nearly as effective as a study of the texts that are used. The great majority of teachers follow the text very closely, for what they consider good reasons. They are exceedingly busy and have little time to formulate an independent treatment even if they had a desire to do so. Moreover, many teachers do not wish to depart from the treatment of the text, because they feel that in a first course the pupil should not be confused with different presentations—one by the teacher in the classroom and another by the text, which the pupil should be encouraged to read. (pp. 57-58)

The hesitation by textbook writers to adopt a functional approach has left its mark on graphical representations of functions. Prior to the 20th century, textbooks published in the United States contained practically nothing on graphical representations. A survey of some of the textbooks used indicates that although the focus on graphing and functions increased throughout the beginning of this century, its role was minor. For example, a 600-page textbook on elementary and intermediate algebra published in

9. IMPLICATIONS OF GRAPHING FUNCTIONS   249

1916 (Slaught & Lennes) contained only 17 pages on graphical representation, a 10-page chapter covering plotting points on a Cartesian coordinate system, and a 7-page chapter on the graphic representation of equations. A footnote to the 10-page chapter on graphical representations stated that this chapter could be omitted without destroying the continuity of the textbook. Upton's 500-page *Practical Algebra* book, published in 1936, contained 6 pages on graphing linear equations and 4 on graphing quadratic equations. He did, however, include a chapter on pictographs, bar graphs, line graphs, and circle graphs.

As the decades of the 20th century passed, textbook writers increased their emphasis on graphing in algebra textbooks. However, even in the 1960s and 1970s textbooks existed that barely addressed graphs. In a 1960 algebra textbook, Grove, Mullikin, and Grove included eight pages on graphing equations, whereas E. I. Stein (1970) did not include a single Cartesian graph in his first-year algebra textbook.

There are two notable features of the use of graphs in most existing textbooks. First, graphs are not integrated with other topics. Instead, they are treated in stand-alone chapters that could be skipped without hampering a student's ability to work through the remainder of the book. Second, textbooks approach graphs as ends in themselves—that is, once the graph has been produced, the problem is complete. When working with parabolas, for example, students might be required to determine the vertex, directrix, focus, axis of symmetry, and finally the graph of the parabola. Seldom are the students expected to use the graph to answer further questions about the function. Metaphorically, this approach to functions seems to be a half circle. One starts with the function and ends with its graph, never returning to the original function, or to the application that spawned the function.

Some recently produced textbooks have begun to change this approach. The University of Chicago School Mathematics Project has recently published a high school textbook that includes the following note to students: "If you plan to buy a calculator, consider buying a graphing calculator. Throughout this book there are questions which are made easier with such a calculator, and you will find such a calculator particularly useful in your future mathematics courses" (Senk, Thompson, & Viktora, 1990, p. 1). Working out of Ohio State, Demana and Waits (1990) wrote a precalculus textbook that requires the use of a graphing device.

In summary, graphical representations of functions have played a minor role in the algebra curriculum of the past, and when they have been included, no attempt has been made to integrate graphical representations of functions with the rest of the algebra curriculum. This situation exists in spite of the recommendations of educators and the pressures of the reform movement of the early part of the 20th century. A possible explanation is that the construction of graphs of functions is a tedious and time-consuming process. Teachers and students would have great difficulty calculating values for trigonometric, exponential, and logarithmic functions without access to present-day calculating devices, making it impractical to graph all but elementary polynomial functions. The amount of effort needed to construct graphs in the past exceeded the benefits that accrued from graphical representations. Whatever the reason, the proposed emphasis on functions or their graphical representations never materialized.

EXAMPLES OF A GRAPHICAL APPROACH IN THE CURRICULUM

Many of the skills currently found in the mathematics curriculum were included because they were necessary to solve specific classes of problems. For example, often a "real-world situation" can be analyzed by finding an appropriate function to model important or significant aspects of the situation. Solutions to the problem are obtained by manipulating the function's representation. Consider the following question: When will an object dropped from some height strike the ground? Neglecting air resistance and working near the surface of the earth, the Newtonian model for gravitational acceleration is a quadratic function relating time, distance, and acceleration due to gravity, and the solution is obtained by finding the zeros of the function.

Students have not been able to construct these functions or to find their zeros until they developed some skills, including the ability to use variables to represent unknown quantities, manipulate polynomial expressions to obtain equivalent forms, and factor expressions to find their zeros. Because of this it has been necessary for students to develop a variety of specialized skills before they can tackle realistic or interesting problems. Graphing utilities now allow teachers to bypass much of the early, unmotivated skill development that was necessary in the past. A graphing utility requires the user to input the closed-form functional representation, which can then be represented graphically. In the traditional curriculum in which the closed-form functional representation has generally been provided, the entire task amounted to plotting the graph of the function. However, when working with real-life problems, composing the closed-form functional representation may not be a simple task. Unlike the graphing utilities, spreadsheet software enables the user to plot the graph of a function without determining the closed-form functional representation. The following example illustrates the use of these two technologies. Unlike the first, more realistic, problem, this problem is unrealistic and contrived. In spite of this, its subject matter may be of more interest to secondary school students than the home loan repayment problem described earlier.

Problem 2. You are the hottest movie star of the 1990s. You have been approached by three separate motion picture studios, each of which would like to hire you to star in one of their upcoming films. All three of the studios plan to shoot their pictures in July, placing you in the position of having to select one studio. All three of the studios have assured you that their movies will require somewhere between 2 weeks (approximately 14 days) and 3 weeks (approximately 21 days) for the shoot. You are pleased with all three of the parts and you want to accept the most lucrative job. The motion picture studios have been experimenting with some rather unusual salary contracts. You have been offered the following contracts:

Universal: A flat rate of $100,000 for each day of work.
Orion: $10 for the first day of work, with your daily salary doubling for each additional day.
Touchstone: One penny on the first day of work, with your daily salary tripling for each additional day of filming.

Each studio guarantees that you will be paid for between 14 and 21 days of work. Which of the motion picture studio contracts would you accept?

Would your answer to the above question depend upon the number of days you worked? If so, then under what conditions (how many working days) would it be most lucrative to work for Universal? for Orion? for Touchstone? If $Y(n)$ represents the total amount of money earned during n days of work, then the following functions correspond to each motion picture studio offer:

Universal: $Y(n) = 100,000n$
Orion: $Y(n) = 10(2^n - 1)$
Touchstone: $Y(n) = .005(3^n - 1)$

The last two equations are exponentials, which are difficult for most people to work with. Graphing utilities can quickly and accurately provide information about the functions, as shown in Fig. 9.6. The region around the points of intersection is particularly important for analyzing this problem. This is easily found graphically by zooming in on the intersection of the three curves, as shown in Fig. 9.7. This view of the graph shows that the linear function representing the offer by Universal Studios provides the best return until somewhere between 17 and 18 days. The doubling exponential becomes more lucrative at that point until close to 19 days, at which time the tripling exponential becomes most lucrative. This display offers an easy comparison between the linear function and the exponential

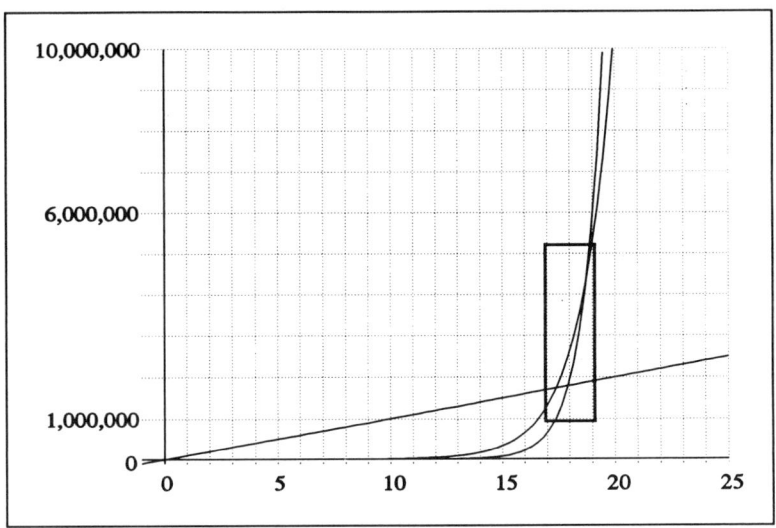

FIG. 9.6. Comparison of payment methods for actor over 22-day period. (Area enclosed by rectangle is enlarged in the next figure.).

functions. One can also observe the effect of decreasing the constant (10 dollars to 1 cent) while increasing the base of the exponential (from doubling every day to tripling). Furthermore, graphing the functions clearly illustrates the fact that although the exponential function may start slowly, it will eventually surpass the linear function.

Graphing utilities offer the ability to quickly graph the functions and zoom in around a specified region. In no more than a couple of minutes, the three functions can be graphed and solutions to any degree of accuracy can be investigated using the zoom feature. The points of intersection represent places at which the value of two studio contracts are equal. For this problem, where the points of intersection were at 17.41 days and 18.75 days, the following pertains:

17 or fewer days: The Universal contract is best.
18 days: The Orion contract is best.
19 or more days: The Touchstone contract is best.

Given the original contract, which guarantees between 14 and 21 paid days of work, there is no one best answer to this problem. In such a case, having a graphical representation may help in more ways than one. In addition to providing a quick answer to any degree of accuracy, the graph offers a visual representation with which a comparison between the three contracts may be made over specified time intervals. For example, the graph indicates

9. IMPLICATIONS OF GRAPHING FUNCTIONS   253

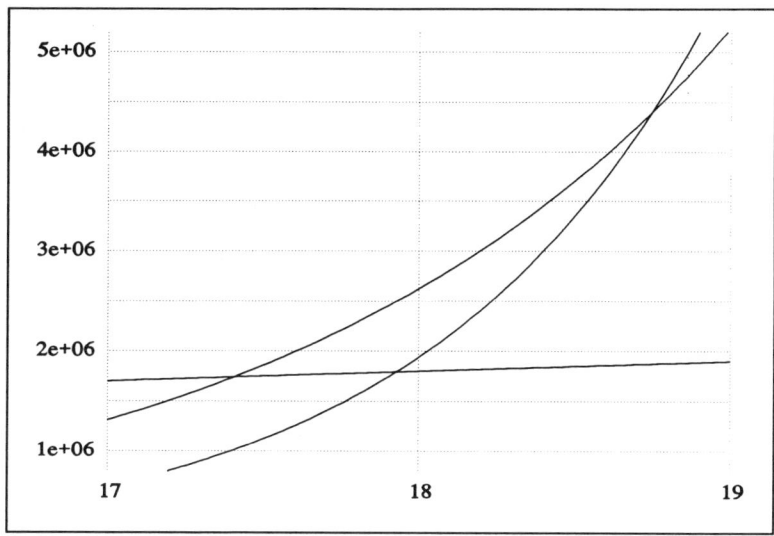

FIG. 9.7. Comparison of linear and exponential functions between 17th and 19th days. (Enlarged from Fig. 9.6).

that the Universal Studio contract, represented by the linear function, offers the most money up until 17.41 days. How much more is it at 16 days? At 15 days? How much better is the Orion contract than the other two contracts over the interval between 17.41 and 18.75 days?

Another approach to this problem involves the use of spreadsheet software, such as Lotus 1,2,3 for DOS or Microsoft Excel for the Macintosh. These applications make it possible to produce graphs directly from tabular information without determining the closed-form functional representation needed for the graphing utility. One of the authors spent one morning teaching himself the fundamentals of Microsoft Excel and then easily produced the worksheet in Fig. 9.8. The column labeled "Univ Total" was produced by entering 100,000 into cell B1 and then writing a formula (=B1+100000) that produced each new cell by adding 100,000 to the cell immediately above. By clicking on cell B2 and dragging down the rest of the column, this iterative procedure was repeated for the rest of the column. Columns C and E were produced using the same approach. The first number was entered and a simple formula was written that produced each new cell by doubling (or tripling) the cell directly above. Again, by clicking and dragging, the rule was repeated for the other cells in each column. Columns D and F were produced by writing a simple formula that produced each new cell as the sum of the cells directly above and directly to the left.

Once the table was created, the graph could be produced by highlighting the appropriate cells. Fig. 9.9 displays the graph of the data from day 14

A	B	C	D	E	F
Day #	Univ. Total	Orion Each Day	Orion Total	Touchstone Each Day	Touchstone Total
1	100000	10	10	0.01	0.01
2	200000	20	30	0.03	0.04
3	300000	40	70	0.09	0.13
4	400000	80	150	0.27	0.4
5	500000	160	310	0.81	1.21
6	600000	320	630	2.43	3.64
7	700000	640	1270	7.29	10.93
8	800000	1280	2550	21.87	32.8
9	900000	2560	5110	65.61	98.41
10	1000000	5120	10230	196.83	295.24
11	1100000	10240	20470	590.49	885.73
12	1200000	20480	40950	1771.47	2657.2
13	1300000	40960	81910	5314.41	7971.61
14	1400000	81920	163830	15943.23	23914.84
15	1500000	163840	327670	47829.69	71744.53
16	1600000	327680	655350	143489.07	215233.6
17	1700000	655360	1310710	430467.21	645700.81
18	1800000	1310720	2621430	1291401.63	1937102.4
19	1900000	2621440	5242870	3874204.89	5811307.3
20	2000000	5242880	10485750	11622614.67	17433922.
21	2100000	10485760	20971510	34867844.01	52301766.

FIG. 9.8. Complete table of the data for the payment methods for actor, using Microsoft Excel Spreadsheet.

through day 21. By highlighting the spreadsheet cells for days 14–19, the software package automatically rescaled the graphs to better illustrate the differences between the three payment methods during this critical period (see Fig. 9.10). The discreet nature of the problem is more evident when viewing the bar graph than when viewing the continuous line graph of the function provided by the graphing utility.

The preceding example illustrates some difference between using graphing utilities and spreadsheet software to represent functions. An advantage of the spreadsheet software lies in its ability to graph a function without first constructing the closed-form algebraic representation. In this problem, the recursive definition of the function used in the development of the spreadsheet solution is more natural and accessible by students than the exponential functions required for the graphing utility solution. In other situations, a spreadsheet can be used to generate graphical representations from tabular data, incorporating a third important functional representa-

9. IMPLICATIONS OF GRAPHING FUNCTIONS

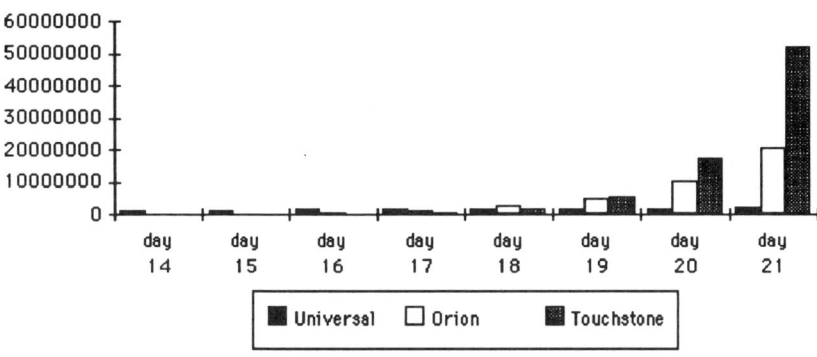

FIG. 9.9. Comparison of payment methods for actor between days 14 and 21, using Microsoft Excel Spreadsheet.

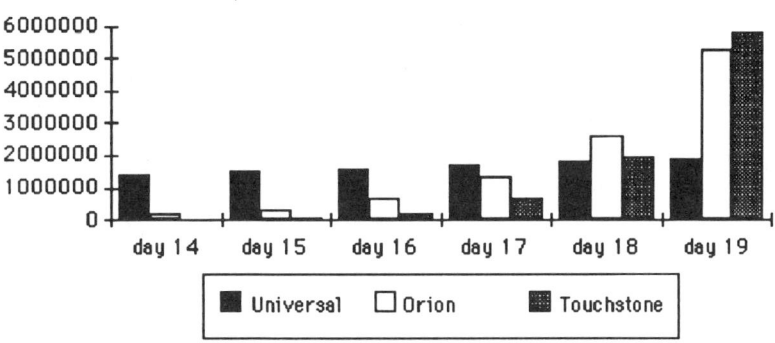

FIG. 9.10. Comparison of payment methods for actor between days 14 and 19, using Microsoft Excel Spreadsheet.

tion. An advantage of the basic graphing utility is its relatively low cost and, in the case of graphing calculators, portability and small size.

Another important contrast illustrated here, which also applies to different implementations of mathematical graphing utilities, is the use of automatic scaling that provides an appropriate view of the relationship between variables. Goldenberg (1988) emphasized the central importance that issues relating to scale had in their work with students' misunderstandings of graphical representations. He also wrote that

> Two conclusions stand out: Simplistic software design or thoughtless use of computer graphing in classrooms may further obscure some of what we

already find very difficult to teach. On the other hand, thoughtful design and use of graphing software presents new opportunities to focus on challenging and important mathematical issues that were always important to our students but that were never so accessible before. (p. 135)

It is apparent that there is a need for research into relationships between characteristics of technologies and student learning. Any implementation of technology in the curriculum needs to match student needs, for example, in terms of the degree of automation versus student control of technology. Other chapters in this volume have examined such issues relating to student thinking in greater depth — in our discussion of curricular issues we have chosen to focus on the use of a basic graphing utility because of its low cost and relatively easy accessibility.

In the past, students who had not developed adequate skills were barred from higher level mathematics due to the fact that the skills were prerequisite to successfully dealing with the problems studied. Furthermore, even students who had gone on to study more advanced mathematics were often prevented from working with realistic and interesting mathematics because they lacked an efficient means by which they might quickly and accurately determine the graph of functions. The examples in this chapter are intended to suggest ways in which graphing technologies might provide a means of remedying these two situations. Another advantage of graphing utilities is that they make it possible for students to tackle problems that in the past could not be solved. Some of these problems come from more advanced courses, such as calculus, whereas others have not been a part of the curriculum at all. Maximum and minimum problems are examples of the former. Such problems are not difficult to set up or to understand mathematically, but differentiation is necessary in order to locate relative extrema. Consider the following example.

Problem 3. Cans are a pervasive part of modern life, though they are rarely given much thought (other than perhaps in terms of the ecological issues of disposal and recycling). They come in a variety of shapes and sizes, and although the size is determined by what they are to hold, what factors influence their shape? For example, an aluminum soft-drink can typically measures about 2½ inches in diameter by 4¾ inches high, whereas a one-gallon paint can has a diameter of 6½ inches and a height of 7½ inches. Both are cylinders, but they are not geometrically similar. Why not?

Discussion. This could be dealt with in various ways and at various levels of detail. The amount of material necessary to contain a given volume is obviously one of the considerations that can be explored with a mathematical model and graphing utility (calculus could also be used, and

9. IMPLICATIONS OF GRAPHING FUNCTIONS

would have been necessary in the past). One question might be, What are the dimensions of a cylindrical can that has the least surface area (approximating material used) for a given volume? Recall that the volume and surface area of a cylinder with radius r and height h can be expressed as follows:

$$V(r,h) = \pi r^2 h$$
$$S(r,h) = 2\pi r^2 + 2\pi r h$$

Using these two formulas and some algebraic manipulation, one can see that a cylinder of volume 1000 cubic units has a surface area given by this function of its radius r:

$$A_c(r) = 2\pi r^2 + \frac{2000}{r}$$

Another question might be, How do different solids compare in the relationship between their volume and surface area? For example, a box with a square base and a volume of 1000 cubic units has a surface area given by this function of w, the side length of the square base:

$$A_b(w) = 2w^2 + \frac{4000}{w}$$

Graphs of both functions are shown in Fig. 9.11. It appears that the cylinder has a smaller minimum possible material, and by zooming in as indicated by the rectangular box, Fig. 9.12 shows that the minimum for the cylinder is about 555 square units, whereas the minimum for the box appears to be about 600 square units (calculus shows that this is exact). These minimums could easily be approximated to any desired degree of accuracy by repeatedly zooming in. Graphs also show that the minimum for the can corresponds to a radius of 5.4 units, whereas the minimum for the box corresponds to a side length of 10 units. It would not be difficult to pursue similar computations for other containers, or even to develop more sophisticated models that incorporate seams and waste material.

Figure 9.13 extends the problem by graphing the relationship between the radius and the height of the cylinder, with the line specifically indicating that a radius of 5.4 units corresponds to a height of 10.8 units. (This could just as easily be found algebraically, possibly with a calculator.) That is, in order to minimize the surface area for a given volume of a cylinder, the diameter ought to equal its height. Furthermore, to minimize the surface area of a rectangular box for a given volume, use a cubic box. Finally, for a given volume, the minimum surface area for a cylinder is always less than

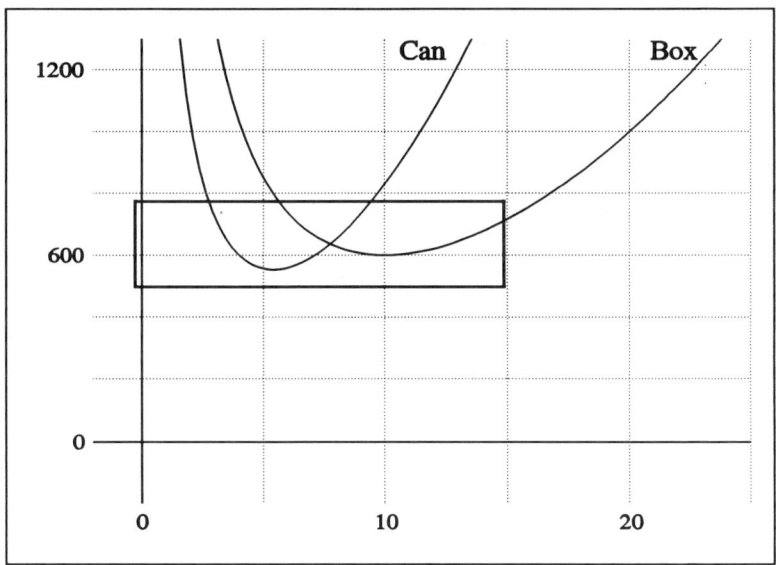

FIG. 9.11. Surface area of a cylindrical can as a function of its radius, and of a square-based box as a function of the base width (each using a volume of 1000 units).

the minimum area for a box. The paint can comes close to using this optimum shape, whereas the soft drink can differs significantly. The graphs can also be used to assess the extent to which variations from the optimum radius lead to increases in the amount of material. Small deviations do not make much difference, whereas larger changes have an increasingly significant effect on the area. This would be difficult to see even with calculus, whereas the graphical information is easy to interpret.

Again, the specific solution is obtained easily, but new questions arise naturally from the graphical representations of the functions. For example, the vertical axis is an asymptote for each of these functions, and it appears that any positive number can be used for the independent variable. How are we to interpret these graphical features in relation to the problem? What is the meaning of the portion of the graph with negative x values? What other considerations might influence design decisions besides the amount of material? For example, suppose purchasers found the shape of the can inconvenient to grasp due to its circumference. Or what about wasted space when packing the containers into boxes or cupboards? The square-based box has a clear advantage in this case! A sphere turns out to minimize surface area for a given volume. This embedding of measurement, volume, area, polyhedra, and polygons in a realistic problem situation illustrates

9. IMPLICATIONS OF GRAPHING FUNCTIONS

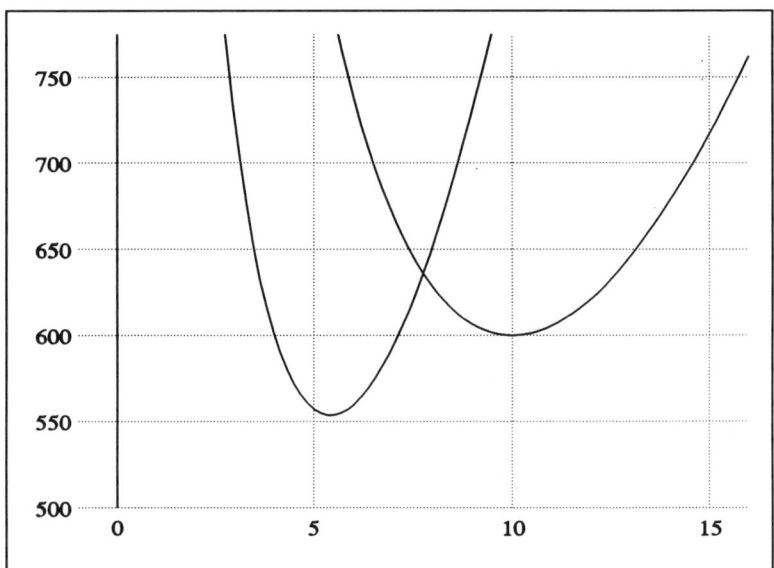

FIG. 9.12. Zoom-in to compare minimum material for can (about 555 square units) and box (about 600 square units). The minimum for the can occurs at about a 5.4 unit radius.

some of the potential benefits of making a graphical approach accessible to students even in the middle grades.

A number of topics currently covered in the mathematics curriculum can be taught more easily using graphing utilities. Curve sketching is the most obvious, but graphing tools make significant contributions to the teaching and learning of topics such as transformations, classification of zeros and factoring, systems of equations, and limits. Each of these topics can be covered in a more unified, conceptual way with the utilities. Some problems become very easy, as the following examples illustrate.

Problem 4. Evaluate

$$\lim_{x \to 0} \frac{\sin x}{x}$$

(This is needed to find derivatives of trigonometric functions)

Discussion. See Figure 9.14. From a rigorous point of view, a graph, because it is based on a finite number of points, does not prove anything. In a widely used college calculus text (Thomas & Finney, 1988, pp. 67, 68)

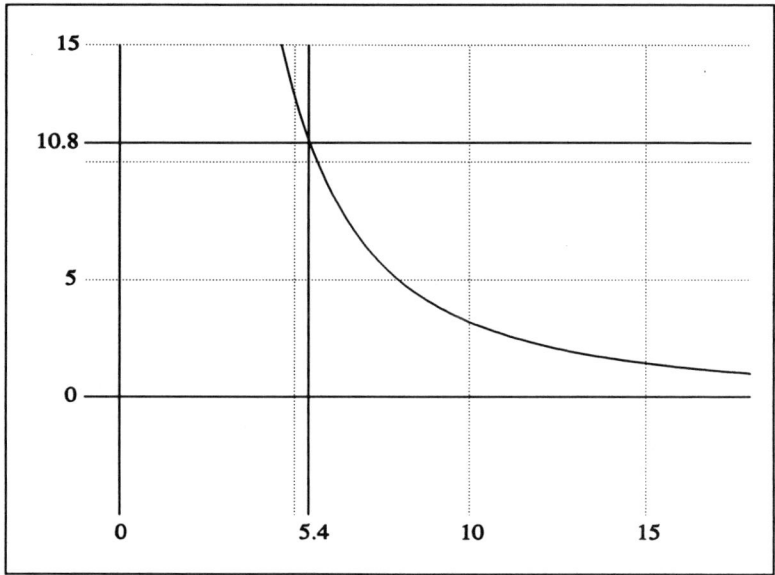

FIG. 9.13. The height of the can (volume=1000) as a function of its radius. Radius 5.4 corresponds to height 10.8, so a can with height equal to its diameter uses the least material.

one page is devoted to proving this limit. In addition to a rigorous proof, this discussion includes a table of values illustrating the behavior of the function for values of x near zero. However, although the authors included many graphs in this chapter, they did not include the graph displayed in Fig. 9.14, which would have illustrated this limit much more clearly than the table of values. The limit is obvious from the graph, which illustrates how graphical information can provide intuitions that may make it easier for students to acquire important mathematical concepts. Many theorems, such as the mean value theorem, or the fundamental theorem of calculus, are suggested and proved using the intuitions provided by a graph as an indication of the direction in which to proceed.

Problem 5. Find the rational zeros of the function

$$f(x) = 2x^4 - x^3 - 38x^2 - 41x + 30$$

Discussion. See Fig. 9.15. Such problems are found in sections of textbooks that cover factoring polynomials using the factor and remainder theorems. Although the theorems are still useful, this particular problem is simple with a graphing utility.

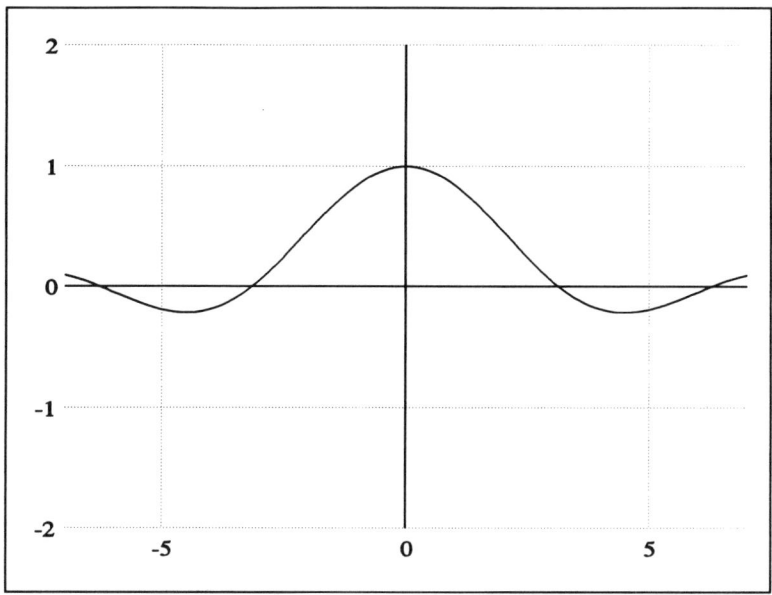

FIG. 9.14. This graph of the function $f(x) = (\sin x)/x$ illustrates that as $x \to 0$, $f(x) \to 1$.

The notion of limit and the nature of zeros of a function are both important mathematical ideas; the following examples illustrate how these concepts can be approached with a graphing utility. In each case, the utility is a necessary tool to obtain information, but the student must already possess important mathematical concepts in order to make effective use of the graphical information.

Problem 6. Classify and find all the zeros of the function

$$f(x) = x^5 + 2x^4 - 3x^2 - 4x - 2$$

Discussion. Fig. 9.16, in conjunction with the rational root theorem, shows that this polynomial has an integer root, two irrational roots, and two nonreal complex roots. This theorem often gives a relatively large list of potential roots, most of which are immediately ruled out by the graph. Contrary to what might be expected, graphing utilities can make the study of factoring and finding the roots of polynomials with integer coefficients an interesting and significant topic. In Chapter 4 of their text, Demana and Waits (1990) illustrated how this traditional topic of factoring can be approached using utilities. The topic hints at some very beautiful mathematics, such as Galois theory and field theory in abstract algebra, which

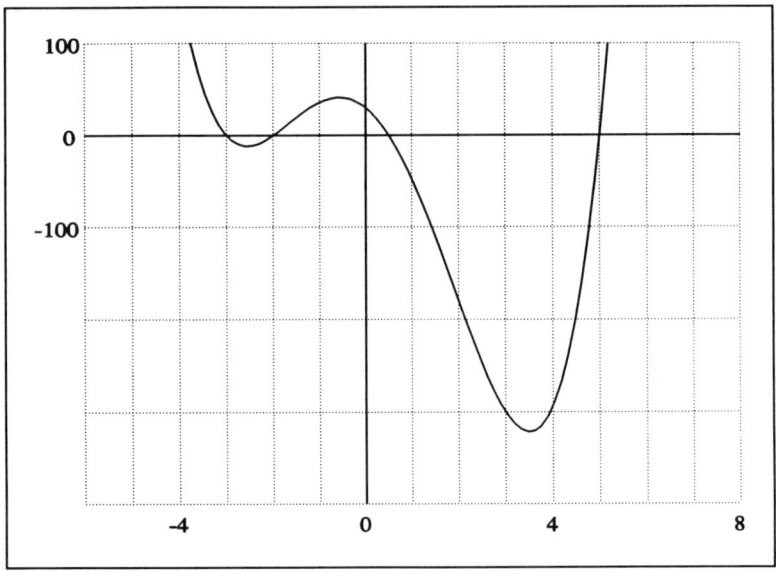

FIG. 9.15. The rational zeros of $f(x) = 2x^4 - x^3 - 38x^2 - 41x + 30$ can be read from this graph.

was partly developed to investigate the problem of finding zeros of functions. The search for solutions of equations has historical and current importance in a number of areas, including linear and abstract algebra, physics, and applied mathematics. The investigation of this problem has spurred development of a great deal of mathematics.

The next two problems are examples of new topics that are now within the reach of students. They are intended to show that areas of mathematics that have been inaccessible can now be investigated using graphing utilities.

Problem 7. What are the *end-behavior models* for the functions in Problem 3, which dealt with the amount of material required to construct containers of a given volume? (That is, what happens to the function values as the independent variable becomes large positively and negatively?)

Discussion. Both functions are already written as a quotient plus a remainder, which makes it easier to see that each behaves like a parabola for values away from the vertical asymptote at $x = 0$:

$$A(x) = 2x^2 + \frac{4000}{x} \quad \text{for the square-based box}$$

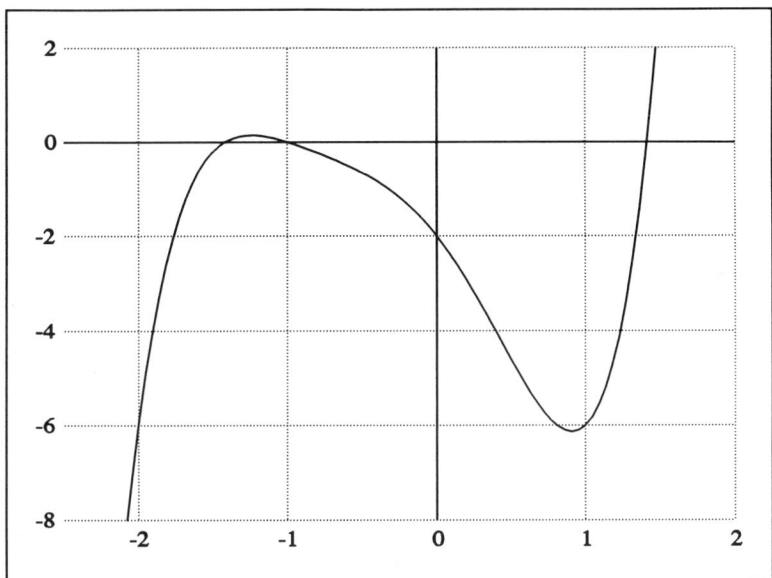

FIG. 9.16. The graph provides useful information for finding and classifying the roots of this fifth-degree polynomial.

$$A(x) = 2\pi x^2 + \frac{2000}{x} \quad \text{for the cylindrical can}$$

The graphical information in Fig. 9.17 plainly illustrates this phenomenon. Students will be able to relate the geometric and the algebraic information as they explore this characteristic of rational functions. Nonlinear asymptotes have not been a part of even the calculus curriculum in the past, but it now becomes natural to consider this more general form of the asymptotic behavior of functions. Students are likely to stumble upon a variety of interesting characteristics of various functions as they seek an appropriate viewing rectangle that displays a complete graph.

Problem 8. Find all solutions to the equation

$$2^x = x^{10}$$

Discussion. Alison's investment in Problem 1, comparing compound and simple interest, involved finding the intersection of a linear and exponential function. Problem 8 is a more difficult situation that looks for the intersection of a 10th-degree polynomial and an exponential function. There are three solutions, two of which are easily found graphically. The

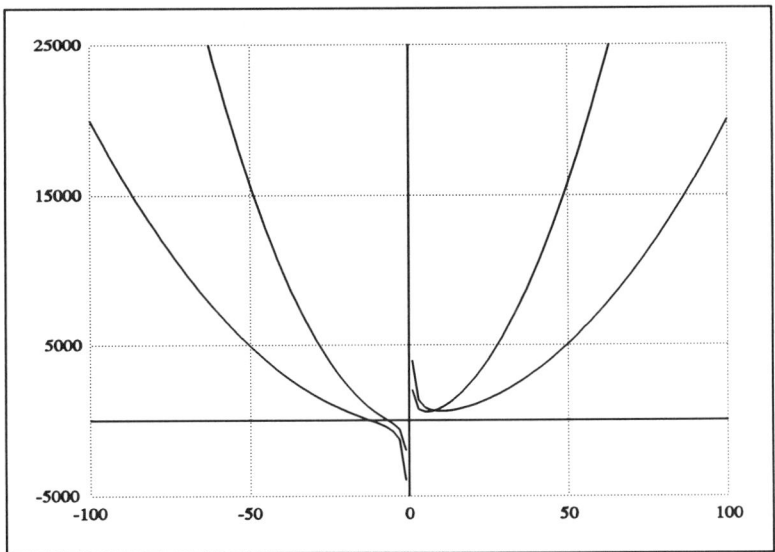

FIG. 9.17. These are the same functions as in Problem 3. The change in scale clearly shows the parabolic asymptotic behavior of the functions as $x \to \pm \infty$.

third solution would probably not be found unless one knew to look for it because "an exponential grows faster than any polynomial." Even when one knows it must exist for some positive x, it is a challenging exercise to find the third solution graphically (the authors are unaware of any algebraic solution). This illustrates that graphing utilities do not do all, or even most, of the mathematics for users. It is still necessary to develop considerable mathematical skill and knowledge to solve problems and to utilize the utilities fully. (Fig. 9.18 provides two solutions; the third is left as a challenging exercise for the interested reader.)

THE ROLE OF GRAPHS IN THE CURRICULUM

The sample problems were chosen to suggest that technological developments provide an opportunity to change the relationship between the algebraic and geometric representations of functions — a change that carries important implications for the mathematics curriculum. Students or teachers can inexpensively purchase or often obtain, free-of-charge, public domain graphing software designed for a personal computer, which interactively allows sketching, scaling, and other manipulations of functions, conic sections, or parametric equations, using both polar and rectangular

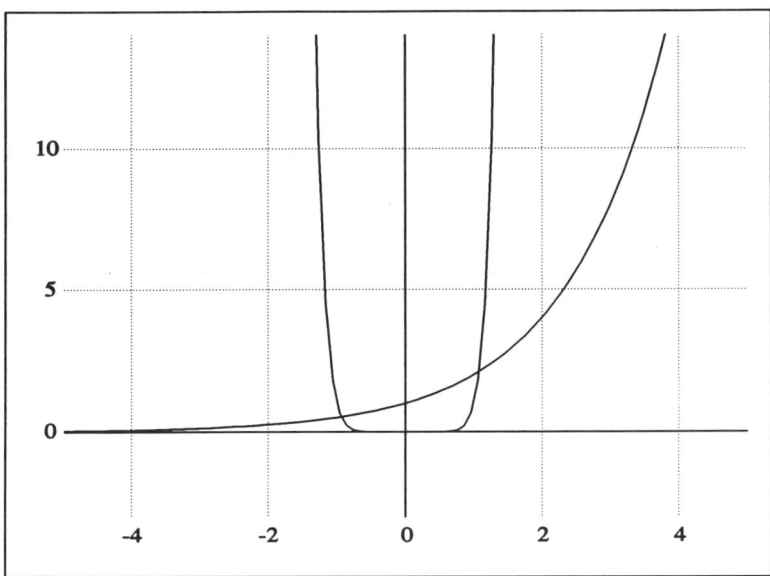

FIG. 9.18. Two solutions to the equation $2^x = x^{10}$ are shown. A third solution occurs for some value of $x > 40$.

coordinates. Pocket calculators that incorporate some of these features cost as little as $60 (for example, from Casio, Sharp, and Texas Instruments). The graphs that can be produced by these tools surpass those that even a skilled person can produce in terms of accuracy, complexity, and speed, providing easy access to the characteristics of a much greater range of functions and relations in mathematics classes. This capability changes what is possible in mathematics classrooms. The amount of effort needed to construct graphs in the past usually exceeded the benefits to be gained from graphical representations. In the next few years, however, as the power of such tools increases and the cost almost surely decreases, more students will have access to such resources. We believe the question will not be whether to use them, but how. That is, how can this technology best be used to influence the teaching and learning of mathematics? What is the role of functions and their graphs in mathematics?

At the same time that technological developments provide an increased capacity to investigate and carry out mathematics, many people and organizations have been criticizing the existing state of mathematics education and proposing changes to improve it. Is there any relationship between technology, functions and graphs, and an improved mathematics curriculum? We believe there is. Furthermore, the graphical approach could be integrated widely into the curriculum, promoting a study of mathematics that in many ways will be closer to the spirit of such important proposals as

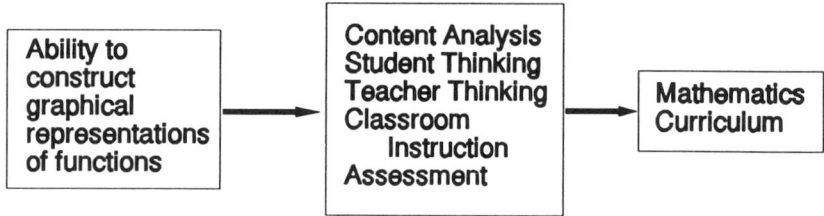

FIG. 9.19. Model of the influence of the graphical representation of functions on the curriculum.

the *Curriculum and Evaluation Standards for School Mathematics* of the National Council of Teachers of Mathematics (1989). To help clarify how the graphical representation of functions could have a positive influence on the mathematics curriculum, offer the model shown above, illustrating its relationship to various components of the mathematics education system (see Fig. 9.19). Technological developments have significantly altered this model. Calculator or computer-based graphing utilities make it much easier to obtain accurate and easily manipulated graphs of a wide variety of functions. The model in Fig. 9.20 suggests how recent changes have influenced the relationships among the three elements in the model, increasing the impact of a graphical approach on the curriculum. It is not our intent to focus on the technology, but rather to show how any graphing utility might influence mathematics education by providing easily manipulated mathematical information. This model is useful as an organizer for the ensuing discussion. We must emphasize that much of what we have to say about curricular implications is opinion. There is little, if any, research support for our ideas, because only recently have developments made graphs easily and widely available. The reader is cautioned to view the following assertions as hypotheses requiring testing and modification. In this sense, most of what follows is a guide and a call for future research into questions relating the graphical representation of functions to their place in mathematics.

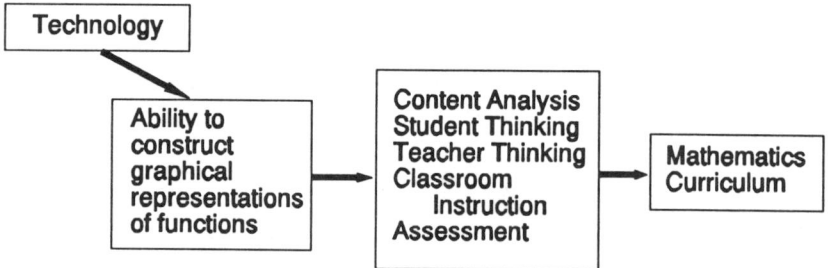

FIG. 9.20. Model of the influence of graphing utilities on the curriculum.

As suggested by the model, there are several components of the educational setting that interact with the curriculum: Graphing utilities will influence the curriculum in different ways through each of those various components. This section describes the impact and suggests research questions related to the five components listed in the figure: content analysis, student thinking, teacher thinking, classroom instruction, and assessment.

Content Analysis

Many of the earlier examples in this chapter illustrated how a graphical approach can influence the content of school mathematics. For example, representing functions with graphs might promote a view of mathematics as the study of a variety of functions rather than, according to the traditional view, as the study of a collection of methods for manipulating symbols. This could in turn lead to several curricular changes. Textbooks might reflect a new organizational structure in which problems are sequenced by the sophistication of their functional representation rather than by their specific type of algebraic representation. For example, it is not uncommon to find second-year algebra textbooks dividing the study of parabolas into many sections, based on the variation of the algebraic representation. The widely used *Merrill Algebra Two* (Foster, Rath, & Winters, 1983) includes a chapter on quadratic functions that covers the following topics: Parabolas (Section 7-2), Graphing $y = a(x - h)^2$ (Section 7-3), Graphing $y = a(x - h)^2 + k$ (Section 7-4), Using Parabolas (Section 7-5), Graphing Quadratic Inequalities (Section 7-6), and Solving Quadratic Inequalities (Section 7-7). Because the graphing utility allows a similar treatment of all functions, the mathematics curriculum could increasingly concentrate on the relationship between the problem and its model, rather than partitioning the study of each class of functions into small chunks as is presently done in many textbooks and courses.

In a discussion of the algebra curriculum, Thorpe (1989) maintained that at least one of three criteria must be met before a topic can be included. These three criteria are: intrinsic value, pedagogical value, and intrinsic excitement or beauty. Intrinsic value refers to topics included in the curriculum because of their potential importance to students' lives. Pedagogical value refers to topics that are important because they form a foundation for topics that have intrinsic value. Intrinsic excitement or beauty refers to topics, such as fractals, that seem to capture the interest of everyone. The graphing utility will change the categorization of some topics. Exponentials, which have great intrinsic value, may become more exciting for students as their useful and unusual properties are more easily and clearly revealed with graphs. Manipulative skills that in the past carried pedagogical value may no longer be necessary prerequisites for the study of topics that have intrinsic value. For example, factoring, which has been taught as a necessary skill for finding the solutions of equations, may lose its importance with the graphing utility. On the other hand, factoring is a basic concept used in many areas of mathematics, such as algebra and analysis. As such it retains pedagogical value. The use of graphs could even increase the intrinsic interest of factoring, as students explore the connections between graphical and algebraic representations of functions. Graphing utilities can help students discover the importance of factoring and enhance their interest in it by greatly reducing the emphasis on an algorithmic, manipulative approach, while revealing the significant information available from the factored form of an expression. Because there will be a greatly reduced need to develop extensive manipulative skills prior to dealing with important mathematical concepts, necessary skills can be acquired in conjunction with the study of interesting mathematical ideas. There is limited evidence from research indicating that skills are readily developed in courses that focus on concepts from the start, rather than on the development of skills first (Heid, 1985, 1988). Further consideration of how technology enabling easy and accurate graphs of functions ought to influence the curriculum is necessary.

Student Thinking

Up until now we have generally focused on the possibilities graphing technology create. The story about the ability of graphing technologies to improve the teaching and learning of mathematics is not all rosy, however, and in no area is this so clear as in studies of students' understanding of graphs (Clement, 1985; Goldenberg, 1987, 1988; Goldenberg & Kliman, 1988; Preece, 1983; Schoenfeld, 1990). Students possess a myriad of misconceptions about graphing, some of which are artifacts of the actual program being used. Moschkovich, Schoenfeld, and Arcavi (Chapter 4, this volume) provide a nice example of an unintended consequence of the use of graphing technology that would likely go unnoticed if not situated in a research climate. In the example, students used a graphing program to graph $y = x + 2, y = 2x + 2, y = 3x + 2, \ldots$ and then stated what they observed. Instead of observing that the lines were increasingly steeper and all passed through (0,2), they commented on the fact that some of the lines were more jagged than others, and furthermore, the bigger the coefficient, the more jagged the lines. This observation is a nice example of students over- or undergeneralizing their observations while using a piece of technology in such a way so as to cause them to see things that were not intended (line jaggedness) and not see things that were intended (point of intersection on the y axis).

An additional example of students' misunderstanding of graphs comes from one of the authors working with a group of college precalculus students solving the following problem:

Determine when $f(x) > g(x)$ for the following functions:

$f(x) = 3x^2 - 2x - 5$
$g(x) = x^3 - 8$

Figure 9.21 provides a graph of the two functions. Even after students generated the graphs using graphing software, many were unable to solve the problem. Some of the questions raised by this example include: Is this an example of students' lack of understanding of the graph itself, or is it instead a lack of understanding of the relationship between the graphical and algebraic representations? To what extent were the difficulties related to the graphing utility? Were the students unable to make sense of the algebraic statement of the problem, managing to graph the given functions without having any reason for doing so?

Much has been written about the distinctions between views of functions as processes and objects (for example, Harel & Dubinsky, 1992). Students need to use both perspectives to solve this problem. Finding coordinates for points of intersection of the graphs draws on the relationship or rule

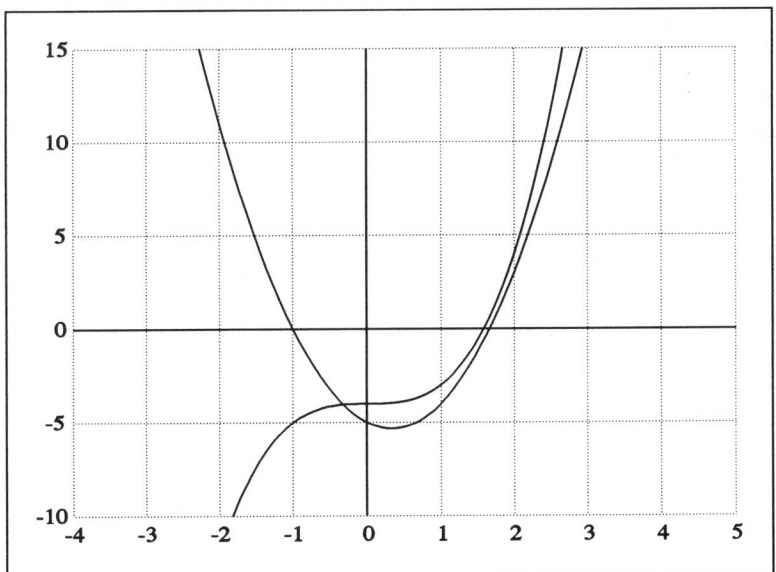

FIG. 9.21. This graph can be used to solve the inequality $x^3 - 4 > 3x^2 - 2x - 5$.

connecting coordinates of points on the graphs, whereas finding where one curve is below the other uses the object perspective inherent in the graphical representation. Goldenberg (1988) and Kieran (1992) are two examples of more detailed discussion of these issues. Moschkovich, Schoenfeld, and Arcavi (Chapter 4, this volume) describe in more detail a situation where a student had identified a graph and its algebraic representation, but seemed unable to make further use of this information to answer related questions about the function. Such difficulties are warnings that the use of graphing technologies will not remove the issues surrounding students' misconceptions; they will at best help to highlight them, and in all probability add a new dimension to them.

There are various research questions related to the links between various forms of representations. For example, will it be sufficient for students to be able to develop the graphical representation from the algebraic representation? Perhaps no amount of staring at an asymptote is going to be meaningful for students unless they are able to manipulate the graph and see its effect on the algebraic representation. To what extent must the link between the graphical and the algebraic be reversible in order for students to gain a better understanding of the connection between them? (R. Lehrer, personal communication, March 1990). Researchers like Goldenberg (1987, 1988), Kaput (1992), and Schoenfeld (1990) are among those who have written extensively about issues relating to how students might come to understand the links between algebraic, tabular, and graphical representations of functions, also approached in other chapters in this volume. The curricular implications are that graphical representations are potentially valuable, but must be used wisely. Research is needed to show ways in which the curriculum can better develop student knowledge in this domain, based on what researchers find about student conceptions and misconceptions.

Teacher Thinking and Beliefs

Many of the questions asked about the thinking of students may also be asked about teachers, who will need to work with the representations offered by graphing utilities. There are some important differences between student thinking and teacher thinking. Students are using the graphing utility while learning the accompanying mathematical concepts, whereas teachers are imposing use of the utility on their preexisting mathematical knowledge. It may be that students who study functions by using graphing utilities develop representations that their teachers, who have already developed effective representations, simply do not understand. The research challenge is to determine the differences between these two representations and then to investigate whether teachers are able to understand and work with their students' different and developing notions. There is evidence that "software graphing utilities tend to hide student misconceptions and lead both teacher and student to believe that the student understands how a graph is constructed, when in actuality the student does not understand the graph as a mathematical representation of the relationship between two variables" (Bohren, 1989, p. 2184).

Teachers possess a variety of beliefs about mathematics that may alter or restrict the way in which the graphical representation of functions influences the curriculum. For example, some teachers may already believe that they use a graphical approach because they include a unit on graphing; believing this, their thinking may not be influenced any further by the graphical approach. Or they may feel that there is no reason to change that which they believe is already effective and appropriate, because the central ideas and methods of mathematics are regarded as unchanging. If either is the case, then graphs are likely to continue to play a minor role in the curriculum.

Just as teachers' beliefs will influence the extent to which they incorporate graphical representations of functions in the curriculum, their use of graphical representations may influence their beliefs. For example, the belief that mathematical problems have unique correct solutions, which developed as an artifact of teachers' mathematical experiences, may be altered through the use of the graphing utility. An example of this is found in the authors' experiences while writing this chapter. The problems used in the text evolved as the authors interacted with the graphs, so that, for example, the aluminum can question (Problem 3), which began as a maximum-minimum calculus problem, became transformed into an open-ended problem situation that could no longer be adequately addressed with a single solution.

It is well documented in the research that teacher beliefs significantly influence curricular implementation in the classroom (Stephens, 1983). It will be important to gain an understanding of how a graphical approach will influence teacher thinking and, in turn, how this affects the implementation of the curriculum. Although little is known about the interaction of teacher beliefs and the curriculum in relation to a graphical approach, anecdotal evidence and the authors' personal experience suggest that teachers who have used graphing utilities in their classes experience significant change in their beliefs and knowledge, change that in turn influences how and what they teach. Farrell (1990) and Rich (1991) carried out studies in mathematics classes that used graphing calculators as a regular tool. Each reported that the nature of classroom interactions and teacher behavior seemed to have been influenced in positive ways—for example, instructors came to take an increased role as consultants rather than lecturers.

The implementation of a curriculum based on these changed beliefs will take time—a commodity that current teachers do not have in abundance. If teachers are to be encouraged to create new instructional environments using new utilities that enhance a changing mathematical epistemology, they must have time to experiment, read, think, and talk to each other.

Classroom Instruction

Graphing utilities have the potential to promote a changed instructional style in the classroom. By making it much easier for students to speculate and then get quick responses, graphing can involve a directed exploratory approach to the introduction of mathematical ideas. Generally, teachers choose a particular problem because its solution involves certain mathematical representations. Instead of presenting a carefully developed solution, the solution can be generated dynamically and interactively by students and teacher, providing a more realistic impression of the ways mathematics is used and developed. This is possible because students have access to a tool that allows them to easily try a variety of approaches and see the results of their conjectures.

In contrast to the algebraic representation, which often has a single most direct solution, a graphical approach usually provides more than one way to find the solution to a problem. The opportunity to select between an algebraic and geometric approach offers a real choice, especially because many problems are still better solved algebraically. For example, finding the zero of a linear function can be done algebraically or geometrically. One of the authors discovered that students rapidly realized that the zeros of linear functions could be derived more efficiently by relying on an algebraic approach. The important point is that a graphical approach provides realistic choices that were not before available.

The dependence on algebraic manipulations to solve most problems was a consequence of the lack of viable alternatives. Graphical representations increase the opportunity for students to make decisions about an assortment of diverse problems. The possibility of such decision making, however, has to be incorporated into instruction. If students are presented polished solutions without seeing the many choices leading to their production, then the full potential of graphing utilities will be squandered. Students need to be encouraged to approach real mathematical problems— that is, problems for which clear solution techniques are not readily available. Furthermore, teachers must increase access to a wide variety of interesting problems requiring varying techniques for their mathematical representation and solution.

Materials, such as textbooks, will need to provide examples that illustrate a variety of problem-solving approaches. It seems unlikely that the traditional pattern of stating definitions and theorems, giving examples of problems utilizing the concepts, presenting carefully worked out solutions, and then asking the students to solve ten similar exercises will be appropriate. The graphical approach may require more open-ended problems that do not have a prescribed or intended means of solution. The first problem in this chapter is an example of one that has many appropriate solutions. Problems illustrating the applications of mathematics to diverse realistic situations will also be needed, reflecting the reduced emphasis on symbolic manipulation and the increased importance of problem representation.

Assessment

Algebra was first introduced into the high-school curriculum because colleges added it to their list of required courses and the material was on entrance examinations. One of the first hurdles in the way of increased utilization of graphs of functions may be the degree to which assessment and entrance tests incorporate the use of technology. If a graphing approach is used in mathematics classes, assessment will need to change in response. Graphing cannot be fully employed in an environment that is strongly influenced by external assessment, especially if that assessment forbids or limits the use of graphing utilities. Existing evidence indicates that many teachers take topics on standardized tests into account when making curricular decisions (Romberg, Zarinnia, & Williams, 1989). It follows that until such tests incorporate functions and their graphs, or at least permit the use of graphing utilities, many teachers will continue to teach a curriculum unaffected by this approach.

CONCLUSION

The ability to accurately represent the graph of a function quickly and easily offers the potential to significantly alter the mathematics curriculum by changing the content sequence, the content selection, and the number of students who have access to functions and graphs. Calls to unify the mathematics curriculum around the function concept date from the turn of this century, but little change has occurred. The graphing utility offers the opportunity to facilitate curriculum restructuring around the function concept. Such radical restructuring will only occur as a result of a variety of separate issues coming together, including a rethinking of the big ideas in mathematics around function; a view of mathematics as emphasizing conceptual understanding over emphasis on manipulative skills; and an opportunity for teachers to rethink their own knowledge and beliefs about these issues. However, we believe that changes in the curriculum not also accompanied by changes in who has access to the curriculum are insufficient.

The graphical approach will affect who has access to mathematical ideas. As noted previously, a graphical approach can bypass the need for the acquisition of extensive manipulative skills before interesting problems can be studied. It is likely that these benefits will be greatest for those who have had difficulty with the traditional approach, simply because they have more to gain. That is, a student who can already deal with a problem may gain intuitions from using the utility, but the student who was unable to do the problem has both the solution and the intuition to gain. At least one anecdote suggests that there are differences between weaker and stronger students in the perceived benefits of a graphical approach. A high-school precalculus teacher in Milwaukee who used the Waits and Demana materials surveyed the students and found that the stronger students reported the graphical information as "very useful," whereas weaker students said that the graphical information was "extremely useful" (Arnie Engebretson, personal communication, September 20, 1989). Further analysis of benefits to students ought to include interactions between high versus low achieving students and conceptual versus procedural knowledge. It is important to ensure that gains in knowledge are conceptual as well as procedural.

Although the graphing utility may enable more students to study higher levels of mathematics than was before possible, we, in the interest of equity, also have another goal. We believe that the graphing utility ought to be available throughout the elementary and secondary curriculum, allowing for interesting mathematics to be pursued by students who are not in the traditionally "rigorous" courses. It is not enough to bring more students up the mathematical mountain—too many are still not making it. Instead, the graphing utility ought to bring the mathematical mountain to more people.

It is conceivable that the use of a graphical approach could exacerbate existing undesirable differences in the rate of participation in mathematics. If a graphical approach becomes an integrated part of the mathematics curriculum, its implementation must proceed in a way that ensures that inequities are reduced. It would be highly undesirable if schools that could benefit from a graphical emphasis lack the necessary financial and human resources to develop it.

As pointed out previously, the psychologist E. L. Thorndike, along with mathematics educators early in this century, believed that less than one third of the population was intellectually prepared to benefit from studying algebra. With the NCTM *Standards* calling for all students to learn meaningful algebra in secondary school and with the increasing availability of graphing utilities, we are at an important point in the history of the learning and teaching of algebra. However, teachers as well as students need to believe that all students are capable of using and benefiting from the representations of functions available with the graphing utilities.

Graphing utilities influence our ability to represent and analyze functions, which in turn influences the curriculum via effects on various components of the educational system. Our discussion of these influences may have conveyed the false impression that this influence proceeds mainly in one direction. In fact, the situation is of a more cyclic nature, involving the complex interaction of many components in mathematics education. To illustrate this, consider the idea that the graphical representation of a function might not provide any information to a student who does not understand what the graph shows. In fact, it might do worse: It might provide false information if the student has misconceptions about graphs. Such results will influence curricular decisions, requiring, for example, a selection of materials and instructional approaches designed to develop an appropriate understanding of graphs. One of these materials is the graphing utility itself. The way in which the utility constructs and presents the graph could strongly influence the user's concept of the graph and its relation to the function it represents. A grapher that instantly displays the complete graph conveys a concept differently from the one that sweeps out a curve, which in turn differs from a grapher that plots individual points. The design of the utility becomes a curricular issue, and the decisions made influence in turn content, thinking, assessment, and so on. It is important that researchers continue to investigate these issues and that they convey their findings to software and hardware designers as well as to those developing and implementing the curriculum.

For most of this century, curricular decisions regarding the use of graphs and functions have been made without the benefit of research. However, with the recent increase in the study of well-defined content domains, the opportunity now exists for better informed research-based curricular decisions. Although we are optimistic that this research-based decision making will continue to increase in the future, only time will tell what the final impact will be on students. In spite of this uncertainty, we believe that a graphical approach will have increasing importance in the mathematics curriculum and that it has the potential to improve mathematics education in the coming years.

REFERENCES

Barber, H. C. (1932). Present opportunities in junior high school algebra. In W. D. Reeve (Ed.), *The teaching of algebra* (NCTM Seventh Yearbook, pp. 155-166). New York: National Council of Teachers of Mathematics.

Betz, W. (1930). Whither algebra? — A challenge and a plea. *Mathematics Teacher, 23*, 104-125.

Bidwell, J. K., & Clason, R. G. (Eds.). (1970). *Readings in the history of mathematics education*. Washington, D C: National Council of Teachers of Mathematics.

Bohren, J. L. (1989). A nine month study of graph construction skills and reasoning strategies

used by ninth grade students to construct graphs of science data by hand and with computer graphing software (Doctoral dissertation, The Ohio State University, 1988). *Dissertation Abstracts International, 49,* 2184A.
Breslich, E. R. (1933). Secondary school mathematics and the changing curriculum. *Mathematics Teacher, 26,* 327-349.
Clement, J. (1985). Misconceptions in graphing. In L. Streefland (Ed.), *Proceedings of the Ninth International Conference on the Psychology of Mathematics Education* (pp. 369-375). Noordwijkerhout, The Netherlands: PME.
Demana, F., & Waits, B. K. (1990). *College algebra and trigonometry: A graphing approach.* Reading, MA: Addison-Wesley.
Farrell, A. M. (1990). Teaching and learning behaviors in technology-oriented precalculus classrooms (Doctoral dissertation, Ohio State University, 1989). *Dissertation Abstracts International, 51,* 100A.
Foster, A. G., Rath, J. N., & Winters, L. J. (1983). *Algebra two with trigonometry.* Columbus, OH: Charles E. Merrill.
Goldenberg, E. P. (1987). Believing is seeing: How preconceptions influence the perception of graphs. In J. C. Bergeron, N. Herscovics, & C. Kieran (Eds.), *Proceedings of the Eleventh International Conference on the Psychology of Mathematics Education* (pp. 197-203). Montreal: PME.
Goldenberg, E. P. (1988). Mathematics, metaphors, and human factors: Mathematical, technical and pedagogical challenges in the educational use of graphical representation of functions. *Journal of Mathematical Behavior, 7*(2), 135-173.
Goldenberg, E. P., & Kliman, M. (1988). *Metaphors for understanding graphs: What you see is what you see.* (Algebra Research report). Newton, MA: Education Development Center.
Grove, E. L., Mullikin, A. M., & Grove, E. L. (1960). *Algebra and its use.* New York: American Book Company.
Harel, G., & Dubinsky, E. (Eds.). (1992). *The concept of function: Aspects of epistemology and pedagogy.* (MAA Notes No. 25). Washington, DC: Mathematical Association of America.
Heid, M. K. (1988). Resequencing skills and concepts in applied calculus using the computer as a tool. *Journal for Research in Mathematics Education, 19,* 3-25.
Heid, M. K. (1985). An exploratory study to examine the effects of resequencing skills and concepts in an applied calculus curriculum through the use of the microcomputer (Doctoral dissertation, University of Maryland, 1984). *Dissertation Abstracts International, 46*(6), 1548A.
Izzo, J. A. (1957). A history of the use of certain types of graphical representation in mathematics education in the secondary schools of the United States. *Dissertation Abstracts, 17*(7), 1506.
Jablonower, J. (1932). Recent and present tendencies in the teaching of algebra in the high schools. In W. D. Reeve (Ed.), *The teaching of algebra* (NCTM Seventh Yearbook, pp. 1-18). New York: National Council of Teachers of Mathematics.
Jones, P. S., & Coxford, A. F., Jr. (1970). Mathematics in the evolving schools. In National Council of Teachers of Mathematics (Ed.), *A history of mathematics education in the United States and Canada* (pp. 11-92). Washington, DC: National Council of Teachers of Mathematics.
Kahn, H. F. (1974). A study of the manner in which selected topics in elementary algebra were presented to students in America between 1900 and 1970 as revealed in selected commercially published textbooks. *Dissertation Abstracts International, 35*(3), 1320B-1321B.

Kaput, J. J. (1992). Technology and mathematics education. In D. A. Grouws (Ed.), *Handbook of research on mathematics teaching and learning* (pp. 515-556). New York: Macmillan.

Kieran, C. (1992). The learning and teaching of school algebra. In D. A. Grouws (Ed.), *Handbook of research on mathematics teaching and learning* (pp. 390-419). New York: Macmillan.

Kliebard, H. M. (1987). *The struggle for the American curriculum, 1893-1958.* New York: Routledge & Kegan Paul.

Lennes, N. J. (1932). The function concept in elementary algebra. In W.D. Reeve, (Ed.), *The teaching of algebra* (NCTM Seventh Yearbook, pp. 52-73). New York: National Council of Teachers of Mathematics.

National Council of Teachers of Mathematics. (1989). *Curriculum and evaluation standards for school mathematics.* Reston, VA: Author.

Osborne, A. R., & Crosswhite, F. J. (1970). Forces and issues related to curriculum and instruction, 7-12. In National Council of Teachers of Mathematics (Ed.), *A history of mathematics education in the United States and Canada* (pp. 155-300). Washington, DC: National Council of Teachers of Mathematics.

Preece, J. (1983). Graphs are not straightforward. In T. R. G. Green & S. J. Payne (Eds.), *The psychology of computer use: A European perspective* (pp. 41-56). London: Academic Press.

Rich, B. S. (1991). The effect of the use of graphing calculators on the learning of function concepts in precalculus mathematics (Doctoral dissertation, University of Iowa, 1990). *Dissertation Abstracts International, 52,* 835A.

Romberg, T. A., & Carpenter, T. P. (1986). Research on teaching and learning mathematics: Two disciplines of scientific inquiry. In M. C. Wittrock (Ed.), *Handbook of research on teaching* (pp. 850-873). New York: Macmillan.

Romberg, T. A., Zarinnia, E. A., & Williams, S. R. (1989). *The influence of mandated testing on mathematics instruction: Grade 8 teachers' perceptions.* Madison, WI: National Center for Research in Mathematical Sciences Education.

Schoenfeld, A. H. (1990). GRAPHER: A case study in educational technology, research, and development. In M. Gardner, J. Greeno, F. Reif, A. H. Schoenfeld, A. Disessa, & E. Stage (Eds.), *Toward a scientific practice of science education* (pp. 281-300). Hillsdale, NJ: Lawrence Erlbaum Associates.

Senk, S. L., Thompson, D. R., & Viktora, S. S. (1990). *Advanced algebra.* Glenview, IL: Scott, Foresman.

Sigurdson, S. E. (1962). The development of the idea of unified mathematics in the secondary school curriculum, 1890-1930. *Dissertation Abstracts, 23*(6), 1997.

Slaught, H. E., & Lennes, N. J. (1916). *Complete algebra.* Boston: Allyn and Bacon.

Smith, D. E. (1927). Mathematics in the senior high school. In National Council of Teachers of Mathematics (Ed.), *Curriculum problems in teaching mathematics* (Second yearbook of the NCTM, pp. 231-241). New York: National Council of Teachers of Mathematics.

Stanic, G. M. A. (1984). Why teach mathematics? A historical study of the justification question (Doctoral dissertation, University of Wisconsin-Madison, 1983). *Dissertation Abstracts International, 44,* 2347A.

Stein, E. I. (1970). *Modern algebra step by step.* New York: American Book Company.

Stephens, W. M. (1983). Mathematical knowledge and school work: A case study of the teaching of developing mathematical processes (Doctoral dissertation, University of Wisconsin-Madison, 1982). *Dissertation Abstracts International, 44,* 1414A.

Thomas, G. B., & Finney, R. L. (1988). *Calculus and analytic geometry* (7th ed). Reading, MA: Addison-Wesley.

Thorndike, E. L., Cobb, M.V., Orleans, J. S., Symonds, P. M., Wald, E., & Woodyard, E. (1923). *The psychology of algebra.* New York: Macmillan.

Thorpe, J. A. (1989). Algebra: What should we teach and how should we teach it? In S. Wagner & C. Kieran (Eds.), *Research issues in the learning and teaching of algebra* (pp. 11–24). Reston, VA: National Council of Teachers of Mathematics.

Upton, C. B. (1936). *Practical algebra: Introductory course.* New York: American Book Company.

WordPerfect Corporation. (1986). WordPerfect 5.0 [Computer program]. Orem, UT: Author.

VII REACTIONS

10 The Urgent Need for Proleptic Research in the Representation of Quantitative Relationships[1]

James Kaput
University of Massachusetts Dartmouth

We argue that research beyond that reported in this volume is urgently needed. This research must abandon basic assumptions: (a) that the future curriculum will be organized in ways that are only local modifications of the current curriculum; (b) that the technology used in research should parallel that available to schools; and, most importantly, (c) that the appropriate representations be restricted to the traditional "big three," namely, symbols, numerical tables, and coordinate graphs. We explore alternative perspectives to all these assumptions and offer particular examples to illustrate the third, which then force rethinking of the first two. These examples suggest ways of using the new representational capacity of computers to bridge the highly persistent gap between the rich experience of living and the sterile formalisms that have dominated our approaches to mathematical learning and knowing. The research questions raised by these alternatives are myriad and deep, and are not yet being explored.

Proleptic? My dictionary defines *proleptic actions* as actions that deliberately anticipate the future. In this chapter, I take the view, not likely to be satisfying to researchers in this field or even among the leaders of the field as represented in this volume, that research in the representation of quantitative relationships should be farther ahead of current practice than it now is. It should be farther ahead in at least three dimensions: technology, curriculum, and representations.

---
[1]The author gratefully acknowledges support for preparation of portions of this chapter: from the Apple Computer, Inc., Apple Classrooms of Tomorrow (ACOT) Program; and from NCRMSE, funded by the Department of Education OERI, grant R117G10002.

Given the large differences between the way functions are treated by the researchers in this volume and the way functions are treated in today's typical school curriculum (if they are treated at all), the reader might legitimately wonder, "What in the world does he want!? Research should reflect practice that must be within reach—and we're not sure that we can get practice to move to where these researchers are now, let alone to more remote possibilities." And given the pressure for school-based research and authentic involvement of practitioners, this reaction carries even greater weight.

The bulk of this chapter is devoted to clarifying what is meant by proleptic research and arguing its necessary place on our agenda. I begin with an attempt to be a bit clearer on the kinds of research that need to be done, from which it is clear that I am not arguing against the type of research discussed in this volume, but rather I am arguing that a new form, a more proleptic research, is urgently needed in addition to what we are doing today.

TIMELINES, THE RATE OF TECHNOLOGICAL CHANGE, AND TYPES OF RESEARCH

What is a reasonable timeline for the application or impact of today's research? In order to be clear on an appropriate timeline, it is necessary to be more specific regarding the types of research we are discussing. We distinguish between two broad classes of research. The first is research that assumes an updated version of the status quo (in terms of curriculum, technology, and representations) and is designed to test the effectiveness or impact of an innovation rooted in current practice and technology. We would describe this research as extending and improving upon current practice and term it *practice-extending research*. A second form of research attempts to test the boundaries of what is possible or appropriate outside the constraints of current practice, technology, and representations. We would term this *future-defining research*.

The timeline for current practice-extending research, assuming that research being planned in the early 1990s will be conducted and reported in the mid 1990s, will exert its impact on curriculum design and perhaps software design in the following two or three years. The actual classroom impact will not occur until late in the 1990s. The total time elapsed is about 8 to 10 years. Future-defining research, on the other hand, examines mathematical content and learning opportunities somewhat independently of current practice and utilizes state-of-the-art university-level technology, that is, technology that might be school-affordable in 5 to 7 years. The timeline for planning, execution, and reporting is essentially the same as for the first form of research, but its impact may well be different, probably less direct: It will tend to influence further research rather than practice. Hence, it is likely to influence research conducted during the second half of the 1990s. This research, assuming that it is close enough to practice to be of the first type, would have its classroom impact 8 to 10 years later, say in 2005. Even if these intervals can be shortened, their order of magnitude is not likely to change.

Given the rate of change of visual-representation technology over the past decade and the customary 10-year migration from laboratory to everyday commercial availability, the representational opportunities available during the early mid part of the next decade are likely to be those just now emerging in research laboratories—real-time video mixed with computer video, high-density television with storable and manipulable dynamic images, hand-held devices embodying today's state-of-the-art workstation computing power, etc. One immediate realization is that research based on today's university-level computer technology, the only future-defining research that will probably be done, is actually aimed at the technological environment of the world targeted by current practice-extending research. But, of course, practice-extending research is essentially oriented to today's technology. The point is that for both types of research, the objective falls short of the technological potential at the time of impact. Thus we see a severe conservative bias in mathematics education research relative to technological change, a bias that is quite likely to endure.

This deeply ingrained habit of aiming below the technological target must be broken before we can expect to see any real breakthrough-level payoff of the new technology. Paradoxically, the inertia of current practice may be more difficult to overcome via incremental reform than by more radical reform. In the last part of this chapter we attempt to illustrate the kinds of opportunities that await research and development testing. We return to these issues after we have provided some examples.

Given our two classes of research, what specifically do we mean by proleptic research? Although future-defining research is more likely to be proleptic research than practice-extending research, practice-extending research is certainly not without value or importance, especially if it is done within a proleptic perspective. Proleptic practice-extending research, for example, recognizes that its real use may be found in curricular, technological, and representational contexts that do not resemble those of today. Thus it takes into account the fact that the experience a future student brings to a particular function-graphing activity may be radically different from that of today's student and that the technological support for that graphing activity might also be substantially different. This "taking into account" can mean that the researcher is in fact helping to define the nature of the appropriate prior experience and available technological support.

Hence, practice-extending research may be carried out in a proleptic manner.

On the other hand, other research may use radically new technology and representations and deal with novel subject matter, but may embody backward-oriented assumptions about the contexts in which it might be utilized, such as a teaching-as-telling, learning-as-witnessing pedagogy. Thus, the term proleptic research refers to research that attempts to anticipate future conditions—the circumstances anticipated at that relatively distant time when the research makes its mark on practice. It takes the long view rather than the short one.

Plan for the Chapter

The basic objectives of this chapter are twofold—first to deal with practical and theoretical barriers to proleptic research, and second to illustrate some new research issues. The investigation of these new research issues simultaneously serves to generate the options that need to be explored in order to define our curricular future and to help us build appropriate experiences for teachers and students when we make our choices among the options. In service to the first objective, and drawing on the research reported in this volume, we examine the character of typical research in the graphical representation of functions. We then offer some alternative assumptions regarding new research with a longer scope of application, together with some fairly specific illustrations of the kinds of new representational opportunities that need to be examined. But first, we review briefly the state of current practice.

THE STATE OF CURRENT PRACTICE

It is first necessary to distinguish between the current practice in schools and the vision of teaching and learning reflected in research activity. Before moving on to details, we must point to the largest difference—that of pedagogy. Although the pedagogy is implicit in most of the research work described or proposed in this volume, one need not be a detective to infer that the preferred approach is guided inquiry. And while this choice reflects deep commitments on the part of the researchers, it likewise reflects the power of the technology that they employ to enhance the empirical side of mathematics.

Computer technology can render experimentation a genuine student activity, with feedback that is immediate and, more importantly, informative—informative in the terms of the experiment. Experiments involving relationships between coordinate graphs and character-string representations of functions yield feedback that is not of the form "Sorry, Sam, that is not the correct answer. Try again." Or even feedback of the form "That's not quite high enough." Rather, the feedback is provided by the computer responding in appropriate representations of interest—the graph appears, or the string representation of the function changes, and the student then responds in the same representational context. Although this does not preclude additional layers of computer interactivity involving natural language, for example, it represents a major difference from activity in static, inert media, which simply do not respond. Further, by providing fluent connections among representations and by handling routine computations, the technology renders experimentation cognitively and practically affordable.

Hence, simply by assuming the use of solid, contemporary hardware and software, current researchers take a major step in the direction of the ambitious inquiry pedagogies that are now the norm in calls for reform. As a result, the distance of these researchers from current pedagogical practice is essentially the same as the distance between current and reformed practice. Although this pedagogical distance is great enough to be intimidating, and its roots and reasons for stability across the years are complex, it is not the direct focus of this chapter (see Chapter 7, this volume; by Norman). Instead, I focus on the curricular, representational, and technological dimensions of current research, ultimately concentrating on the latter two. The reason for this choice is that these are the dimensions where the greatest change is possible and where the opportunities may be the greatest.

The Current Curriculum

As Philipp, Martin, and Richgels point out (Chapter 9, this volume), there is a substantial difference between the character-string-based manipulation of expressions that dominates algebra, where functions play a minor role until precalculus courses, and the graphical and numerical approaches that computers facilitate so well. They also document that some recommendations for using functions as a central organizing concept in the curriculum have accumulated at least a hundred birthday candles. More generally, the notion that algebra is a secondary-school topic (whether or not it is built around the idea of function) seems deeply entrenched.

An articulate critique of this highly alienating practice, and its associated heavy emphasis on syntactic maneuvers of empty symbols apart from any wider meanings, has been offered by many reformers, most notably Usiskin (1988). Suggestions for a very early approach to algebra were offered by Robert Davis (1972, personal communication) 20 years ago. He suggested that algebra is actually a subject appropriate to the upper elementary and lower middle school level. In any case, we can safely say that the current curriculum treats arithmetic and algebra as independent networks of concepts and skills. It assumes that arithmetic has only incidental relationship to the algebra, based on the fact that the manipulation of literals requires some acquaintance with the basic operations of arithmetic, of the arithmetic of pure numbers. To "simplify" the string "$2x+3x$" it helps to know that $2+3 = 5$. I make some suggestions on arithmetic/algebra connections later in the chapter.

Current Uses of Technology

Let us review where we are technologically (for a comprehensive account, see Kaput, 1992). As of this writing, in the early 1990s, electronic computing has been available in some form for four decades. It has been available in the service of education for almost three decades. But it has been available in forms that would serve the graphical representation of functions for only about a decade and a half, and for most of that time it was in the hands of pioneers, not in the hands of practitioners.

Becker (1991) reported that, on the basis of the U.S. portion of the 1989 IEA Computers-In-Education survey, only 42% of all secondary mathematics teachers used computers at all and, of these, only 40% used them with some regularity. My calculations indicate that only about 17% of mathematics teachers used computers with regularity. But then, on exploring how they used them (based on Becker's data), we see that teachers used computers for programming 32% of the time, drill and practice 10% of the time, and for graphing only 9% of the time. Computer-based graphical approaches to functions are certainly a long way from common in schools!

We should also note that the technology and curriculum issues are closely joined, even if we factor out pedagogical matters. One basic reason for this is that the graphing software currently requires a functions approach. In the design of the software, one can only plot quantitative relations if they are introduced as closed-form rule-defined functions (or relations). This closed-form rule approach to quantitative relationships turns out to be an important interlocking feature of the technology, the curriculum, and the underlying representations that lies behind much of the mathematics that is regarded as appropriate or even possible.

Although the types of bit-mapped graphics systems that allow true manipulation of high quality graphics have been available in school-accessible forms for about five years, relatively few schools have such computers. However, these are essentially the only computers commercially available at this time. Thus, the percentage of teachers using the current generation of commercially available computers is quite small. It should be no surprise then that our store of reliable experience in using computers to teach and learn functions using contemporary technology, especially in graphical contexts, is so limited.

The Current, Historically Inherited Representational Context. Since ancient times, to assume that a quantitative relationship exists among items in one's experience has meant assuming that numbers and units of measure (sometimes the latter were suppressed) could be assigned to those items, and they could be related to one another using the formal tools available. Among the Greeks, the tools were mainly ratios (not actually numbers as we know them, as made clear by Fowler, 1987), operations on ratios, and geometrically based formulas admitting unknown quantities (but definitely not variables, as argued by Klein, 1968), and these were used to describe static relations, not dynamic situations (Boyer, 1959). Apart from the allowable operations on ratios and the "exhaustive" techniques embodied by the work of Archimedes and a few others, the actual formal operations on these formal systems were very limited. And they were devoted essentially to the description of static relationships.

Descriptions of dynamic relationships were not attempted until the advent of the Scholastics in the 13th and 14th centuries, when the initial language used was Latin, and when protographical techniques were introduced by Oresme (Kaput, in press). Algebra as we know it was not used and, in fact, was not fully available as a modeling tool until about 300 years later, in the century up to and including Newton and Leibniz. It was during this period that the fundamental idea of using the formal language of algebra to build models came to prominence. (Indeed, the whole idea of building testable theories about nature was quite new at the time of Oresme [Kline, 1972].) The subsequent 300 years since Newton can be regarded as a grand elaboration of the formal tools that were developed at that time. This elaboration was a complex dialectic involving observation, theory building, and algebraic formalization of theory—and then formalization of that formalization. As of the turn of the 20th century, the results in the field of mathematical analysis and its applications in physics were one of the great achievements of humankind.

However, given the code in which it was developed and is currently available to us, the beauty and power of this achievement can be appreciated and utilized by only a miniscule fraction of humankind. Why? These formal tools were developed by and for a knowledge-producing elite in static, inert media. Although coherence and comprehensibility of formal systems among the knowledge producers was an ongoing concern, their learnability by the wider population was never a significant issue. There were occasional exceptions; for example, Robert Recorde, who was among the first textbook writers, concerned himself with the learnability of ideas and notations, such as the equal sign (Boyer & Merzbach, 1989). And of course, Leibniz made notation design an important part of his life's work (Edwards, 1979). But even he was concerned only with the usability of notations by other mathematicians and scientists.

Thus the representation of general quantitative relationships has been restricted to formal algebraic notations and sometimes their coordinate graphs (treated in an ancillary manner); particular quantitative relationships have been represented primarily using numeration systems. Thus, it is natural that these same numeric and algebraic systems have been the primary notational contexts in which we have attempted to teach quantitative mathematics across the years. The question of how best to teach and to make learning possible has assumed these systems as a starting point and has centered on how to introduce these systems so that students become fluent in their use. Little attention has been given to the building of more learnable systems.

I return to the question of learnability shortly.

School Change and Technology Integration

It is surely unnecessary to dwell on the slow pace of school change or on how schools assimilate innovation into highly stable patterns of practice. Although many factors in today's reform-oriented climate are challenging the status quo, most thoughtful observers agree that substantial change will require systemic reform, which implies a timeline of at least a generation.

Combining this forecast with the technological forecast presented in the introduction leads to the conclusion that the gap between school mathematical practice and technological possibility will continue to widen for the foreseeable future. The forces responsible are partly practical and economic, but to a greater extent they are political and cultural. However, one factor that may serve to narrow the gap is the gap itself—that is, as it grows larger, it becomes more conspicuous and harder to ignore. It is but another facet of the gulf between the world of school and the world outside of school.

The rates of integration of computer technology into practice have been in large part governed by factors not directly involving the technology itself. But this may change because the technology is rapidly becoming easier to integrate. We have left the Model T stage and are now moving past the Model A stage of technology use (see Kaput, 1992, for extended treatment of this analogy). Vast numbers of people—teachers, students, administrators, parents, and the public at large—find the computational medium part of their daily lives because as computers become powerful, they become easier to use not merely for traditional activities, but for entirely new ones. As the technology becomes easier to use, more familiar, and more pervasive, it becomes invisible. This will eventually happen even in schools. But our research patterns, with a few exceptions, like the Apple Classroom of Tomorrow Program (in which all students and teachers get computers for both home and for school), assume patterns and density of computer use similar to the "technologically ideal" classroom of today—at most a few computers, frequently but not universally available, for activities of a fairly traditional sort, akin to seatwork or homework assignments.

Given all of the constraints we face, economic and otherwise, assumptions about the level of technology integration in classroom practice are perhaps the most difficult to overcome in research practice. We are in a double bind.

THE BACKGROUND ASSUMPTIONS OF TODAY'S PRACTICE-EXTENDING RESEARCH

Most mathematics education research in the United States assumes an overall curriculum configuration similar to today's standard. And this research is carried out using the current level of school technology, which focuses on traditional representations. I now want to examine these research assumptions in more detail. Although there is variation among different researchers' adherence to the assumptions listed next, the assumptions listed represent something of a "mean" across most research activity, particularly that described in this volume.

Curricular Assumptions, Global and Local

By and large, researchers represented in this volume assume that, globally, the organization of the curriculum is fixed and beyond reach and that significant changes are to be made locally, within existing courses. Hence, questions about the integration of graphical representations with other subject matter at the cross-course level are not raised. Similarly, the existence of courses with the title "algebra" is assumed, and it is further assumed that such self-contained courses are the vehicle for the delivery of any new material. These appear to be political decisions, and in the case of Schwartz are quite explicitly so.

Related to the assumption of global curricular stability is a more specific assumption that algebra, however conceived—as character-string manipulation activity, or as the study of the properties and uses of functions, or as the process of symbolic generalization and abstraction from quantitative instances—begins relatively late in the student's school mathematical experience and is primarily a domain for secondary school study. Nonetheless, there seems to be general agreement that the study of algebra should begin in the upper middle school, a year or two earlier than it now does. This view, voiced most incisively by a nonresearcher, Usiskin (1988), carries with it a corollary, that algebraic thinking, in its several aspects, should be deliberately promoted only when arithmetic, in its several aspects, is mastered. This assumption is sharply questioned by the work of Thompson (1991), who deliberately and systematically examines the integration of algebraic and quantitative thinking among younger students.

An important assumption of virtually all researchers is that algebra is about functions and their application to meaningful problems. Although this approach is not new, its realization may be far more likely in today's graphing, computational context than in the past because of two very simple and interrelated facts. First, what is one to graph? The quantitative relationship under consideration must be described as a relation, and, more likely, as a function (of a single variable). Second, essentially all graphing software requires such a description. This automatically narrows the objects of study to functions. The extent to which applications are central seems to vary across researchers, however.

Technological Assumptions

The research studies included in this volume utilize current school software and hardware and primarily consist of practice-extending research. With the possible exception of beginning experiences, this research assumes that the appropriate medium for function graphing is the computational medium—which nowadays means microcomputers or hand-held graphing devices (supercalculators). Clearly, the typical supercalculators of today cannot accommodate the variety of activities supported by appropriately programmed microcomputers. For example, one cannot act directly on supercalculator graphs. One acts strictly on algebraic representations—but this is likely to change. To the extent that the actions require large amounts of display space, supercalculators will never make up the difference. However, nothing in principle seems likely to hold back their growth in processing power. And if the curriculum assumptions described in the foregoing hold constant, then the choice between microcomputers and supercalculators will be based primarily on the chosen balance between accessibility and ease of use and display. Indeed, a compromise seems likely wherein activities requiring large computers will yield data that will be processed further on hand-held devices, and vice versa.

However, as argued, some of the key new uses of the computational medium will undoubtedly involve new representations and linkages among them, which in turn are likely to require as much display space as can be mustered.

Representational Assumptions

The assumptions regarding appropriate representations are quite uniform among the researchers reporting in this volume. Linkages among representations are confined to the traditional "big three," graphs, tables, and closed-form formulas. Quantitative relationships are expressed primarily through these formal representations, chief among which is the closed-form algebraic representation of functions. This representation maintains a privileged ontological status in that the others are considered to be representations of it. A notable exception to this view is provided by Dugdale, who compares student learning of trigonometric identities using what amounts to alternative ontologies: In one case, the algebraic representation was the thing and the graphical was treated as the representation of it; the alternative treated the graphical representation as the thing and the algebraic as a representation of it. The asymmetry lies in the locus of the primary actions that were supported. In either case, the thing was defined as the place where one's actions were centered—the thing one acted upon—and the secondary representation was what followed along in response to one's actions.

Relative to character-string and coordinate graphical representations of functions, Yerushalmy and Schwartz (Chapter 3, this volume) provide a reasonably symmetric approach to supporting actions, including actions that involve comparison of functions. However, lying behind all these actions is the closed-rule form of the function, which is the way that the function is entered initially. Nonetheless, Schwartz reports (personal communication, May 1991) that by acting directly on the graphical representation of functions to construct or modify functions, fundamental changes occur in what properties are observable and what kinds of inferences are "natural." Furthermore, experienced and highly competent teachers find a graphical approach to functions a significant challenge and show difficulty in breaking away from reasoning in a strictly symbolic, or character-string, mode. The challenges in coming to grips with this level of novelty have not been reported among researchers dealing with the more traditional activities associated with graphing calculators or traditional graphing utilities, which primarily automate standard graphing activities and, more generally, facilitate actions on the traditional representations.

Another, primarily tacit, assumption of standard research is that situations giving rise to functions occur off-line, usually in the form of text, so that functions are entered into the computer via closed-form rules by the user. A notable exception, not reported on in this volume, occurs in the work of McArthur (1990; McArthur, Stasz, & Hotta, 1987). He has constructed microworlds where quantitative relationships (and, in fact, inferences based on these) derive directly from actions, especially measurements, on objects in the microworld. Another option is to pull real (numerical) data into the computer from some source, either controlled by the student or not, and deal with that data in some way in the computer environment. We consider this option further.

SOME (POTENTIALLY) PROLEPTIC ALTERNATIVES

Here I examine some alternatives to the three sets of assumptions listed and discussed in the previous section. I do not discuss directly some of the issues raised in the introduction relating to future school circumstances or the kinds of prior experience that a teacher or student might bring to a particular activity in the future. I begin the discussion of proleptic assumptions and alternatives with the technological, because the technological seems to be the simplest. You may notice that each subsequent assumption builds on its predecessor.

Some Baseline Technological Assumptions

Basic Computer Power and Data Storage. We can safely assume that by the mid to late 1990s, today's university workstation processing and memory capability will be available on portable computers at the current cost of portables, and perhaps even available on hand-held computers. The major differences and cost variation will involve display size and color availability — unless some new display technology appears on the scene. The bottom line is that for all standard purposes, computer power will not be a key issue — no more than engine power is an issue today for normal automobile driving. A similar statement applies to memory storage capacity, with some form of data storage providing copious information for student and teacher alike. The major issue will be what to do with that capability and how to integrate it with historically received curricula and the established economics of schools and publishing.

Teacher workstations will be relatively standardized to include a large display for classroom viewing and a high quality printer. And hand-held devices will be available for student use away from the classroom, with up- and down-loading capability for software and data. These will open the possibility for two classes of machines differing mainly in display and memory capability, one stationary in school and the other portable and inexpensive, costing perhaps $200 in 1992 dollars. Although the former may have fancy display capability to handle realistic representations and probably video display, the latter might be restricted to more traditional mathematical representations, which might even be built into ROM (read only memory).

Networks. Most prior work has taken place with a tacit assumption that individual students or small groups of students use their computer in a stand-alone mode. But networked configurations are likely to predominate in the future, although the extent to which cross-computer linkage will have a significant pedagogical function beyond its initial managerial function is unclear. The most obvious pedagogical function involves communication between teacher and student. Although student-to-student communication across machines for purposes of collaboration has not been seriously examined within mathematics classes, it has been studied in project-oriented writing and multimedia activity (Scardamalia & Bereiter, 1991). With new mathematics curricula taking a more project-oriented approach and with the possibility for shared data and access to remote data, opportunities for meaningful collaboration increase (Hunter, 1992).

However, full communication across computers on a network opens the opportunity for rich "cross-representation" activities of the following general sort. Imagine physically separated students working on two computers, I and II, where each is involved with two types of linkable representations, A and B. There are then four representations available to work with, I^A, I^B, II^A, and II^B, as depicted in Fig. 10.1. It becomes possible to engage students on the respective computers in challenge activities. Students on Computer I can create an object in Representation A, but whose counterpart in Representation B is sent (1) to Computer II, where the representations are *not* linked (see Fig. 10.1). The goal is for students working on Computer II to reconstruct the A version of the object (2), and (3) send it back to the students working on Computer I, where their construction is critiqued and compared with the original A version.

This could take place at various levels of difficulty and complexity. For example, the students on Computer I may or may not have A and B linked on their computer (4). And the communication between Computers I and II might be controlled to allow only messages in text, or only in the languages of one of the respective representations. Or, if the A construction received by students on Computer I is incorrect, they might send back either the B

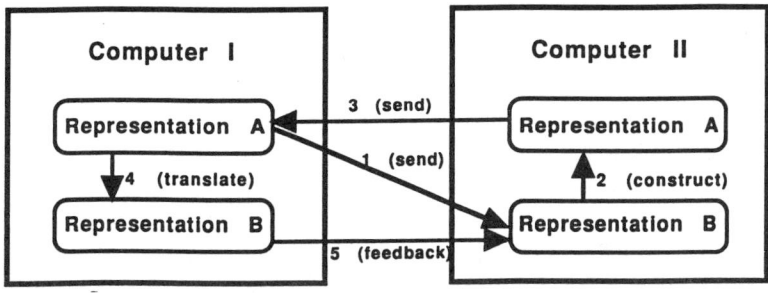

FIG. 10.1. Cross-representational challenges on a computer network.

version of what they received (5), so the students on Computer II could compare it with their originally received B, or they might send back the A version of what they received (not indicated in the diagram), so the students on Computer II could compare it with their A construction. Intuitively at least, it seems that such cross-representational activities could be quite engaging as students attempt either to stump one another, or, as one might imagine in an assessment activity, to do exactly the opposite—students would be graded on both sides on the basis of how well they communicated across representations. In effect, this type of activity is a standard (single-computer) two-representation target-matching activity broken apart to include student constructions in the feedback loops.

Video and Animation. Full merging of computers and video will be well underway by the mid 1990s, which will increasingly make possible measurements and mathematical activity based on realistic visual data—that is, data that vary continuously in time. Perhaps even more importantly, computer animation will become easy to construct and manipulate, and hence to integrate into mathematical activity. A significant step has been taken in this direction by *The Geometer's Sketchpad* (Jackiw, 1992). Animation seems to have some of the same advantage over real video as computer manipulatives have over physical ones: better control over representational features and more direct linkage possible with formal representations.

Curricular Alternatives

The Need for a Broader View and Clarification of the Role of Algebra. Although algebra, including the study of functions, has typically been seen mainly as a secondary-school domain—constituting several courses—I propose that it be thought of in a more integrated way, both vertically integrated with the curriculum of the earlier grades and horizontally integrated with other mathematical domains. In particular, we need to ask questions regarding the kinds of experiences in the elementary grades that build conceptual foundations for the various faces of algebra and how these earlier experiences relate to later work, especially how early graphical representations of quantitative relationships can serve as the means for reasoning about those quantitative relationships. This requires clarifying the forms of reasoning that graphical representations can support and clarifying how these relate to the forms of reasoning supported by character-string representations. These clarifications seem to be needed before we can begin to make judgments about how either should best be learned. The work in this direction that has been done by Dugdale (1990, and Chapter 5, this volume), Confrey (1991), and Yerushalmy (1991a, 1991b; Yerushalmy & Gafni, in press) shows strong potential for linked representations where the (coordinate) graphical representation is given prominence. Further, there is now a growing body of software that is deliberately designed to support learning of linkages between the graphical, symbolic, and numerical, most notably *Function Probe* (Confrey, 1992), and the *Function Supposer* (Schwartz, Yerushalmy, & Harvey, 1991). Data indicates that students are put in the position of gaining a view that is more than the simple sum of its parts. Indeed, earlier worries (Ward & Sweller, 1990) about the increased complexity involved may not be warranted. It is this author's attitude that we should not be overly cautious in seeking alternatives to the standard way of doing things in algebra—we needn't fear the destruction of a well-functioning system of instruction! We are not tampering with success, but with failure.

Nonetheless, clarifications relating algebra and arithmetic (where by "arithmetic" we mean something other than the study of representing and operating on pure numbers) seem necessary regarding the relationships between quantitative reasoning involving particular quantities with particular values versus quantitative reasoning involving general quantities and variables (Kaput & West, in press; Thompson, 1991), and we must also get a better handle on the shift, if it exists, between numeric and algebraic reasoning.

A Scaling Curriculum. Not to be ignored are relations between algebra and geometry, especially as connected to scaling, and especially in the earlier grades. Graphs of functions that do not represent meaningfully experienced quantitative relationships are graphs of abstract objects—akin to the abstract objects of pure arithmetic—and rescalings of such graphs may not be easily understood or appreciated. In particular, the crucial distinction between a change in view of a fixed object and a change in object may not be understandable unless the object is strongly anchored to one's experience. If the graph represents something that one knows directly, say, the velocity of a vehicle that one has driven, or one's own velocity as recorded by a motion detector, then changing the view of that graph by rescaling, or even by changing the viewing window, is much more likely to be understood than is the case when the graph is merely another representation of another abstract object, such as a polynomial.

Indeed, a major effort needs to be invested in building and testing a scaling curriculum, or perhaps a scaling strand in the curriculum, one that begins in the early grades and continues throughout. Such an effort would probably begin with the simplest of activities with measurements of physical objects using ideosyncratic and then more conventional units (as is already done in some curricula), and then would focus on the graphical representations of these, to be followed by variations in those representations done in a systematic way, such as first rescalings in a single dimension (perhaps with one-dimensional graphs, i.e., number lines), before rescalings in more than one dimension. (Such was actually begun in the 1960s in the MINNE-MAST Project.) We also need to engage students in experimentation that compares rescaling of representations of objects with remeasurings of those same objects with different units. When are we changing units and how are units accounted for in our graphs? This aspect of the curriculum draws heavily on and adds meaning to the proportional reasoning curriculum.

A great deal of understanding is required to comprehend a non-square zooming of a coordinate graph, for example, or the "auto-scaling" done by most graphing utilities. I recall an entire in-service class of secondary-school teachers totally at sea in a discussion of the graphing of inverse functions when the expected reflection across the diagonal of an exponential function disappeared—the auto-scaling of the exponential function for the given domain rendered the diagonal visually horizontal and obliterated the expected picture. The fact that the understanding of scaling is normally very limited among today's students and teachers may simply be an artifact of our very limited use of graphs and especially our very limited assessment consciousness regarding this topic. We have not yet recognized that it is important to assess the understanding of scaling. But such assessment is growing drastically in importance as graphical representations proliferate in daily life due to the ease with which they can be created and published.

Although visual aspects of single-variable functions have been examined (Janvier, 1981), to my knowledge no one has systematically investigated scaling issues in three dimensions, or in the context of level curves of two-variable functions. Nor have we investigated nonlinear scaling, although logarithmic scaling has been used by specialists for generations to straighten out exponential graphs. Perhaps nonlinear polynomial scaling should be investigated, which might deepen understanding of the whole process. What does a cubic look like when the vertical axis is scaled quadratically? And vice versa: What does a quadratic look like when the vertical axis is scaled cubically? Or quadratically? Much needs to be done, and it requires stepping out of the constraints assumed by most of today's research.

Widening the Classes of Meaningful Functions and Actions

Although it is generally agreed that functions should be at the heart of the quantitative core curriculum, and although we have suggested that they should occupy this central place from the early grades onward (so extended experience with functions should not be reserved for the narrow group intending to take calculus courses at the late secondary level), we further argue that functions need not be studied strictly as algebraic rules given in closed form. Rather, we suggest that a much wider class of functions modeling many different dependency relationships should be experienced in a variety of representational forms and applications, including data imported via interfaces with measuring instruments. After all, the class of functions describable in closed form is quite small compared to the class of functions that occur in the contexts of empirically generated data, particularly when discrete measurements are allowed. Yet all these latter functions can be directly represented and studied as coordinate graphs—graphs of outdoor temperature during the day as a function of time, velocity of a student as measured by a Microcomputer Based Laboratory (MBL) motion detector, time or distance olympic records as a function of time, the inflation rate as a function of time, and so on. And of course, all such graphing raises issues relating discrete or continuous phenomena to discrete or continuous graphs—when are our measurements representing continuous phenomena as discrete, and when are our graphs representing discrete data as continuous? And vice versa. These issues, unimportant to the traditional core curriculum, become crucial when the curriculum gets serious about admitting and representing real data, including inherently irregular real data.

We certainly do not mean to imply that closed-rule descriptions of functions are unimportant. Quite the contrary: they represent the most efficient and potentially powerful of all types of data descriptions. Further, they play a key role in normalizing and understanding by supporting comparison activities for the more irregular functions that abound in natural experience. They and their graphs act as categories of relationships (e.g., linear, quadratic, exponential) that can be used to interpret the more irregular on a piecewise basis (e.g., "It's growing linearly here and then it peaks out like a parabola before decreasing linearly for a while. . . ").

As is now widely understood among researchers, the computer medium can change the character of traditional representations from display representations to action representations (Kaput, 1987, 1989; Yerushalmy & Schwartz, Chapter 3, this volume). Coordinate graphs and data tables, which once were used strictly to display quantitative relationships, now can be acted upon directly, either to change the function being represented (e.g., by translating or stretching a graph) or to change its form (e.g., by changing the coordinate graph's viewing window or rescaling within a given window, or by changing the step size in a data table). This complicates the question of scaling, although I suggest that this matter has a natural place in the context of a larger scaling curriculum. As Yerushalmy and Schwartz (Chapter 3, this volume) and Goldenberg (1991a) pointed out, the software itself may play an important pedagogical role in this activity, where, for example, the software would not render the scale of the graph visually implicit, but rather would show, via a grid or otherwise, what is taking place as one rescales. And it should make it especially clear when one is changing the function via a graphical action. It is not likely that a conspicuous change in the algebraic representation will suffice to make matters clear if both the rule and its graph are poorly understood as abstract objects. Indeed, care must be taken to assure that the facile transformations and associated linked representations that contemporary software makes possible do not contribute to a meaningless game in the same way that, previously, the manipulation of character strings was a meaningless game—but now with fancier pictures. One should apply to the study of functions the same criteria for significant meaning that seem, finally, to be gaining acceptance in the arena of arithmetic and quantitative reasoning. Not only should the functions be experienced as meaningful, but actions upon them should likewise have meaning. This meaningful quality need not always be based on referential integrity, where the function and actions on it stand for some phenomena, but can be based on a rich aesthetic experience of pattern, just as appropriate play with abstract numbers can sometimes be experienced as meaningful. However, the fact that an action may be computationally possible is not solely the grounds for it being supported by software and/or researched, let alone being required of students.

Historically, complicated functions were built up termwise using series approximations based on such simple functions as polynomials, which are defined on large, typically unbounded, domains. In particular, the improvement in the approximation given by, say, increasing the number of terms of a Taylor's series, takes the form of widening the interval over which the approximation holds within a given error tolerance. This is a very different style of construction than the piecewise building of a function across a domain of numbers. As is well known (Dreyfus & Eisenberg, 1983; Vinner & Dreyfus, 1989), our curricular concentration on continuous (and usually analytic) closed-form functions, defined for all or almost all real numbers, narrows students' image of what constitutes a legitimate function. Building-up processes based on patching piecewise across a domain deserve careful attention as an alternative and cognitively accessible means of understanding phenomena in a local fashion. It is graphically quite simple to understand, although most graphing utilities are not designed to support it. (Schwartz and Yerushalmy are currently building such software, however.) Such an approach, with a strong graphical orientation, may deserve a place alongside traditional globally oriented build-up processes in the core curriculum. It may also require more flexible curve-fitting software than is now available, where one may designate several subintervals and use different classes of curves for each in a way that better corresponds with our visual examination of a graph.

All the attention here to graphs and the associated visual experience may mislead the reader into believing that we are leading toward a kind of trivialization of the functions curriculum, where all actions are based on a direct match or action on what we see graphically. But we must recall a basic lesson that the history of mathematics and science has shown us—the real payoffs in the form of abstractions and deep principles often are found by looking beneath the surface appearances, formally represented or otherwise. But the recommendations above are made precisely because students need to be better equipped to move beyond visual epiphenomena to deeper regularities when this is required, and they should have some sense of when this move is required.

REPRESENTATIONAL ALTERNATIVES—ATTACKING THE MATH/EXPERIENCE LINKAGE PROBLEM

Moving Beyond the Big Three—Linking Experience With Formal Mathematics

Perhaps the greatest unsolved problem in school mathematics to which technology may contribute a solution is linking authentic experience with formal mathematics. As suggested in Fig. 10.2, there is a "high-capacitance" gap between the rich worlds children experience in daily life and the world of school mathematics—a spark of connection between them is difficult to generate by working from only one side.

Countless attempts have been made over the years to create meaningful activities for children. Sometime the attempts take the form of enriching activity with formalisms, such as work with number patterns. Sometimes the attempts take the form of providing rich contexts for the use of formal mathematics. More recently, attempts have taken the form of building bridges between the naturally occurring reasoning patterns arising from everyday experience and the more abstract conceptual activity possible when such patterns are extended to more formal systems of representation.

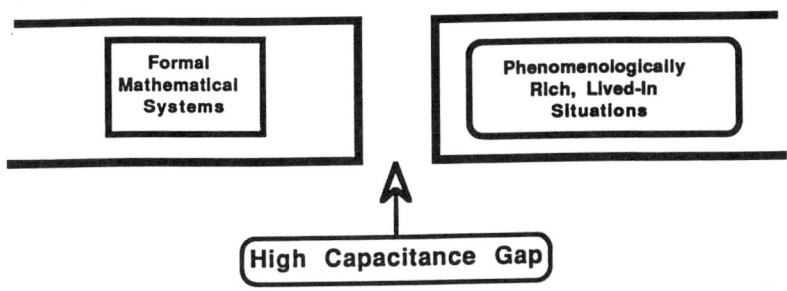

FIG. 10.2. The formal-experiential gap.

Convincing successes have been achieved in the conceptual field of additive structures in the Cognitively Guided Instruction Project at the University of Wisconsin, and more preliminary but promising results have been achieved for multiplicative structures (Kaput & West, in press). And the promise of activity-based approaches, apart from research, has motivated curriculum design for many decades, reaching back before even Dewey and Montessori.

The message of this work is that the powerful, but context-bound, situated reasoning patterns that develop normally in children and adults can be extended and generalized to more formal systems of reasoning. However, the success of this strategy seems to depend upon at least two factors: (a) a structural adjacency or congruence between the structure of the reasoning in the situation and that of the reasoning in the mathematical system, and (b) a relative simplicity of the reasoning activity overall. (These factors may also contribute to the communicability of this overall approach to teachers and students.)

When one moves to reasoning about nonlinear functions that in turn represent dependency phenomena (perhaps hypothetical) among quantities in some situation (perhaps hypothetical), and to reasoning about relations among such functions and actions upon the functions, these similarity and simplicity factors no longer hold. The mathematical experience is more distant from the experience of the phenomena, requiring much more abstraction away from the phenomena. And usually both are much more complex. In fact, the phenomena are usually not experienced at all, but rather some representation of the phenomena serves as a surrogate, where the representation is often in the form of static, inert text, such as a word problem, perhaps augmented by static, inert graphics. Here the linkage problem is much more difficult. And our only real educational solution to date has been demographic—to eliminate most students from responsibility for learning the more complex mathematics. Even for the minority who are expected to learn it, our expectations in terms of competence have been egregiously limited.

Between the simple mathematics of grade school and the more complex and abstract mathematics of secondary school, we have seen an alternative linking strategy that shows some promise as a way to connect previously established quantitative reasoning patterns with algebraic reasoning. It is based on two changes from traditional practice. First, it expands the types of situations to be modeled to include more realistic ones (Fey, 1989). Second, it exploits calculators or simple computer software to help students generate tables of paired values based on specific numerical computations from which function-based models might be abstracted and formally represented. This linking strategy is optimally executed by the Ohio State group (see Chapter 2 by Demana, Schoen, & Waits in this volume and the

Transition to Algebra materials [Demana, Leitzel, & Osborne, 1988] aimed at middle school students; see also Philipp, Martin, & Richgels, Chapter 9, this volume). Nonetheless, the numbers in these situations are inevitably the result of an explicit numeric computation based on some closed-form rule; the situations themselves are yet represented only weakly, off-line, by text; and the representations are limited to the traditional, and formal, big three—tables, formulas, and graphs. In other words, the experiential link between the mathematics and the situation being mathematized is still relatively indirect.

We now explore two types of graphical uses of new technologies that have yet to be seriously employed to attack the linking problem. Each involves directly manipulable simulations, but they differ in their degrees of phenomenological richness.

Dynamic, Interactive Diagrammatic Representations

Using a relatively simple problem that is widely used as an optimization problem in beginning calculus courses, I illustrate what I mean by a dynamic, interactive diagrammatic representation. The problem is as follows:

> A farmer has 120 m of fencing with which to enclose a rectangular pasture along a straight river, where no fence will be needed. How should the farmer lay out the fence in order to enclose a pasture with the largest area, and what is this area?

The standard approach is to draw a diagram of the typical pasture, label it with variables, write the function that we wish to maximize in terms of a single variable (using an auxiliary equation if needed), and then maximize that function using derivatives. Of course, the first issue with many students is whether there is a problem at all. That is, does it really make any difference how you lay out the fence? Won't the area stay the same? The best response is to try a few different layouts to reveal that the size of the area does, in fact, depend on how you use the fence—as illustrated by the four diagrams in Fig. 10.3, where the corresponding area is given inside the rectangle.

Suppose that one were able to draw a variable diagram, constrained by the given condition— namely, that it be three-sided, rectangular, with a fixed perimeter of 120 units. Suppose further that one were able to label part of this diagram with its measure so that as the diagram varied in response to the user's control, dependent measures could be applied to other, more formal representations.

The usual figure is given in Fig. 10.4, together with its linked represen-

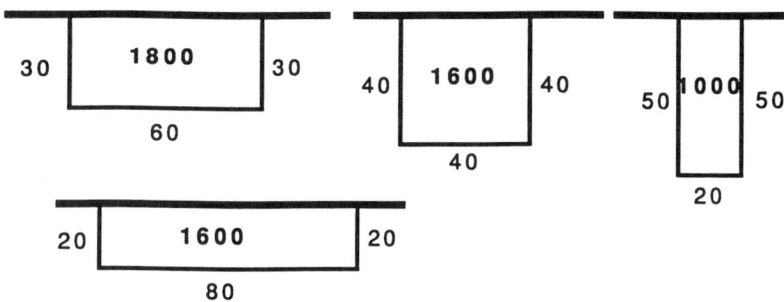

FIG. 10.3. The fence problem diagrams.

tations—namely, a table and a coordinate graph. The dashed box contains the original construction, while the rectangle to its right is its "hot" version, which can be dragged continuously through allowable values by pulling the lower right corner. The table and coordinate graph in Fig. 10.4 were created by setting the X-step to be 10 and then dragging the hot corner of the figure through its allowable values. After creating them, we can point and click at respective points in the table or graph to recover the associated rectangle, which we have done to highlight the $X=40$ instance in each of its four representations. (Note that the algebraic representation is at the top of the table as well as inside the construction rectangle.)

This type of dynamic diagram-constructing software is now available commercially (Jackiw, 1992; Laborde, 1992; Richards, 1992; Schwartz & Yerushalmy, 1992). McArthur (1990) also built geometry-oriented microworlds sharing some of these characteristics, as did Grant and Borovoy (1990), where the domain is population dynamics and where the population levels and densities are concretely represented on the screen together with more formal data about the characteristics of the population. Work by Kaput and colleagues involving concrete representations of multiplicative structures (Kaput, Luke, Poholsky, & Sayer, 1987; Kaput & West, in press) and by P. Thompson and A. Thompson involving computer-based Dienes-type manipulatives (Thompson, P. 1992a, 1992b; Thompson P. & Thompson, A. 1990) are of a similar genre in that they are based on dynamic linkages between concretely manipulable representations and formal representations. It is also possible to add a dynamic, direct manipulation feature to traditional formal graphics without regard to external modeling, such as that made possible by CABRI (Laborde, 1990), as when one takes the traditional definition of a conic as a locus of points and actually drags a constrained point to generate the conic, such as a parabola (Klotz, 1991).

This general approach adds an important new set of teaching and learning opportunities that supplement the traditional representations. But research is needed to determine the balance between the amount of structure

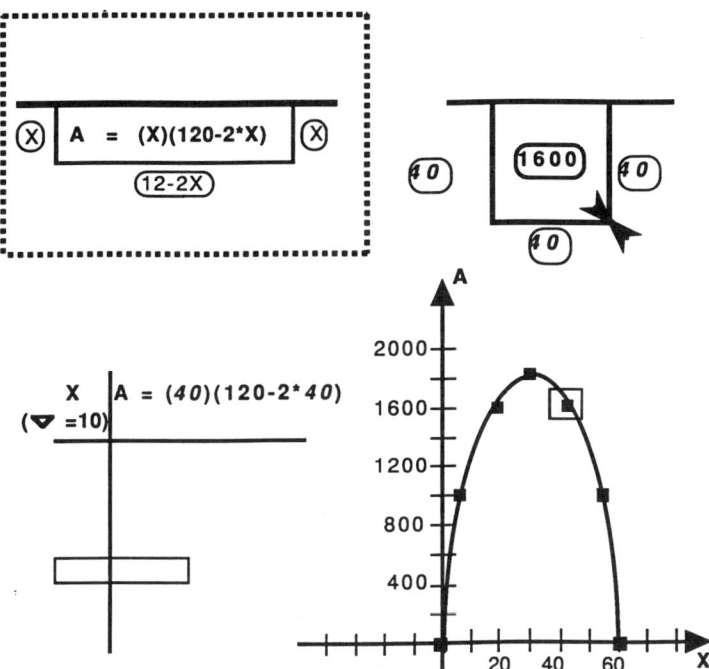

FIG. 10.4. Linked diagram, graph, table, and equation.

that is provided by the software (possibly with the help of the teacher) and how much should be the direct responsibility of the student. A finer version of this question is, how should the balance shift as a function of student experience and problem complexity? The reader has undoubtedly noticed that the acts of building and labeling a dynamic diagram are quite similar to the standard initial steps of setting up a function, although in the computer environment, constraints and supports with immediate feedback can be provided to guide the process.

Lived-In Representations

I have used the word "representation" in the title of this section where most would have used the word "simulation." But a simulation is a representation in the sense that I have defined it elsewhere (e.g., Kaput, 1987, 1989) — in fact, it is an action-representation, a representation that one acts on directly. The intent of simulations is to capitalize on student prior knowledge about the situation being modeled and to enable the student to integrate that knowledge with or transform it into mathematical knowledge.

Historically, most simulations have taken the form of parameter adjusting followed by observation: One sets parameters and then runs the simulation and observes the effects of one's choices—there is a gap between one's action and the results. A potentially more powerful approach involves direct manipulation simulations that provide continuous, real-time feedback in several experiential dimensions while simultaneously providing data that feeds into more formal representations, as suggested in Fig. 10.5.

A useful image is provided by contemporary commercially available flight simulators. However, although these flight simulators provide real-time data in dials and gauges, they do not provide the kinds of formal representations that need to be learned in mathematics classes, such as coordinate graphs. Furthermore, a system designed specifically for mathematics learning should enable a student to treat any formal object generated by the simulation, such as a graph, as an object of study in its own right, either within the system or as an object exported to another, more analytic system (as is the case with *The Explorer's Toolkit:* Richards, 1992). Pictured in Fig. 10.6 is a typical screen from one part of MathCars, a system designed, but not yet implemented, by the author (Kaput, in press). Its intent is to provide strong linkage between the phenomenologically rich experience of motion in a vehicle and the more formal representations of velocity and position as functions of time.

Shown in Fig. 10.6 is the one-car mode with the velocity versus time graph, a digital clock, and an odometer displayed on the dashboard, below a windshield view. At the lower right is the accelerator by which one controls the essentially linear motion of the simulated vehicle. By holding the pointer on the lower part, where it is shown, acceleration is negative, indicated by the minus sign in the display at the center of the accelerator. This fact is also apparent from the graph, which is tending downward. Note that the graph is actually being generated as the trace of the tip of an analog speedometer that slides from left to right as time elapses, thereby revealing the velocity graph as a record of the velocities at all the instants of the trip.

Across the top is a menu palette that contains blocks of items for three different types of parameters describing a MathCars trip—time, velocity, and position. In the time block are three display modes, from left to right first an ear-icon indicating sound (a metronome beep, whose default frequency is one beep-per-second), and an analog and digital clock (chosen

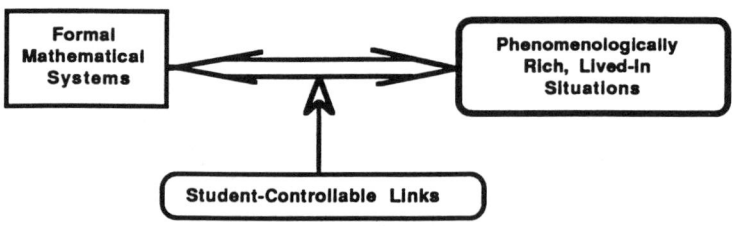

FIG. 10.5. Student-controllable formal-experiential linkages.

10. URGENT NEED FOR PROLEPTIC RESEARCH   303

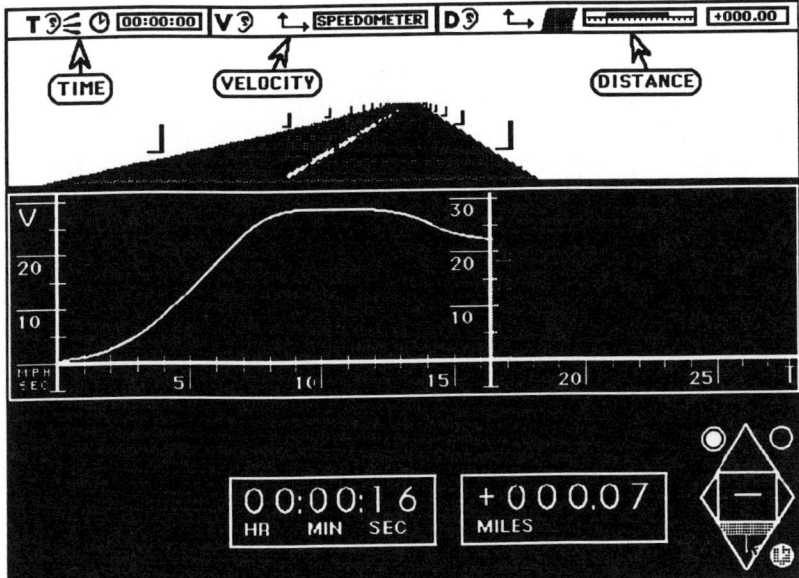

FIG. 10.6. A MathCars screen. (Source: Screen from a video-computer animation. [1992]. Developed under the auspices of the Apple Classrooms of Tomorrow Program, jointly funded by Apple Computer and the National Center for Research in Mathematical Sciences Education.)

for display), respectively. Note that the ear has "sound marks" emanating from it to indicate that time sound is turned on. In the velocity block are again a sound display (here it is a continuous sound whose pitch varies with velocity), a coordinate graph display of velocity versus time (which is the one chosen), and an analog speedometer (which can be displayed horizontally apart from the graph if desired). In the position block are sound (an echo from the "posts" along the road), a coordinate graph display of position versus time, a map (essentially a vertical view), and an odometer (which can be paired with the digital clock to generate a table of position versus time data, which can further be keyed to the beep interval to coordinate the table's step size with beep rate).

Clearly, this is a rich environment for displaying the many different ways we experience motion in a vehicle. Although we have not included kinesthetic aspects of the experience, it is not difficult to imagine Math-Bikes, where one rides a stationary bike whose rate-of-spin data feeds into a MathCars type of display via a Microcomputer Based Laboratory (MBL) interface. In addition, the MathCars design has provision for two cars, where the operator has control of one and the other has its motion specified in advance. Furthermore, provision has been made for target activities, in which, for example, one attempts to "drive" a given position versus time graph where one has feedback only on one's velocity versus time. In this case, one is, in a very real sense, "driving" a version of the Fundamental Theorem of Calculus. In more advanced options of the system, one can specify the motion algebraically, as well as model other types of quantities such as fluid flow and accumulation by replacing the windshield view with other views—but with most of the other representational aspects of the system deliberately held constant to help reveal the underlying mathematical structure across such situations.

The basic notion of this simulation and of extensions not described here is to provide a very rich "lived-in" experience that is under the student's control and that provides a carefully designed web of connections between that experience and more formal mathematical representations. Although we have not yet had opportunity to implement and test this system, evidence exists from MBL research (Rubin & Nemirovsky, 1991), as well as work by Swan (1985) at the Shell Centre, that dramatic improvements are possible in student understanding of velocity and position functions as well as understanding of relations between them. It is important to note that the functions dealt with here are not defined by closed-form rules (at least not in students' initial experience with the systems). However, students would gradually move in the direction of specifying motion by using formulas and would be involved with linear functions rather early, experimenting with constant and linearly increasing velocity, for example, as did the Scholastics (Kaput, in press).

A very important research question involves the study of how the complex experience and associated knowledge of vehicle motion can be abstracted toward increasingly formal knowledge. It is not difficult to imagine a form of "arcade knowledge" of, say, the Fundamental Theorem of Calculus embodied in a skillful ability to match target velocity or position curves with feedback only in their counterparts. But how does this ability, instantiated almost at the level of one's fingers, come to be the basis of deeper understanding? It is important not to devalue this level of skill, because it appears that most reliable, genuine personal knowledge has this characteristic, that it has many levels (Polanyi, 1958), and further, these levels are intimately connected, often tacitly connected (Polanyi, 1966). It may be that this level of knowledge may be an important missing ingredient, not only in calculus knowledge of even advanced mathematics students, but also physics students (McDermott, Rosenquist, & van Zee, 1987). Moreover, it may be possible to build a significant edifice of understanding and competence in the graphical realm without complex algebraic representations. Consider, for example, the process of analyzing a position versus time graph to determine the approximate velocities associated with the motion that gave rise to that graph. In Fig. 10.7 are depicted two graphs, in which the position graph on the top was generated by a short trip with the

10. URGENT NEED FOR PROLEPTIC RESEARCH   305

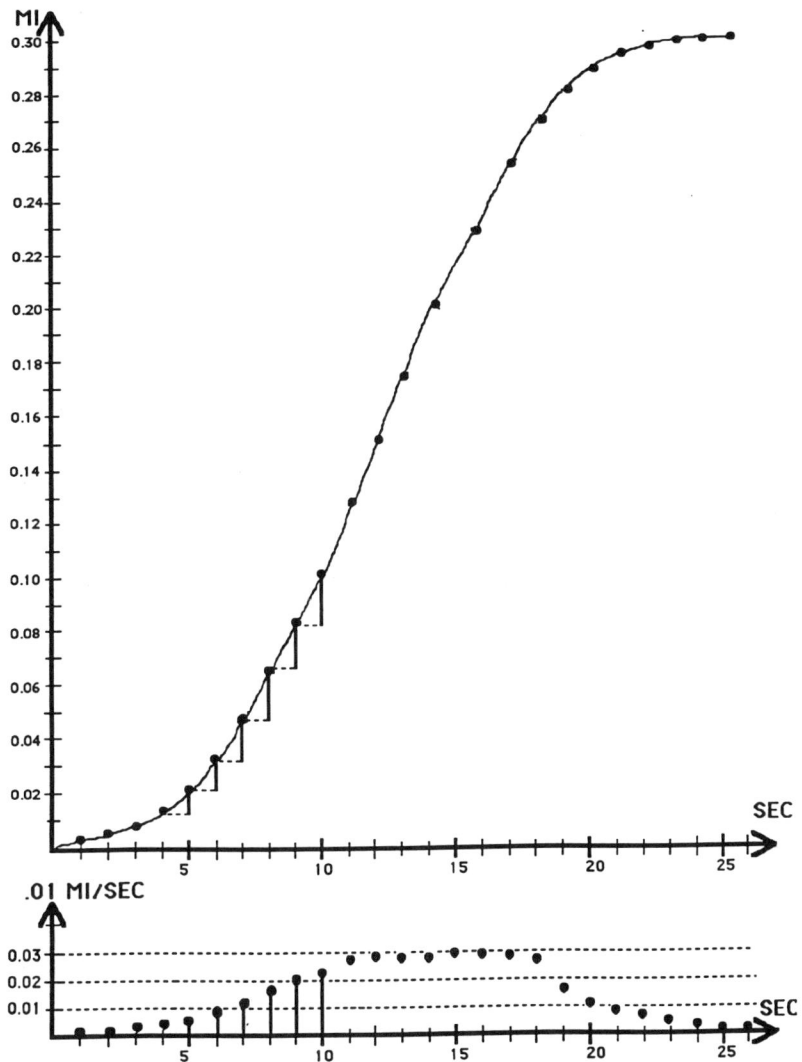

FIG. 10.7. Building velocity graphs from position graphs.

time beeper set at its default value of one beep per second, and in which a dot was deposited at each beep. This discretizes the graph for purposes of analysis.

As mentioned earlier, any formal object generated by a simulation ought to be available as an object of further analysis afterward. Here we assume that the position graph (as a set of ordered pairs) has been put into an analytical environment where we have set up another time axis and asked the question, What is the interpretation of the vertical distance between each dot? This, the length of the vertical line segment constructed for some of the dots (yielding a "lollipop"), is, of course, the change in position over the 1-sec interval between dots. Suppose now that we copy those vertical segments onto the new time axis below as indicated and ask what the interpretation of the vertical axis should be? The answer is given by the conspicuous label, namely, a distance unit per second (where in this case the distance unit, derived from the scale of the position graph, is .01 mi). After all, if the length of the vertical line segment is a change in position for each 1-sec interval, then we can think of this distance as an average velocity over that 1-sec interval. We further ask, What is the meaning of the entire set of dots? The answer, given readily here, is that it is an approximation to the velocity of the vehicle on a one per second sample of position change. More abstractly, it is an approximation to the derivative of the position function.

If we decrease the beeping interval, we get more dots and a better approximation to the derivative (or velocity function). The reader will note that this process makes almost no use of algebra and does not depend on the function being given as a closed-form rule. Further, the same process could be applied to a velocity function to yield a change-in-velocity graph, namely, an acceleration approximation, that is, a second derivative. The process can also be reversed, using an area interpretation, to yield antiderivatives. In fact, the elements of differential equations can be approached in this manner (Kaput, in press). Although we are sweeping some detail under the rug, particularly issues of scale (recall the discussion of a scaling curriculum earlier), the reasoning involves simple ratios, which in turn require only a modest amount of symbolic representation to be dealt with efficiently (Kaput & West, in press).

An ancillary issue that this highly graphical approach raises concerns the appropriate grade level for the introduction of these ideas. My preliminary hypothesis is that middle school is probably a good place to start. But this then challenges some very fundamental assumptions about the role of algebra as a language for the expresssion of important ideas and the usual prerequisite relation between algebra and calculus. It also puts a much heavier burden on understanding graphs and manipulating graphs—understanding that is well known to require specific instruction. Other aspects of this experience do not involve graphics per se, but involve what Williams (Chapter 11, this volume), drawing on the work of Winograd and Flores (1986), refers to as "thrown-ness." The degree of immersion and style of interaction of a "lived-in simulation" are very different from those of a parameter-adjusting simulation, where feedback, while it may be prompt, is separated from one's choices as a different event.

More reflections on the implications of such departures from standard practice are offered in the next section. Put briefly, there is much to be explored, and most of it is not being explored today.

OTHER UNCOVERED TERRITORY

Language for Describing Graphical Phenomena: Change in View Versus Change in Function

A concern deserving fuller attention is the matter of the language we use to talk about graphical representations, with students and with one another. As of this time, there is no standard language for discussing graphs and the actions on them, or for the features of software that support these actions. For example, we need good terms for dealing with the various types of scaling and "zooming" — very strange things happen, for example, if one does repeated zooms (rescales by a given magnification factor in each dimension) with a non-square zooming window. Screen aspect and its relation to the experience of slope even on square viewing windows is another matter. We need means for discussing these actions, features, and events systematically.

It would be nice if we could build a language that is as consistent as possible across all representations. Given a functions-based approach, it makes sense to speak of transformations of two basic types — those that change our view of the function and those that change the function itself. If one performs algebraic simplifications or factoring, then one is changing one's view without changing the function, just as occurs when one performs scale or other viewing window changes on a graph, or when one changes the step size in a table. On the other hand, if one adds a constant to the function or slides a graph vertically, then one is changing the function, whether or not one is changing view (note that we did not say "the" graph, which would imply that the function is the same before and after the transformation). The differences between change in view and change in function are a source of much confusion among students. However, with character-string and graphical representations linked, there is opportunity to clarify the difference. This is especially the case in the context of function comparisons, such as equations, examined in a linked representation environment.

Functions of More Than One Variable and Alternative Graphical Descriptions

We have not discussed functions of more than one variable with graphs and level curves in two and three dimensions. As graphics power and plotting techniques evolve, these will become more commonplace. All the factors that are driving change in our approach to single-variable functions will extend to functions of several variables. Again, the curriculum will be challenged. Although solid geometry slipped out of the standard secondary school curriculum in the 1950s and 1960s, it may, and probably should, return in the context of graphing functions of more than one variable (Banchoff, 1991). A similar statement may apply to functions of a complex variable. Increasingly we will see opportunities for real cross sections of complex functions of complex variables and will get an opportunity to see roots of quadratics, for example, in the broader context of complex variables, which is their natural mathematical context. Phase diagrams for real-valued functions of a real variable are likely to become increasingly useful as alternative representations of phenomena. Student interpretations of phase descriptions of motion and fluid flow using real-time data display devices are now being studied by Nemirovsky at TERC (Rubin & Nemirovsky, 1991), and there is reason to believe that when high-school students themselves control the data being generated, they are capable of learning the phase-description approach. Further, even elementary students can make sense of change of quantity and its different graphical representations in situations that are real to them (Nemirovsky, Tierney, & Oganowski, 1992).

Other Function-Based Phenomena

We have not discussed recursively or iteratively defined functions that give rise to fractals and that are used to describe dynamical systems. These are not especially complicated algebraically, but their graphics are enormously rich, especially when viewed dynamically (Devaney, 1990, 1991; Devaney & Keen, 1989; Goldenberg, 1991b; Mandelbrot, 1982; Peitgen & Richter, 1986; Sandefur, 1991). Currently, these are seen mostly as curiosities or, especially by Devaney, as powerful motivators to get students excited about mathematical discovery, including the discovery of an entirely new mathematics.

Although we cannot discuss these new mathematical adventures further here, their exploration as candidates for inclusion in the school curriculum of the next century is an absolute necessity. If we are to bring the turn-around time for new mathematics in the school curriculum to less than a century (currently, it seems to be several centuries), then we must immediately begin determining the nature of the environments and the student experiences that are appropriate for meaningful student learning. Then we can get on to determine the appropriate teacher education and assessment consequences of potential curricular changes. It has become relatively clear that as new curriculum development at the elementary and middle school levels proceeds, the traditional secondary curriculum will require even greater modification than was imagined in the late 1980s. By the turn of the century, most students will be familiar with most of the traditional core ideas of algebra and geometry by the time they enter 9th or 10th grade, and they will be ready for mathematical experiences that are not at all available today.

REFLECTIONS

In this chapter, we have attempted to outline the character of current research in contrast with the kinds of future-defining research that would expand the envelope of current practice to new possibilities offered primarily by the availability of new technological power. This is not to say that the kinds of practice-extending research described earlier in this volume are without importance. Quite the contrary! This is some of the most important work taking place today. Rather, the contrast is offered in the spirit of pointing to additional work that needs to be undertaken. Although, as suggested earlier, there is room for making practice-extending research more proleptic by anticipating explicitly the new conditions under which it might be applied, the future-defining work deserves special advocacy, for it is the hardest to do. Not only does it require expensive technology and special expertise and support, it requires a good dose of imagination and daring, not only among the researchers but among the patrons who would support this activity. It requires resistance to the pressures to provide quick payback, pressures felt everywhere in the education establishment.

Other pressures likewise act as constraints, chief among these being the widely expressed feeling that research should be school based and involve practitioners at all levels. Although forward-looking and imaginative practitioners are surely as abundant as university research and development people, the circumstances of school-based work make major departures from established norms quite difficult. Hence, it seems quite likely that highly proleptic work is likely to remain university based, although mathematics practitioners should be part of the effort at every stage.

Perhaps it is merely my personal bias, but from my vantage, it appears that the most substantial changes in student graphical experience of quantitative relationships will derive from changes in the representational forms and actions on these that students will have available to them. These changes will surely drive other changes, especially those in curriculum. To get a sense of the impact of representational innovations on curriculum, imagine what our quantitative curriculum would look like if the standard system for representing numbers and the associated algorithms had not been invented. We would inhabit a very different world indeed. A fully exploited set of graphical representations may make a similar difference in the long run. But we must get started on exploring those graphical opportunities — proleptically.

REFERENCES

Banchoff, T. (1991). Student-generated software for differential geometry. In W. Zimmerman & S. Cunningham (Eds.) *Visualization in teaching and learning mathematics* (pp. 165-172). MAA Notes 19. Washington, DC: Mathematical Association of America.

Becker, H. (1991). Mathematics and science uses of computers in American schools, 1989. *Journal of Computers in Mathematics and Science Teaching.*

Boyer, C. (1959). *The history of calculus and its historical development.* New York, NY: Dover.

Boyer, C., & Merzbach, U. (1989). *A history of mathematics* (2nd ed.). New York: John Wiley & Sons.

Confrey, J. (1991). *Constructing and linking functions.* Final Report to the National Science Foundation. Available from the author, Department of Education, Cornell University, Ithaca, NY.

Confrey, J. (1992). *Function probe.* [Software] Santa Barbara, CA: Intellimation.

Demana, F., Leitzel, J., & Osborne, A. (1988). *Transition to algebra: Seventh and eighth grade units.* Lexington, MA: D.C. Heath.

Devaney, R. (1990). *Chaos, fractals, and dynamics.* Reading, MA: Addison-Wesley.

Devaney, R. (1991). The orbit diagram and the Mandelbrot set. *College Mathematics Journal, 22,* 23-37.

Devaney, R., & Keen, L. (1989). Chaos and fractals: The mathematics behind the computer graphics. Providence, RI: *American Mathematical Society.*

Dreyfus, T., & Eisenberg, T. (1983). The function concept in college students: Linearity, smoothness, and periodicity. *Focus on Learning Problems in Mathematics,* 5(3&4), 119-132.

Dugdale, S. (1990). Beyond the evident content goals. Part III: An undercurrent-enhanced approach to trigonometric identities. *Journal of Mathematical Behavior, 9,* 233-287.

Edwards, C. (1979). *The historical development of the calculus.* New York: Springer-Verlag.

Fey, J. (1989). School algebra for the year 2000. In S. Wagner & C. Kieran (Eds.), *Research issues in the learning and teaching of algebra* (pp. 199-219). Hillsdale, NJ: Lawrence Erlbaum Associates; Reston, VA: National Council of Teachers of Mathematics.

Fowler, D. (1987). *The mathematics of Plato's Academy: A new reconstruction.* Oxford: Clarendon Press.

Goldenberg, E. P. (1991a). The difference between graphing software and educational graphing software. In W. Zimmerman & S. Cunningham (Eds.), *Visualization in teaching and learning mathematics* (pp. 77-86), MAA Notes 19. Washington, D.C.: Mathematical Association of America.

Goldenberg, E. P. (1991b). Seeing beauty in mathematics: Using fractal geometry to build a spirit of mathematical inquiry. In W. Zimmerman & S. Cunningham (Eds.), *Visualization in teaching and learning mathematics* (pp. 39-66). MAA Notes 19. Washington, DC: Mathematical Association of America.

Grant, W., & Borovoy, R. (1990). *Simulation environments to support the construction of mental models.* Cupertino, CA: Apple Classroom of Tomorrow Program, Apple Computer, Inc.

Hunter, B. (1992). Linking for learning: Computer and communications network support for nationwide innovation in education. *Journal of Science Education and Technology.*

Jackiw, N. (1992). *The geometer's sketchpad II* [Software]. Berkeley, CA: Key Curriculum Press.
Janvier, C. (1981). Difficulties related to the concept of variable presented graphically. In C. Comiti & G. Vergnaud (Eds.), *Proceedings of the Fifth International Conference on the Psychology of Mathematics Education* (pp. 189-192). Grenoble, France.
Kaput, J. (1987). Toward a theory of symbol use in mathematics. In C. Janvier (Ed.), *Problems of representation in mathematics learning and problem solving* (pp. 159-195). Hillsdale, NJ: Lawrence Erlbaum Associates.
Kaput, J. (1989). Linking representations in the symbol systems of algebra. In S. Wagner & C. Kieran (Eds.), *Research issues in the learning and teaching of algebra* (pp. 167-194). Reston, VA: National Council of Teachers of Mathematics; Hillsdale, NJ: Lawrence Erlbaum Associates.
Kaput, J. (1992). Technology and mathematics education. In D. Grouws (Ed.), *Handbook on research in mathematics teaching and learning* (pp. 515-556). New York: Macmillan.
Kaput, J. (in press). Democratizing access to calculus: New routes to old roots. In A. Schoenfeld (Ed.), Mathematical thinking and problem solving. Hillsdale, NJ: Lawrence Erlbaum Associates.
Kaput, J., Luke, C., Poholsky, J., & Sayer, A. (1987). Multiple representations and reasoning with intensive quantities. In J. Bergeron, C. Kieran, & N. Herscovics (Eds.), *Proceedings of the 11th Annual Meeting of the PME-NA* (Vol. II, pp. 289-295). Montreal: University of Montreal.
Kaput, J., & West, M. M. (in press). Missing value proportional reasoning problems: Factors affecting informal reasoning patterns. In G. Harel & J. Confrey (Eds.), *The development of multiplicative reasoning in the learning of mathematics* (Research in Mathematics Education Series). Albany, NY: State University of New York Press.
Klein, J. (1968). *Greek mathematical thought and the origins of algebra.* Cambridge, MA: MIT Press.
Kline, M. (1972). *Mathematical thought from ancient to modern times.* New York: Oxford University Press.
Klotz, E. (1991). Visualization in geometry: A case study of a multimedia mathematics education project. In W. Zimmerman & S. Cunningham (Eds.), *Visualization in teaching and learning mathematics* (pp. 95-104), MAA Notes 19. Washington, DC: Mathematical Association of America.
Laborde, J.-M. (1992). *CABRI geometry.* New York: Brooks-Cole.
Mandelbrot, B. (1982). *The fractal geometry of nature.* San Francisco: W. H. Freeman.
McArthur, D. (1990). *Overview of object-oriented microcomputer microworlds for learning mathematics through inquiry.* Technical Report. Santa Monica, CA: Rand Corporation.
McArthur, D., Stasz, C., & Hotta, J. (1987). Learning problem-solving skills in algebra. *Journal of Educational Technology Systems, 15*(3), 303-324.
McDermott, L., Rosenquist, M., & van Zee, E. (1987). Student difficulties in connecting graphs and physics: Examples from kinematics. *American Journal of Physics, 55*(6), 503-513.
Nemirovsky, R., Tierney, C., & Oganowski, M. (in press). *Children additive change, and calculus.* TERC Working Paper No. 6-92.
Peitgen, H., & Richter, P. (1986). *The beauty of fractals.* New York: Springer-Verlag.
Polanyi, M. (1958). *Personal knowledge.* Chicago: Chicago University Press.
Polanyi, M. (1966). *Tacit knowledge.* New Haven, CN: Yale University Press.
Richards, J. (1992). *The explorer's toolkit* [Software]. Sunnyvale, CA: Wings for Learning, Sunburst Communications.
Rubin, A., & Nemirovsky, N. R. (1991). Cars, computers, and airpumps: Thoughts on the roles of physical and computer models in learning the central concepts of calculus. In G. R.

Underhill (Ed.), *Proceedings of the 13th Annual Meeting of the PME–NA* (Vol. 2, pp. 168-174). Blacksburn, VA: PME.

Sandefur, J. (1991). Discrete dynamical modeling. *College Mathematics Journal, 22,* 13-22.

Scardamalia, M., & Bereiter, C. (1991). Higher levels of agency for children in knowledge building: A challenge for the design of new knowledge media. *Journal of the Learning Sciences, 1*(1), 39-68.

Schwartz, J., & Yerushalmy, M. (1990). *The function supposer* [Software—part of the Visualizing Algebra series]. Pleasantville, NY: Sunburst Communications.

Schwartz, J., & Yerushalmy, M. (1992). *The super-supposer* [Software]. Sunnyvale, CA: Wings for Learning, Sunburst Communications.

Schwartz, J., Yerushalmy, M., & Harvey, W. (1991). *The Algebra Toolkit* [Software]. Pleasantville, NY: Sunburst Communications, Inc.

Swan, M. (1985). *The language of functions and graphs.* England: University of Nottingham.

Thompson, P. (1991). *A theoretical model of quantity-based reasoning in arithmetic and algebra.* Revised version of a paper presented to the Annual Meeting of the AERA, San Francisco, CA. Available from the author, San Diego State University, Department of Mathematics.

Thompson, P. (1992a). *Blocks microworld.* Santa Barbara, CA: Intellimation, Inc.

Thompson, P. (1992b). Notations, conventions, and constraints: Contributions to effective uses of concrete materials in elementary mathematics. *Journal for Research in Mathematics Education, 23*(2), 123-147.

Thompson, P., & Thompson, A. (1990). Salient aspects of experience with concrete manipulatives. In G. Booker, P. Cobb, & T. Mendicuti (Eds.), *Proceedings of the 14th PME Conference,* Oaxtepec, Mexico.

Usiskin, Z. (1988). Conceptions of algebra and uses of variables. In A. Coxford & A. Schulte (Eds.), *The ideas of algebra, K-12* (1988 NCTM Yearbook). Reston, VA: National Council of Teachers of Mathematics.

Vinner, S., & Dreyfus, T. (1989). Images and definitions for the concept of function. *Journal for Research in Mathematics Education, 20*(5), 356-66.

Ward, M., & Sweller, J. (1990). Structuring effective worked examples. *Cognition and Instruction, 7,* 1-39.

Winograd, T., & Flores, F. (1986). *Understanding computers and cognition.* Norwood, NJ: Ablex.

Yerushalmy, M. (1991a). Student perceptions of aspects of algebraic function using multiple representation software. *Journal of Computer Assisted Learning, 7,* 42-57.

Yerushalmy, M. (1991b). Effects of computerized feedback on performing and debugging algebraic transformations. *Journal of Educational Computing Research, 7.*

Yerushalmy, M., & Gafni, R. (1992). Syntactic manipulations and semantic interpretations in algebra: The effect of graphic representation. *Learning and Instruction.*

11 Some Common Themes and Uncommon Directions

Steven R. Williams
Washington State University

I begin a commentary on the foregoing chapters by assuming the uncomfortable role of academic curmudgeon and engaging in a little critical hyperbole. Specifically, I will voice three brief (and somewhat heretical) reactions to this book as a whole, then return to each in order to clarify and to fine-tune the criticism.

Heresy 1. Research, as represented in the body of this book, is being driven by available and promised technologies; it seems, if anything, too sanguine about the role that technology might play in classrooms. Having caught a vision of what we *can* do, we too often uncritically assume not only that it should be done, but that it will in fact solve long-standing problems.

Heresy 2. Research has failed to deal effectively with what I will call the content domain of the graphical representation of functions. Specifically, we as researchers have failed to provide a comprehensive analysis of the subject matter with an eye toward informing research on learning or teaching.

Heresy 3. Research has also failed to deal effectively with understanding the graphical representation of functions. Specifically, we have failed to engage in the sort of careful analysis of tasks and structures that would allow for modeling the understanding, learning, and teaching of graphs and functions.

These reactions seem all the more churlish because these problems are neither egregious, nor unexpected. It is obvious that the study of functions and graphs, with an eye toward informing teaching and learning, is in its infancy. A great deal of work remains to be done. Far from being able to see the big picture, we are only now learning what the puzzle pieces look like. Still, given the intent of this book, I am left with the feeling that critical pieces of the puzzle are not only missing, but are not being sought; or at least they are being skipped over in favor of easier or more appealing puzzle pieces.

I was asked recently by a mathematical colleague for a list of questions that would do for mathematics education what Hilbert's 23 *Paris Problems* of 1900[1] did for mathematics: generate a great deal of interest and research, by a large number of people, on problems of mutual interest and acknowledged importance. In a sense, this book attempts to provide a local set of *Paris Problems* for the domain of functions and graphs. As discussed in the introductory chapter by Romberg, Carpenter, and Fennema, this book is the result of a conference that brought together researchers in mathematics education with a common interest in the graphical representation of functions. The goal of the group was to delineate and to clarify the important issues in the field, to summarize what has been done, and to set a course for future work. The historical success of this approach in the Georgia Conference, the Wingspread Conference, the Research Agenda Project, and other similar efforts (see Sowder, 1989) is reason for optimism that an analogous approach would work in this content area. Indeed, Hiebert and Wearne (1991), following Belmont and Butterfield (1977), put forward a similar sequence of projections as a model for how learning might be studied so as to inform teaching. They suggest that the process involves:

> (1) selecting the content domain and defining it clearly; (2) identifying the cognitive processes that are critical for successful performance in the domain; (3) designing instruction to promote the acquisition and use of the key processes; and (4) examining the relationship between instruction and cognitive change and assessing the extent of cognitive change. (Hiebert & Wearne, 1991, p. 161)

In a general sense, this volume represents a beginning point for following the aforementioned course for the graphical representation of functions. It is partly this common-sense view of the task before us that provides clues as to the nature of the puzzle pieces, and that gives rise to my concerns. The obvious differences between the domain of, for example, early number concepts and skills, on the one hand, and functions and graphs, on the

---

[1] In 1900, David Hilbert presented a list at the Paris International Congress of 23 unsolved mathematical problems that have stimulated a great deal of research and discovery in the field of mathematics.

other, are reason enough to believe that the path ahead for researchers in this area will present very different and very difficult challenges.

Moreover, there are considerations that lie beneath the four steps outlined and that are brought to light by my colleague's challenge to pose a new set of *Paris Problems*. As I struggled in the ensuing weeks to provide for myself and for my colleague an intelligent response, the value-laden nature of mathematics education as a profession became more apparent. My colleague's request took for granted a particular view of what it means to learn mathematics or, at least, it took for granted that settling on such a view was unproblematic within our discipline. Instead, I came to realize that any response I gave to his request would of necessity be relative, both to a particular culture of mathematics users, and to a view of competent performance within that culture. The real problem in providing an intelligent answer for my colleague lies in the fact that there is no one accepted view of what learning mathematics is. There exist several mathematical subcultures that students might reasonably wish to enter in their lifetimes, including those of research mathematicians, of technical users of mathematics such as scientists and engineers, of mathematics teachers, and of more casual or occasional users of mathematics such as Lave's (1988) "just plain folks." There are thus many societies of users of mathematics that our students may reasonably want to join, and each will engage in practices that relate differently to the graphical representation of functions. The answer to the seemingly straightforward question, "How can we better teach mathematics?" really depends in large part on *which* mathematics we wish to teach.

Of course, there is no one culture or collection of cultures that intrinsically deserves a privileged position. What we have instead is a description of a uniquely ethereal culture that, it can be argued, does not currently exist. It is well described by the *Curriculum and Evaluation Standards for School Mathematics* of the National Council of Teachers of Mathematics ([NCTM], 1989; hereafter called the *Standards*), a document that is indeed proleptic in describing the mathematical behavior of the well-educated, just plain folks of the future. From the viewpoint of the *Standards*, the work represented in this volume is tautologically on target. Specifically, the NCTM Standards suggest that high school graduates should be able to "represent and analyze relationships using tables, verbal rules, equations, and graphs; translate among tabular, symbolic, and graphical representations of functions; . . . analyze the effects of parameter changes on the graphs of functions; [and for college-bound students] understand operations on, and the general properties and behaviors of, classes of functions" (1989, p. 154).

Elsewhere, the *Standards* suggest that middle-school students be able to "represent numerical relationships in one- and two-dimensional graphs" (p.

87), and to "analyze tables and graphs to identify properties and relationships" (p. 102). It is clear and not surprising that the great majority of the work reported in this volume fits comfortably within the set of goals made explicit by the *Standards*. Moreover, such work has met with a promising amount of success. Part of my purpose in this chapter is to suggest ways in which a broader set of goals can be examined and what a research agenda might look like within this broader context.

As is evident, I take seriously the position that learning mathematics is profitably viewed in terms of enculturation into a society of users of mathematics. From this viewpoint, successful learning is learning that enables students to participate in a reasonable way both in the practices and in the language games of a mathematical community. This view is compatible with current work that draws on situated cognition (Brown, Collins, & Duguid, 1989), social constructionism (Gergen & Davis, 1985), and cognitive anthropology (Carraher, Carraher, & Schliemann, 1985; Lave, 1988; Saxe, 1991) for theoretical grounding. The growing body of work undertaken from this general viewpoint has drawn our attention to the situated and cultural nature of knowledge and the importance of everyday practice to defining mathematical behavior. From this viewpoint, functions and graphs are seen as cultural artifacts, endowed with meaning only as they are used within a culture, specifically, within a culture of users of mathematics. And, as just noted, this is not in fact one culture, but several: "Just as carpenters and cabinet makers use chisels differently, so physicists and engineers use mathematical formulae differently" (Brown, Collins, & Duguid, 1989, p. 33).

In the remainder of this chapter I assume a view of learning as enculturation. I believe it to be workable, sensible, and practical in the domain of the graphical representation of functions. Moreover, I believe such a view both pays homage to the complexity of graphical representations of function and also provides enough structure to support worthwhile research efforts. This view of the cultural dependence of mathematical learning will run as an undercurrent through my treatment of each heresy in turn.

HERESY 1: THE INFLUENCE OF TECHNOLOGY: WHAT CAN BE AND WHAT OUGHT TO BE

It is hardly controversial to suggest that this book is being written because of recent advances in graphing technology. To be sure, there have always been policy makers, researchers, and teachers with particular interests in graphs as a content area. It is conceivable that, had computer and calculator-assisted graphing capabilities not become readily available, sufficient interest in the field could have been generated for this book to be written. In reality, however, it is clear that a major reason for the current interest in graphs is that technology now allows us to do a number of things we could never do before. As is pointed out in virtually every chapter, such capability is reason to stop and evaluate current practice. We have entered a necessary phase of exploring what can be done, but it will soon be time to critically examine what ought to be done, not only in terms of how we can teach, but what we should teach.

I hasten to point out that even I will not allow my curmudgeonly persona to deny graphical representations a prominent place in the curriculum. As one of several possible ways to represent the numerical relationships in a wide range of phenomena, graphical representations provide a compact and efficient means of communicating and conceptualizing for students. At the same time, a reader of this volume might share with me the sense that aspects of this conceptualizing have not been critically examined. It is not simply a question of whether graphical representations of functions should be taught. The more important questions are: What kinds of expertise with such representations are really needed? How much time should reasonably be spent in creating such expertise?

Some of the research foci discussed in this volume represent areas of clear importance for our students. Most of us would agree with Thorpe (1989) that "an understanding of graphs is required to appreciate fully the weather report on television or in the newspaper and to understand trends in such diverse areas as air quality and stock market performance" (p. 18). More generally, students should be able to interpret a graph globally—that is, to discern from a graph general properties of the relationship it represents as a whole. Representative of these tasks are those problems discussed in this volume by Dugdale (Chapter 5) and Kieran (Chapter 8), which call for students to interpret graphs of power consumption or population over time. This seems like a reasonable and valuable skill for intelligent living in our society; it is important for just plain folks. Visual (graphical) presentation of such data is compact, efficient, and of obvious value, and it is sometimes the only symbolic access available to important relationships in data. At the same time, it is worth noting that much of this kind of information is, in fact, given in bar or pie charts rather than in the kinds of graphs discussed in this volume, so links to functions in any but the most formal sense are missing. Moreover, it calls for a sort of graphing technology that allows for the plotting of data, either gathered through "probeware" or from other sources. This represents a comparatively straightforward use of graphing technology, one that seemingly approximates the use of graphing tools outside the classroom.

Part of the work done in this area attempts to help students make the connection between the phenomenological and representational world.

Eureka (Phillips, Burkhardt, & Swan, 1982) is a simple example, as is Kaput's *MathCars* (Chapter 10, this volume). This is a reasonable and promising approach. Here technology is used as a tool to capture aspects of a phenomenon in terms of a graph; students then need to make the correspondence meaningful (one obvious gap in our knowledge is the means by which this is or can be made meaningful to students, an issue discussed under Heresy 3). Making the relationship between graphs and the phenomena they represent meaningful does not require the forging of especially strong links between a graph and an algebraic representation. In fact, it is clear that the links between real-world data and the graph that represents them most often cannot be made via an intermediary algebraic representation. For this area of expertise, at least, multiple-linked representations seem unnecessary.

Thorpe (1989) made another statement that most of us would agree with as well: "Functions and graphs should go hand in hand, each reinforcing the understanding of the other" (p. 18). This statement is as good a summary of the underlying curricular assumptions guiding work in this area as can be wished for. Similarly, Leinhardt, Zaslavsky, and Stein (1990) pointed out that the different representations of function exemplify a level of abstraction significantly higher than, for example, the use of manipulatives to represent numbers:

> Although much of the prior mathematical work in the student's life may have dealt with concrete representations as the basis for learning more abstract concepts, functions and graphs is a topic in which two symbolic systems are used to illuminate each other. . . . It thus can serve as an interesting bridge topic between reasoning from the concrete to the abstract and reasoning among abstractions. (p. 3)

Thus, as discussed in previous chapters, the purpose of using graphing technology to provide multiple-linked representations for students seems to be to encourage the production of a concept image (Vinner, 1983) of function that is more flexible, more powerful, and more compelling than is currently available. In so doing, it is hoped that students can build the bridge between the concrete and the abstract more effectively. Much of the work that uses multiple-linked representations concerns flexibility and choice of representations. Traditionally (Philipp, Martin, & Richgels, Chapter 9, this volume; J. Kaput, personal communication October, 1989), the typical instructional path through the representations of functions has been from algebraic expression through ordered pairs, to graphs. Any linking that occurred among the representations tended to follow this path. This has given rise to students' beliefs that graphs were something extra, something appended to functions but not a fundamental representation of

11. SOME COMMON THEMES AND UNCOMMON DIRECTIONS

the function (Yerushalmy & Schwartz, Chapter 3, this volume). With the advent of new technology, all three representations can be thought of as a starting place for translation among representations. In addition, as Moschkovich, Schoenfeld, and Arcavi (Chapter 4, this volume) point out, the ability to begin with a graphical, as opposed to a symbolic, representation may encourage a view of function as an object, rather than as a process (this point is discussed in more detail later).

In the days before the ready availability of graphing software, I introduced the concept of *function* by dividing the chalkboard into three regions for the algebraic, tabular, and graphical representations. I felt (and feel) it is vital for students to understand that all three representations are useful, important, and related. In the same way, I feel that it is vital to understand that the long-division algorithm (or the proper sequence of key punches on a calculator) accomplish what might otherwise be accomplished by repeated subtraction or by physically dividing objects into piles. I am not yet convinced that for many students, proficiency with the links between the three representations is worth the time it might take to establish. At a time when the emphasis is on the use of division (whether algorithm or calculator-based) to solve problems, it does not seem sensible to spend too much time developing the cognitive links between two representations of that act. At a time when many scholars are questioning the need for competence in algebraic manipulations, it makes little sense to invest additional effort in forging the links to graphs that provide an intuitive feel for the correctness of manipulations.

As an illustration, suppose a clever computer program had been developed that provided a representation of the long-division algorithm using base-10 block icons linked to actions in the division algorithm itself (such a program may very well exist). As I interactively make guesses as to the number of times N would go into M, the computer would group and regroup the icons to demonstrate the current status of the division problem. That the program could be used to make the long-division process more meaningful is granted; perhaps it could even improve performance on long division tasks and give a better intuitive feel for what each step of the procedure accomplished. This is a laudable goal, viewed from the perspective of some mathematical subcultures. Still, we would naturally ask ourselves if gaining such knowledge is time efficient in a world of $5 calculators.

The analogy is not perfect (e.g., algorithms can only loosely be considered as representations, in the sense that graphs or formulas are), but the issue is much the same. We would agree that students should understand the division process in terms of a physical model; we would agree that they should be able to perform some simple algorithm, or use a calculator, to do a division problem. Should the links be made? Are the links critical to successful performance of mathematical tasks or to the accomplishment of mathematical thinking? In the same sense, we might question whether it is worthwhile for students to understand, for example, the effects of parameter shifts in functions (replacing x by nx) and their effects on graphical representations. The technology exists to help students achieve proficiency in this area. Although no one would argue that simply being able to do something is sufficient reason to begin doing it, it underscores the fact that in the absence of a comprehensive and accepted view of what it means to understand graphs of functions — one that is grounded in the practice of those who use such functions — understanding might come to be seen in terms of what technology is capable of doing.

It is arguable that other manifestations of our current preference for the graphical as opposed to algebraic representation of function is due simply to our new-found ability to produce and "zoom in on" graphs quickly and easily, rather than a reasoned judgment based on social or psychological considerations. Having such wonderful capabilities available, we tend to use them to their fullest. It is true that we can use graphical representations to solve equations quickly and easily that before could not be solved or that took a great deal of work; certainly "zoom-in" features of graphing technology allow us to do this. Students using this technology have access to problems they did not have access to before. But so do students using symbolic algebra systems and technology-based numerical methods. Granted that there are other aspects of graphs that make such a solution method appealing (and Philipp, Martin, and Richgels do a good job of describing this in Chapter 9, this volume), it seems important to distinguish what is inherently and uniquely graphical (such as the ability to see functional behavior as a whole) from that which is a result of technology (quick solutions to messy problems). It remains to be seen whether using graphing technology to gain an overall view of the functions involved in a problem, followed by other computer-aided techniques used to obtain needed answers, might be just as effective (and more efficient for many students) as relying solely on graphing technology. To me this seems to be a more reasonable approach to maintaining the links between graphical and algebraic symbol systems.

Problems Inherent In Technology

It is by now considered a truism that the introduction of graphing technology will engender its own set of misconceptions. Most of us readily admit that there will be problems introduced by graphing technology, some of which have substantial research interest. Goldenberg's (1988) often cited work suggests a few of these, and suggests that dealing with graphing technology calls for increased sophistication in interpreting visual informa-

11. SOME COMMON THEMES AND UNCOMMON DIRECTIONS 321

tion. The work reported by Moschkovich, Schoenfeld, and Arcavi in Chapter 4 certainly speaks to the complexity of the tasks involved in dealing with information on a computer screen. Beyond these perceptual and conceptual issues lie epistemological problems. The power of technology may also shift the location of intellectual authority from the teacher (which is good) to the technology itself (which is not so good). As an example, one college student who was asked to prove that the angle bisectors of a triangle were concurrent elected to use the *Geometric Supposer* (Schwartz & Yerushalmy, 1983, 1985, 1988, 1990) software to aid his exploration. As he systematically drew in the first two angle bisectors, the computer labeled the point of intersection. As the third bisector was drawn in, the computer renamed the point of intersection. The student concluded that this proved that they were the same point because "the computer wouldn't have renamed it if it weren't the same point." Other students in the same class felt that because the computer was able to generate a "truly random" figure, a demonstration that used such a figure constituted a proof. A considerable amount of effort was expended by their teacher in dislodging these notions.[2]

Of course, the problem was not in the software, nor was it insurmountable. It may even be, as Dugdale (Chapter 5, this volume) suggests, an excellent learning opportunity. Certainly the student can and should deal with what constitutes a proof, with how the software itself works, and with interpreting visual information in the context of using that software. But it underlines two important points about technological tools and the reason that I argue for one type of technological tool over another. First, it is important for such tools to be seen as *tools* and not as creators of knowledge. Students never accuse blackboards of doing mathematics; students who essentially use computers as sophisticated blackboards might. The tendency of humans to anthropomorphize computers is well documented. It is often a mildly amusing and insignificant habit. It may prove to be more significant in the context of a classroom, where students expect teachers to know and do mathematics, and may come to expect their computer to do so as well. Indeed, students may more readily expect it of their computers, because the computer may seem closer to the external, immutable world of mathematics than the more human teacher. Second, inherent in any combination of software and hardware is what Winograd and Flores (1986) called *blindness*. This refers to the inescapable fact that no tool will be infinitely flexible—tools are inevitably designed to perform certain functions and will be blind to others. A tool created to "teach" links between symbol systems, for example, may be blind to or even inhibit other kinds of understanding.

For this reason, I am less excited about technology designed for specific

---

[2]I am indebted to a colleague, Verna Adams, for this example.

teaching purposes than about that designed as a tool for problem posing or problem solving. Graphing calculators and other function plotters, data plotters, and curve-fitting software all seem more useful (perhaps because they more closely duplicate tools in the workplace) than software designed with specific educational goals in mind. They will, of course, be open to both misconceptions and blindness of their own; being tied to specific practices makes that blindness more evident and less dangerous. To the extent that educational graphing technology is designed to approximate operation as a tool for the workplace, both teachers and students will find it more flexible and ultimately more useful.

The ubiquitous importance of graphing technology in this volume presses the issue of how we might best design both tools and instruction to empower our students. And, just as software design is inherently subject to blindness, so is instructional design. I believe the answer lies first in recognizing the problem and second in having the most comprehensive possible view of what we wish to accomplish: What is it that we wish students to be able to do? In what practices should they be able to engage? In short, within which cultures of users of mathematics should they be able to enter and work comfortably? This suggests again the importance of the question of mathematical culture; it also suggests a reason why the design of instructional technology should be as sensitive to practice as possible—that is, in order to cut across the numerous cultures in which it may be used.

Prolepsis and the Problem of Design

Although it may be premature to say that the battle to get calculators into the classroom has been won, it is increasingly true that calculators are becoming part of the educational background. The questions and concerns voiced by my fellow curmudgeons of 20 years ago are rapidly receding as calculator use becomes pervasive. Thus in the NCTM *Standards* (1989), calculators are seen more as an everyday context for instruction than as either a special problem or a special opportunity. Part of the reason is that calculators are becoming part of the cultural background. They are on their way to being, in Heidegger's (1962) words, "ready-to-hand." That is, they are present as a tool, just as a carpenter's hammer is present. Very seldom are we explicitly aware of the characteristics of a hammer, and even less seldom do we debate its efficacy for the task of pounding nails. It is in this direction that graphing technology will inevitably go. As with calculators, we will eventually begin to focus not on what technology can do per se, but on how it can help us meet more important goals, such as those of the NCTM *Standards*: those that affect student's attitudes and beliefs about mathematics, their confidence, and their ability to reason and to communicate mathematically.

To carry the analogy further, mathematics educators are in the somewhat enviable position of helping to design the hammers of tomorrow. Although some insights can be gained by studying the effects of other innovative technologies through history (for example, textbooks, paper and pencil replacing chalk—see Cohen, 1987), we have little in the way of general theory to guide us. There is little question but that beliefs about the nature of mathematics, the structure of the content area, and the nature of the learner will help provide direction to this process. It seems reasonable, therefore, to begin as does Pea (1987):

> My perspective on the functions necessary for cognitive technologies thus has two vantage points. First, students are purposive, goal-directed learners, who have the will (on any given occasion or over time) to learn to think mathematically or not. Then once they have embarked on mathematical thinking, they may be aided by technologies in mathematical thinking. (p. 100)

If we take this view of the learner, what does it imply about our design of cognitive technologies?

The Importance of a Focus on Human Activity. Little is known about how the function concept as a written mathematical artifact relates to the function concept as used in practice, and still less about how graphing tools might affect this. Any speculation regarding the effects of technology, and indeed any attempt to do research in the context of technology, is compounded by the rapidity with which technology advances. We are in a very real sense in the position of trying to design what is constantly changing; we are constantly defining students' knowledge in terms of technological contexts that are rapidly growing obsolete. There is no way around this difficulty, but it does underscore the importance of a proper view of what technology can do. It is for this reason that an understanding of human activity as it relates to the function concept is so vital. Winograd and Flores (1986) treated the problem of computer system design extensively and conclude that such design is always inextricably linked with human activity and human concerns. To return to the example of the hammer, it is vital to understand the action of driving nails as we design and redesign new hammers. For this reason, it is important to understand how both functions, and their graphical representations, are conceived and used, not only in school situations but in the work place; we need more careful study of everyday cognition as it relates to the use of graphs by both teachers and students, and by the user of mathematics outside the school context.

Dealing With Breakdown. Winograd and Flores (1986) made another point that is worth emphasizing in this context. By nature of the way technological tools are designed, the potential for *breakdown* is always present. Here, breakdown means an interruption in the flow of our experience that causes us to stop and look more closely at the tools we use, just as when the wrench we are using is too small. Tools have their inherent limitations, and the problem of selecting among tools so as to avoid breakdown, as well as the process of recovery from such breakdowns, must be a part of the circle of design, implementation, and redesign. Dugdale (Chapter 5, this volume) and others (Goldenberg, 1988) have provided evidence of a number of misconceptions that students develop when using graphing tools. Certainly we can and should attempt through design and instruction to minimize these misconceptions. However, from the point of view of Winograd and Flores (1986), these "breakdowns" are inevitable. The issue then becomes, as Dugdale suggests, teaching students how to recover from these breakdowns—how to negotiate meaningfully and sensibly with the world of mathematics. A large part of this may be, as Moschkovich, Schoenfeld, and Arcavi (Chapter 4) suggest, helping students to see what to focus on. Every representation is imperfect, and every analogy is also a disanalogy. Much as students must learn when to adopt one representation and let go of another, so they must learn to attend to useful aspects of a graphing tool and to let go of irrelevant aspects.

Summary

I have argued that, partly because we as a discipline are in an exploratory phase with regards to graphing technology, decisions are by default being made based on what we *can* rather than what we *should* do. At the same time, what can be done rapidly changes, and there seems to be little if any guidance in deciding what should be done. I have suggested that the problem of designing instruction for the future involves a careful study of human activity and concerns and that it will in fact never be complete or even find itself on solid ground. However, keeping the design process constantly within the dialogue of what the various users of mathematics do provides what I feel is a reasonable and workable means to inform educational practice.

HERESY 2: THE IMPORTANCE OF DEALING WITH CONTENT

Efforts to define a mathematical *content domain* can range from a simple delineation of a subset of the traditional curriculum (so that specifying *graphical representation of function* is adequate) to more involved analysis of the structures, interrelationships, and applications surrounding the

11. SOME COMMON THEMES AND UNCOMMON DIRECTIONS 325

subset in question. A simple delineation of the first type is sufficient to gather together a group of teachers and researchers with a common interest. It is also sufficient for performing an historical analysis of what graphs and functions are. Certainly this has been done in the current volume. Reviews in this work have outlined the evolution of the function concept in history, specifically the more recent return from a view of function as a set of ordered pairs to a view of function as a relationship or an expression of dependence. Current educational thought has in general tended to move away from the logical structure of mathematics as a guide to provide meaning for mathematical content, focusing instead on its usefulness in expressing regularities and patterns (National Council of Teachers of Mathematics, 1989; Steen, 1990). Such a change has been recommended strongly in recent years (Thorpe, 1989), and is probably a welcome one.

A second aspect of mathematical function addressed by several authors in this volume—Philipp, Martin, and Richgels, (Chapter 9); Cooney and Wilson, (Chapter 6); and Demana, Schoen, and Waits (Chapter 2)—has been the evolution of the function concept as portrayed in school textbooks. Here again, the sorts of changes in the view of function just discussed have taken place, but in addition there is reported a nearly ubiquitous treatment of function from the symbolic rather than the graphical point of view. This tendency seems to have been stronger in the past than is currently the case and modern textbooks contain substantially more material dealing with graphs. However, with some few exceptions, this material is often poorly connected to the symbolic treatment of function. Thus, graphs are treated, as suggested by Yerushalmy and Schwartz (Chapter 3), as something external to the function, as extra baggage that the function carries, and not as a primary means of accessing functional information. The treatment of this belief using available graphing technology is a common theme throughout the current literature on graphs and functions.

Sprinkled throughout the chapters are important and illuminating aspects of function that become even more cogent in the context of a graphical representation. End behaviors, extrema, intersection points, and so forth, all help to describe why function is a core mathematical concept and why learning to use functions is a worthy goal. The use to which functions are put as described by Philipp, Martin, and Richgels (Chapter 9) is enough to convince even a skeptic like myself that graphical representations have something to offer and might even have primacy over algebraic representations in certain cases. But the treatment of function and graph, viewed as mathematical or even as phenomenological objects, remains sketchy in the research as a whole.

An example of the second, more careful way of specifying a content domain is the treatment given to the structure of addition and subtraction word problems in, for example, Carpenter and Moser (1983). This analysis linked word problems with the phenomena they model, providing a common ground for discussion and a framework that was both accessible and applicable to the bulk of the domain. It is obvious that a careful analysis of this type provides more fertile ground for research; it is equally obvious that it entails a great deal of careful and thoughtful work. Kaput (in Chapter 10) points out that such an analysis is critical to the success of programs aimed at helping teachers to provide instruction based on student cognition. He also rightly points out that the success of this approach with addition and subtraction word problems and, more recently, with rational number concepts depends on the comparative simplicity of the domains — the direct reflection of lived experience in the mathematical structures. That this is not true of graphical representations of functions is also obvious. This is a result of the comparative complexity of the functional domain and the fact that, unlike more everyday forms of mathematical thinking, the use of graphical representations of function is a comparatively esoteric skill, without obvious analogues in everyday practice. Thus, where addition, subtraction, and even proportional reasoning can be fairly rich phenomenologically, graphs seem to be substantially removed from their phenomenological base (Kaput, Chapter 10).

Still, it is likely a more in-depth analysis of the content domain would prove useful (albeit difficult). One possible first step in dealing with this content area might be to follow the lead of Vergnaud and attempt to specify the *conceptual field* for functions and graphs. According to Vergnaud (1983), this constitutes a "set of problems and situations for the treatment of which concepts, procedures, and representations of different but narrowly interconnected types are necessary" (p. 127). This approach, once taken in large part to the domain of addition and subtraction of whole numbers and, more recently, to the multiplicative domain, forms the basis for much of the current work in these fields. As Kaput pointed out in the initial meeting of the group out of which this book evolved, however, the range of situations to which we can apply even simple linear proportions, let alone functions in general, is extremely large. It is probable that the sheer magnitude of ways in which functional thinking might be used in problem solving, mathematizing, exploring, or communicating precludes the sort of analysis of the content area of functions and their graphs that so simplified the work in addition and subtraction. This is to be expected if function is, as is often claimed, the foundational concept of so much of modern mathematics (Yerushalmy & Schwartz, Chapter 3, this volume). Virtually every chapter of this volume provides a number of examples of how functions may be applied to a wide variety of interesting problems; however, there seems to be no comprehensive framework by which to classify and categorize the uses of function and graphs, and nothing akin to the specification of problem types, which was foundational to the work of

Carpenter and Moser (1983), likely to be forthcoming for the functional domain.

In summary, at the same time I recognize the extreme difficulty, or even impossibility, of providing a content analysis for the graphical representation of functions, I also recognize the need for something like it to inform research. One alternative approach to this content domain would be careful anthropological studies of how functions and their graphs are used in everyday practice by the various cultures that employ them. Of course, such studies would still face the dizzying number of possible applications of functions in the everyday practice of many mathematical subcultures. It is also true that scholars hoping to build explicit models of expert (or novice) behavior may be unsatisfied with the methods and underlying assumptions basic to such studies. Still, the insights gained would certainly be of value and may well be the most reasonable approach to answering the question of what constitutes the content domain of functions and their graphs.

What I am suggesting here is more than a traditional laundry list of the "when will we ever use this" type, and more than an appeal to what Kliebard (1987) termed the *social efficiency* tradition of curriculum design; it is not an appeal to provide for students skills specific for extant professions. What I have in mind grows more from a tradition of Continental philosophy[3], and has an impact on our study of understanding as well. A first step might be a more in-depth analysis following the lead of Freudenthal's *didactic phenomenology* (1983). Further steps would follow the suggestion (Heidegger, 1962) that understanding graphs involves disclosing graphs *as* something, with this *as-structure* inextricably linked to the practical, everyday concerns of humans as they go about their business: (a) graphs *as* a means of communicating my plan to my employer, (b) graphs *as* a means of visualizing probable growth of my company, (c) graphs *as* a way of convincing the court of illegal hiring practices. Again, I am calling for an analysis of graphs based on practices that involve graphs within the various mathematical cultures that use them. It would focus on the meaning graphs have for those who use them. Rather than focusing on the potentially large number of component skills, it would attempt to have the ways in which graphs are used reveal themselves. It is in harmony with the way Putnam, Lampert, and Peterson (1990) described understanding as situated cognition:

> In terms of what it means to know and understand mathematics, this perspective suggests that the nature of a person's knowledge of mathematics is inextricably tied to the situations in which that knowledge was acquired. . . . It also suggests that meaningful knowledge of mathematics

---
[3]The tradition represented, for example, by Hegel, Heidegger, Gadamer, and Wittgenstein.

cannot be crystallized and made entirely explicit. Important aspects of knowing mathematics will inherently remain implicit and intertwined with situations in which is it used. (p. 94)

Thus, to understand (as researchers) the content domain of the graphical representation of functions, we need to understand the practice of those using them. In this view, such practice is inherently tied to how graphs are understood. This again is in keeping with a Continental tradition, because it supposes that graphs are not best seen as objects out there that need to be interpreted, but rather as tools that present themselves as already meaningful to a user, as already useful in everyday practice. This is the essence of the graph, or of any mathematical object, and hence a proper starting place for a study of subject matter.

HERESY 3: LEARNING AND UNDERSTANDING: A COGNITIVE VIEW OF UNDERSTANDING

As will now be clear, problems not unlike those that arise in describing the content domain also arise as we attempt to identify the cognitive processes critical for successful performance (Hiebert & Wearne, 1991) in the functional domain. Partly because of the complexity of the functional domain, it is difficult to describe exhaustively the constellation of procedural and conceptual understandings that underlie competent performance. Much of what we do know about students' cognition is from isolated studies done in differing theoretical contexts. And, as will be noted, much of it lacks the specificity that has been the hallmark of cognitive science research in general and that has made such research of great value in other domains. In part, this dearth of comprehensive research programs is the very situation that this volume hopes to ameliorate, by delineating what has been done and what remains to be done.

Let us pause for a moment to ask one version of what is a very popular question: What does it mean to know and do mathematics *as it relates to the graphical representation of functions*? To provide an answer as I have in terms of practices within a mathematical culture begs the question. What are the mathematical practices, relevant to functions and graphs, in which we hope our students will be able to engage? Three answers predominate in the current research and emerge from the chapters of this volume, though these are certainly not the only answers. The first deals with the qualitative interpretation of graphs and was discussed under *Heresy 1*. This seems very amenable to the kind of analysis I suggested in the last section. Interpretation seems very much embedded in most reasonable uses of a graph and can probably be effectively teased apart by more cognitively focused studies.

11. SOME COMMON THEMES AND UNCOMMON DIRECTIONS   329

A second theme is that students should be able to move between viewing function as a *process* and viewing function as an *object* (Sfard, in press; Moschkovich, Schoenfeld, & Arcavi, Chapter 4, this volume). It is certainly true that expert users display this ability, and it is clear from the work reported in Chapter 4 that it is vital even for students performing comparatively simple tasks to be able to move comfortably between these two perspectives. It is less clear that such an ability is necessary for the practice of just plain folks, although it should be within their mathematical reach. Here again, researchers have come to hope that graphing technology, which can capture a function as a graph and render it able to be manipulated, will help to facilitate the movement from process to object (and vice versa, presumably). Graphing technology provides one among many methods that would aid this ability (see Ayers, Davis, Dubinsky, & Lewin, 1988, for another approach). Again, it seems important to separate the inherently graphical from what is essentially technological. There may be hope that visual processing will somehow make a difference. But it is not clear why a graph is any more "objectified" than a closed algebraic formula, or why operations on graphs will be any more meaningful than operations on formulas. Two different formulas may indeed represent the same function, just as two different fractions may represent the same rational number. To me this does not mean that either representation is less an "object." Rather, it suggests that symbolic notation, whether representing functions or rational numbers, is in some sense richer than graphical representations in its ability to encompass both process and object perspectives. Granted, graphs may be another means by which to make this distinction; there may be value in a second more visual treatment, particularly for some students. But it will take considerable research effort to show why twice, or even three times, will be the charm, or why graphical representations, more than others, will aid the process. Yerushalmy and Schwartz's research (Chapter 3, this volume) leads in this promising direction. Their work, and that reported by Moschkovich, Schoenfeld, and Arcavi (Chapter 4), is a significant first step in unpacking the cognitive demands made in using both perspectives in appropriate ways. In bringing to the forefront the Cartesian connection, they offer a compelling argument that competent behavior requires flexibility in moving between representational modes and perspectives. A great deal more work is needed to establish the role graphing technology can play in this process.

A third and final theme emerging from this volume is the importance of being able to move comfortably between and among the three different representations of function: algebraic, graphical, and tabular. Indeed, it is almost axiomatic in the literature that understanding of functions and graphs entails this skill. Yerushalmy and Schwartz (Chapter 3, this volume) and Moschkovich, Schoenfeld, and Arcavi (Chapter 4) make compelling arguments that such skill is needed for competent behavior at even fairly elementary tasks. I have argued that, although this particular type of understanding is valuable, it is instantiated in ways that are less universally important than is sometimes supposed (e.g., examining the results of replacing x by nx or by $x+n$). Moreover, although such understanding is ultimately valuable, it is not clear that the only, or even the best, way to arrive at such understandings is by way of graphing tools that forge specific links. The work reported in Chapter 4 seems to rely in large part on tutors guiding a process of reflection on examples. The power of graphing technology comes in providing those examples easily, not in dynamic linkages between symbol systems.

Supporting these three emerging themes are a number of interesting and important research studies on student cognition (for reviews, see Dugdale, Chapter 5, this volume; Norman, Chapter 7; and Kieran, Chapter 8. At least part of the value of studies such as those of Dreyfus and Eisenberg (1982), Dugdale (1987), Janvier (1981), Karplus (1979), Kerslake (1977), and Vinner and Dreyfus (1989) is that they bring to light ways in which students' basic knowledge is meager or remains inaccessible. Thus we know, for example, that students confuse graphs with pictures of physical situations (Kerslake, 1977), are unable to interpret graphs arising from physical situations (Dugdale, 1984), have concepts of function strongly tied to linearity (Markovits, Eylon, & Bruckheimer, 1983), and have problems dealing with scale (Vergnaud & Errecalde, 1980). Such knowledge of students' conceptions forms the basis for designing instruction that both enriches their knowledge base and enables them to make better use of alternate representations of functions. They form a corpus of important first steps in describing student understanding.

Attempting to gain a more complete understanding of the cognitive processes underlying functions and graphs almost certainly calls for careful analysis of behavior (both expert and novice) on mathematical tasks, for analysis of typical developmental growth in understanding, and often for analysis of the tasks themselves for clues regarding needed procedural knowledge (Hiebert & Wearne, 1991). The ability to model observed behavior (either expert or novice) in terms of explicit rules, often embodied in programs or production systems, is seen as the fundamental proof that such knowledge has been captured. As Wenger (1987) states:

> Such "production rule" descriptions, even incomplete ones . . . , make the forms of understanding sought significantly more explicit than other types of descriptions do. This explicitness can be exploited to develop text (and computer-based) materials which support students' attempts to infer salient patterns and their use in planning strategies and recalling relevant procedures for carrying them out. (p. 238)

Such a view of understanding is in harmony with the analysis of tasks in Leinhardt, Zaslavsky, and Stein (1990) and Swan (1982) mentioned in this volume, as well as some of the literature on students' conceptions and misconceptions regarding functions and their graphs. What seems to be lacking in the literature on knowledge structures is the sort of careful analysis of behavior that could give rise to computer simulations of graph-related behaviors. Whatever the ultimate value of such simulations might be (see, e.g., Cobb, 1990), the sort of thoughtful analysis that makes such simulations possible has proven valuable in other domains. The lack of such work in the domain of functions and graphs may be in part due to the complexity of the domain, as previously mentioned. Still, careful analysis of graphing tasks, graph comprehension tasks, translation tasks, and so forth, have yet to be synthesized in even a preliminary model of graph-related behaviors; a great deal of work remains to be done toward this end. One way of beginning this next step would be to do a careful analysis of what might be called a graph interpretation schema. The components of such a schema are conspicuously lacking and provide a wide open field for further research. Such analysis may proceed through the analysis of behaviors of experts and novices as they interpret graphs, or through studies of the development of expertise in students over time. Such research will certainly prove to be as interesting and valuable as it will be difficult.

A final caveat regarding the type of research I have suggested seems to be in order. It is true that careful analysis of tasks, in terms of students' actual performances, is of great theoretical interest. It seems intuitively reasonable that such analysis should lead to prescriptions for instruction. This second assumption seems less defensible. To borrow the analogy from Moschkovich, Schoenfeld, and Arcavi (Chapter 4), it may be of great theoretical interest to understand the processes involved in riding a bicycle. It does not follow that theoretical knowledge of these facts instantiated in a series of algorithms will make learning to ride any easier. Indeed, the work of Dreyfus and Dreyfus (1986) suggested that use of this knowledge in a rule-governed way, as is usually done in production systems, yields performance at only novice or at best advanced beginner levels. Truly expert behavior is characterized instead by more situation-dependent, intuitive involvement, which moves away from the rules needed by novices. In short, although I have called for careful studies of the ways in which both experts and novices "know that," I would encourage with equal intensity careful studies of experts' "know how."

A Phenomenological View of Understanding

In addition to the more traditional cognitive research just discussed, there seems to also be a need for research that sheds some light on how students make sense of functions and graphs in their own personal world. Such issues are, I believe, important because they bring to light the meanings that graphs and functions have for students and reflect the ways in which linguistic, experiential, and informal knowledge are brought to bear upon mathematical thinking. These meanings are distinct from those given to graphs and functions within the mathematical subcultures just discussed, being based on different practices and cultural understandings. The kinds of cultural and experiential predications that students bring forth to lend meaning to graphs need more careful attention than they have received.

The recent study by Schoenfeld, Smith, and Arcavi (in press) followed the development of one student's understanding of linear functions. The study reflected a careful analysis of the student's production of transitional conceptions between a less- and a more-complete understanding of the relation between slope, intercepts, and equations of linear functions. The authors may be surprised at my characterization of their study as shedding light on how students extend meaning (in the sense I have discussed) to a mathematical concept. However, the level of detail at which the concepts were studied is one that is appropriate for such analysis. The work exhibits care in attempting to study meaning, taking into regard the student's phenomenal world as well as the mathematics involved. It also demonstrates that the kind of research I am calling for can be done from numerous methodological and theoretical perspectives, as long as care is taken to uncover the understandings that students call on in making mathematical practice meaningful to them.

A second example is represented by recent work on students' understanding of what might be called the topology of the real line (and hence of graphs): the relationship between curve and point, and the connectedness and continuity of curves (Williams & Walen, 1992; Williams, Walen, & Cockburn, in preparation). In this study, students in the fifth month of a graphing calculator-based algebra curriculum were interviewed as to their understanding of graphs, points, the process of "zooming in," and other aspects of graphs relating to connectedness. Students displayed some expected alternative conceptions: ". . . there is like a point, and in that point there are many points, and in that point there are more points" or ". . . everything is infinite space between two points." One intriguing aspect of this work is its relation to expert understanding of these concepts. Leaving aside the formal definitions of connectedness or continuity, it is not clear that an expert would have any less difficulty describing the relationship between points and lines.

> Investigator: So when you graph, do you think of it more as plotting points or more as drawing a curve?

Student: It's more of a curve. That's kind of a trick because they are points, each one are [sic] points but it's a continuous line because the points are infinite, you can't really make points.

It indeed appears to be extremely tricky cognitive territory. Although it is somewhat removed from the understandings used in applying graphs, the study of such territory requires a kind of analysis that is more deeply rooted in the rich phenomenological world on which students draw to make mathematics meaningful. This is a level of analysis that I believe is both valuable and different from that commonly seen.

Learning and Teaching

Many of the studies that shed light on students' knowledge structures occurred in situations where the researchers were primarily interested in learning; of these, many employed graphing technology. In general, learning is viewed in these studies as both the assimilation of new knowledge into the extant knowledge structure and as the making of connections to enrich the knowledge structure (Rumelhart & Norman, 1978); these connections might include connections between the representations already discussed. As Kieran (Chapter 8, this volume) suggests, the outlook for such learning is in general optimistic. At the same time, the kinds of learning situations and teaching that give rise to the successful learning of functions and graphs do not seem specific to this content area, nor startlingly new. Indeed, most of the studies seem to echo the National Council of Teachers of Mathematics' *Professional Standards for Teaching Mathematics* (1991), which emphasizes the importance of choosing worthwhile mathematical tasks, providing opportunities for discourse, and using the appropriate tools to encourage learning. It is often reported in these studies that fairly straightforward classroom tasks (e.g., worksheets), which provide experience with making links among different concepts and procedures, are effective in enriching student understanding (Dugdale, 1987; Schwarz, Dreyfus, & Bruckheimer, in press). Dreyfus and Halevi (1988), for example, emphasize the importance of the tasks the teacher chooses, the role of the teacher as guide and resource, and the technological environment in promoting learning. Although it is difficult to tell, based on the information available in Chapter 4, the tutors in the work of Moschkovich, Schoenfeld, and Arcavi seem to offer a straightforward kind of assistance to their students, helping to focus attention, or asking students how they decided on an answer. They appear to assume a role more or less compatible with descriptions of cognitive apprenticeship in the literature (Burton, Brown, & Fisher, 1984; Collins, Brown, & Newman, 1989).

In some sense, little attention has been given to the topic of teaching graphical representations of function. The excellent reviews provided in this volume by Cooney and Wilson (Chapter 6) and by Norman (Chapter 7) make it clear that there is very little research evidence dealing specifically with the knowledge or beliefs of teachers in this content domain. Aside from what is skillfully gleaned in both of these chapters from related work, little can be said, and little can be gained by further belaboring the point. What can be said is that good teaching—that is, the careful choice of mathematical tasks coupled with efforts to establish an environment for exploration, conjecture, and dialogue—has been shown in a number of studies to be successful in leading students to understanding functions and their graphs. This is not revelatory, but it is of some comfort.

SUMMARY

I have enjoyed the role of iconoclast is this chapter, sometimes overstating my case in an effort to make a point (but it has not been nearly as much fun as playing with computers). I have suggested that our research has been driven in large part by a desire to explore the possibilities of technology, without sufficient attention to what is being taught and why. I have suggested that in the absence of theories to guide us, we have tended to define understanding in the domain of functions and graphs in terms of what computers allow us to do. And, I have suggested that we have neglected careful analysis of the ways in which functions and graphs are disclosed and put to use in the mathematical cultures our students will need to join. The common thread running through my three heresies is that, as we engage in proleptic research (see Kaput, Chapter 10, this volume), we must be sure it is grounded in everyday, lived, phenomenological experience, viewed from the cultures we feel are most important for our students. We have, in the NCTM *Standards*, a good beginning at describing a sort of common culture for mathematics. My belief is that, as we flesh out the research implications from within the dynamic culture described by the NCTM *Standards*, we will benefit from more careful attention to the practices of lived experience. In this I agree with Lave (1988) that "If everyday practices are powerful it is because they are ubiquitous. If ubiquitous, they are synomorphically organized and sites of the direct, persistent, and deep experience of whole-persons acting. These seem to be crucial conditions for efficacious human activity" (p. 190).

As we attempt to provide our students access to mathematical power, a view of whole persons acting seems to me to be both useful and necessary.

REFERENCES

Ayers, T., Davis, G., Dubinsky, E., & Lewin, P. (1988). Computer experiences in learning composition of functions. *Journal for Research in Mathematics Education, 19*(3), 246-259.

Belmont, J. M., & Butterfield, E. C. (1977). The instructional approach to developmental cognitive research. In R. V. Kail, Jr. & J. W. Hagen (Eds.), *Perspectives on the development of memory and cognition* (pp. 437-481). Hillsdale, NJ: Lawrence Erlbaum Associates.

Brown, J. S., Collins, A., & Duguid, P. (1989). Situated cognition and the culture of learning. *Educational Researcher, 18*(1), 32-42.

Burton, R. R., Brown, J. S., & Fisher, G. (1984). Skiing as a model of instruction. In B. Rogoff & J. Lave (Eds.), *Everyday cognition: Its development in social context* (pp. 139-150). Cambridge, MA: Harvard University Press.

Carpenter, T. P., & Moser, J. M. (1983). The acquisition of addition and subtraction concepts. In R. Lesh & M. Landau (Eds.), *Acquisition of mathematics concepts and processes* (pp. 7-44). New York: Academic Press.

Carraher, T. N., Carraher, D. W., & Schliemann, A. D. (1985). Mathematics in the streets and in schools. *British Journal of Developmental Psychology, 3*, 21-29.

Cobb, P. (1990). A constructivist perspective on information-processing theories of mathematical activity. *International Journal of Educational Research, 14*, 67-92.

Cohen, D. K. (1987). Educational technology, policy, and practice. *Educational Evaluation and Policy Analysis, 9*(2), 153-170.

Collins, A., Brown, J. S., & Newman, S. E. (1989). Cognitive apprenticeship: Teaching the crafts of reading, writing, and mathemataics. In L. B. Resnick (Ed.), *Knowing, learning, and instruction: Essays in honor of Robert Glaser* (pp. 453-494). Hillsdale, NJ: Lawrence Erlbaum Associates.

Dreyfus, H. L., & Dreyfus, S. E. (1986). *Mind over machine: The power of human intuition and expertise in the era of the computer.* New York: The Free Press.

Dreyfus, T., & Eisenberg, T. (1982). Intuitive functional concepts: A baseline study on intuitions. *Journal for Research in Mathematics Education, 13*, 360-380.

Dreyfus, T., & Halevi, T. (1988, July). *Quadfun—A case study of pupil computer interaction.* Paper presented to the theme group on Microcomputers and the Teaching of Mathematics at the Sixth International Congress on Mathematical Education, Budapest, Hungary.

Dugdale, S. (1984). Some computer applications for the pre-college mathematics and science curriculum. *Technology in Education and Training: Planning and Management*, Information Dynamics, Inc.

Dugdale, S. (1987). Pathfinder: A microcomputer experience in interpreting graphs. *Journal of Educational Technology Systems, 15*(3), 259-280.

Freudenthal, H. (1983). *Didactical phenomenology of mathematical structures.* Dordrecht: D. Reidel.

Gergen, K. J., & Davis, K. E. (Eds.) (1985). *The social construction of the person.* New York: Springer-Verlag.

Goldenberg, E. P. (1988). Mathematics, metaphors, and human factors: Mathematical, technical, and pedagogical challenges in the educational use of graphical representation of functions. *Journal of Mathematical Behavior, 7*, 135-173.

Heidegger, M. (1962). *Being and time.* (J. Macquarrie & E. Robiuson, Trans.). New York: Harper & Row. (Original work published 1926)

Hiebert, J., & Wearne, D. (1991). Methodologies for studying learning to inform teaching. In E. Fennema, T. P. Carpenter, & S. J. Lamon (Eds.), *Integrating research on teaching and learning mathematics* (pp. 153-176). Albany: SUNY Press.

Janvier, C. (1981). Difficulties related to the concept of variable presented graphically. In C. Comiti & G. Vergnaud (Eds.), *Proceedings of the Fifth International Conference for the Psychology of Mathematics Education* (pp. 189-192). Grenoble, France: Laboratoire I.M.A.G.

Karplus, R. (1979). Continuous functions: Students' viewpoints. *European Journal of Science Education, 1*(3), 397-415.

Kerslake, D. (1977). The understanding of graphs. *Mathematics in Schools, 6*(2), 22-25.

Kerslake, D. (1981). Graphs. In K. M. Hart (Ed.), *Children's understanding of mathematics: 11-16* (pp. 120-136). London: John Murray.

Kliebard, H. M. (1987). *The struggle for the American curriculum, 1983-1958.* London: Routledge & Kegan Paul.

Lave, J. (1988). *Cognition in practice.* New York: Cambridge University Press.

Leinhardt, G., Zaslavsky, O., & Stein, M. K. (1990). Functions, graphs, and graphing: Tasks, learning, and teaching. *Review of Educational Research, 60*(1), 1-64.

Markovits, Z., Eylon, B.-S., & Bruckheimer, M. (1983). Functions—linearity unconstrained. In R. Herschkowitz (Ed.), *Proceedings of the Seventh International Conference for the Psychology of Mathematics Education* (pp. 271-277). Rehovot, Israel: Weizmann Institute of Science.

National Council of Teachers of Mathematics (1989). *Curriculum and evaluation standards for school mathematics.* Reston, VA: Author.

National Council of Teachers of Mathematics (1991). *Professional standards for teaching mathematics.* Reston, VA: Author.

Pea, R. D. (1987) Cognitive technologies for mathematics education. In A. H. Schoenfeld (Ed.), *Cognitive science and mathematics education* (pp. 89-122). Hillsdale, NJ: Lawrence Erlbaum Associates.

Phillips, R. J., Burkhardt, H., & Swan, M. (1982). *Eureka.* [Computer-based software]. St. Albans, Herts., UK: Crown.

Putnam, R. T., Lampert, M., & Peterson, P. L. (1990). Alternative perspectives on knowing mathematics in elementary schools. In C. Cazden (Ed.), *Review of research in education* (Vol. 16, pp. 57-150). Washington, DC: American Educational Research Association.

Rumelhart, D. E., & Norman, D. A. (1978). Accretion, tuning, and restructuring: Three modes of learning. In J. W. Cotton & R. A. Klatzky (Eds.), *Semantic factors in cognition* (pp. 37-53). Hillsdale, NJ: Lawrence Erlbaum Associates.

Saxe, G. B. (1991). *Culture and cognitive development: Studies in mathematical understanding.* Hillsdale, NJ: Lawrence Erlbaum Associates.

Schoenfeld, A. H., Smith, J. P., & Arcavi, A. (in press). Learning: The microgenetic analysis of one student's evolving understanding of a complex subject matter domain. In R. Glaser (Ed.), *Advances in instructional psychology* (Vol. 4). Hillsdale, NJ: Lawrence Erlbaum Associates.

Schwarz, B., Dreyfus, T., & Bruckheimer, M. (in press). The Triple Representation Model curriculum for the function concept. *Computers and Education.*

Schwartz, J. L., & Yerushalmy, M. (1983, 1985, 1988, 1990). *The Geometric Supposer* [Computer software and teachers' guide]. Sunburst Communications, Pleasantville, NY.

Sfard, A. (in press). On the dual nature of mathematical conceptions: Reflections on processes and objects as different sides of the same coin. *Educational Studies in Mathematics.*

Sowder, J. T. (1989). *Setting a research agenda.* Hillsdale, NJ: Lawrence Erlbaum Associates.

Steen, L. A. (1990). Pattern. In L. A. Steen (Ed.), *On the shoulders of giants: New approaches to numeracy* (pp. 1-10). Washington, DC: National Academy Press.

Swan, M. (1982). The teaching of functions and graphs. In G. van Barneveld & H. Krabbendam (Eds.), *Proceedings of the Conference on Functions* (pp. 151-165). Enschede, The Netherlands: National Institute for Curriculum Development.

Thorpe, J.A. (1989). Algebra: What should we teach and how should we teach it? In S. Wagner & C. Kieran (Eds.), *Research issues in the learning and teaching of algebra* (pp. 11–24). Hillsdale, NJ: Lawrence Erlbaum Associates.

Vergnaud, G. (1983). Multiplicative structures. In R. Lesh & M. Landau (Eds.), *Acquisition of mathematics concepts and processes* (pp. 127–174). New York: Academic Press.

Vergnaud, G., & Errecalde, R. (1980). Some steps in the understanding and use of scales and axis by 10–13 year-old students. In R. Karplus (Ed.), *Proceedings of the Fourth International Conference for the Psychology of Mathematics Education* (pp. 285–291). Berkeley: University of California.

Vinner, S. (1983). Concept definition, concept image, and the notion of function. *International Journal of Mathematics Education in Science and Technology, 14*(3), 293–305.

Vinner, S., & Dreyfus, T. (1989). Images and definitions for the concept of function. *Journal for Research in Mathematics Education, 20*(4), 356–365.

Wenger, R. H. (1987). Cognitive science and algebra learning. In A. H. Schoenfeld (Ed.), *Cognitive science and mathematics education* (pp. 217–251). Hillsdale, NJ: Lawrence Erlbaum Associates.

Williams, S. R., & Walen, S. B. (1992, August). *Conceptual splatter and metaphorical noise: The case of graph continuity*. Paper presented at the meeting of the International Group for the Psychology of Mathematics Education, Durham, NH.

Williams, S. R., Walen, S. B., & Cockburn, C. (in preparation). *Informal cognitive models of graph continuity*.

Winograd, T., & Flores, F. (1986). *Understanding computers and cognition: A new foundation for design*. Norwood, NJ: Ablex.

Author Index

A

Appel, M., 103, *130*
Arcavi, A., 45, *67*, 73, 74, 75, 86, *99*, *100*, 124, *130*, 233, *236*, 332, *336*
Assessment of Performance Unit (APU), 204, *234*
Austin, G.A., 161, *183*
Ayers, T., 329, *335*

B

Ball, D.L., 6, *8*, 131, *156*, 162, *182*
Banchoff, T., 308, *310*
Barber, H.C., 248, *275*
Barclay, W.L., 113, 127, *129*
Barrett, G., 146, *156*
Beckenbach, E.F., 143, *157*
Becker, H., 284, *310*
Bednarz, N., 43, *67*
Begle, E.G., 161, *182*
Behr, M.J., 162, 163, 177, *185*
Belanger, M., 43, *67*
Bell, A., 168, *182*
Belmont, J.M., 314, *335*
Bereiter, C., 291, *312*
Berliner, D.C., 161, *182*
Berman, S.L., 142, *156*
Bernoff, R., 103, *130*
Betz, W., 141, *156*, 248, *275*

Bidwell, J.K., 245, *275*
Bittinger, M.L., 144, *158*
Bohren, J.L., 271, *275*
Boileau, A., 231, *235*
Borovoy, R., 300, *310*
Boyer, C., 285, 286, *310*
Breidenbach, D., 72, *99*
Breslich, E.R., 140, *156*, 246, 247, *276*
Brophy, J.E., 7, *8*
Brown, C.A., 162, *183*
Brown, J.S., 107, *129*, 316, 333, *335*
Brown, S.I., 131, *156*
Browning, C., 30, *38*
Bruckheimer, M., 19, *39*, 167, *184*, 196, 216, 217, *236*, 330, 333, *336*
Bruner, J.S., 161, *183*
Buck, R.C., 139, *156*
Burkhardt, H., 104, 106, 107, *130*, 199, *234*, 318, *336*
Burton, R.R., 333, *335*
Butler, C.H., 140, *156*
Butterfield, E.C., 314, *335*

C

Carey, D.A., 131, *156*
Carpenter, T.P., 3, *8*, *9*, 131, *156*, 160, 173, 174, 175, 180, *183*, *185*, 203, 204, *234*, 248, *277*, 325, 327, *335*

339

AUTHOR INDEX

Carraher, D.W., 316, *335*
Carraher, T.N., 316, *335*
Charles, R.I., 144, *158*
Chiang, C.-P., 173, *183*
Clark, C.M., 7, *8*, 132, 148, 149, 154, *156*, 161, *183*
Clark, J.R., 141, *158*
Clasen, R.G., 245, *275*
Clement, J., 196, *237*, 268, *276*
Cobb, M.V., 246. *277*
Cobb, P., 331, *335*
Cockburn, C., 332, *337*
Cohen, D.K., 323, *335*
College Entrance Examination Board (CEEB), 139, 145, *156*
Collins, A., 107, *129*, 316, 333, *335*
Comstock, M., 22, *38*
Confrey, J., 292, 293, *310*
Cooney, T.J., 131, *156*, 132, *156*, 170, 172, *183*, *184*
Corbitt, M.K., 203, 204, *234*
Coxford, A.F., 243, *276*
Crosswhite, F.J., 140, *157*, 243, 244, 246, *277*

D

Davis, G., 329, *325*
Davis, K.E., 316, *335*
Davis, R.B., 107, 122, *129*, *130*, 146, *156*, 191, *234*
Demana, F., 12, 21, 22, 24, 28, 35, *38*, 145, *156*, 227, *237*, 249, 261, 264, *276*, *278*, 199, *310*
Devaney, R., 308, *310*
Dick, T.P., 225, 226, *234*
Dion, G., 28, *38*
Dolciani, M.P., 142, 143, *156*, *157*
Dossey, J.A., 70, *100*, 144, *158*
Douglas, R.G., 182, *183*
Dowker, A., 98, *99*
Dreyfus, H.L., 331, *335*
Dreyfus, S.E., 331, *335*
Dreyfus, T., 43, 44, 46, *66*, *67*, 115, 128, *130*, 146, *158*, 166, 167, 168, 169, 171, *183*, *187*, 193, 196, 197, 216, 217, 218, *234*, *235*, *237*, 296, *310*, *312*, 330, 333, *335*, *336*, *337*
Dubinsky, E., 45, 65, *67*, 72, *99*, 163, *183*, 233, *234*, *235*, 269, *276*, 329, *335*
Dufour-Janvier, B., 43, *68*

Dugdale, S., 45, *67*, 109, 110, 114, 117, 120, 124, 126, 127, 128, 129, *130*, 169, *183*, 207, 208, 220, 221, 225, 226, *234*, 292, *310*, 330, 333, *335*
Duguid, P., 107, *129*, 316, *335*
Dunham, P., 30, *38*

E

Education Development Center, 45, 48, 49, *68*
Edwards, C., 286, *310*
Eisenberg, T., 43, 46, *66*, *67*, 115, 128, *130*, 162, 166, 168, 169, *183*, 196, *234*, 296, *310*, 330, *335*
Ellis, J.R., 163, *184*
Elstein, A.S., 161, *186*
Errecalde, R., 330, *337*
Even, R., 6, *8*, 71, *99*, 163, *183*
Eylon, B.-S., 19, *39*, 167, *184*, 196, 216, 217, *236*, 330, *336*

F

Farrell, A.M., 30, *38*, 271, *278*
Fehr, H.F., 140, *157*
Feiman-Nemser, S., 162, *183*
Fennema, E., 3, *8*, 131, 150, *156*, *157*, 160, 173, 174, 175, *183*
Ferrini-Mundy, J., 166, 167, 170, *184*, *185*
Fey, J.T., 146, *157*, 298, *310*
Finney, R.L., 259, *277*
Fisher, G., 333, *335*
Fisher, L.C., 162, *183*
Floden, R., 12, *39*, 162, *183*
Flores, F., 306, *312*, 321, 323, 324, *337*
Foerster, P.A., 143, *157*
Foster, A.G., 267, *276*
Fowler, D., 285, *310*
Franke, M., 150, *157*
Fraser, R., 219, *235*
Freeman, D., 12, *39*
Freilich, J., 142, *156*
Freudenthal, H., 5, *8*, 190, 191, 201, 202, *235*, 327, *336*
Friske, J.S., 220, *235*

G

Gafni, R., 292, *312*
Gage, N.L., 160, *183*
Gamoran, M., 74, 86, *100*

AUTHOR INDEX   341

Garaçon, M., 231, *235*
Gergen, K.J., 316, *335*
Glover, R., 170, *184*
Goldenberg, E.P., 14, *38*, 45, 65, *67*, 76, 99, 123, 129, *130*, 169, *183*, 198, 223, 224, *235*, 255, 256, 268, 270, *276*, 295, 308, *310*, 320, 324, *335*
Goldin, G.A., 43, *67*, 163, *184*
Good, T.L., 7, *8*, 170, *184*
Goodnow, J.J., 161, *183*
Graeber, A.O., 170, *184*
Graham, K.G., 166, 167, *184*
Grant, W., 300, *310*
Green, T.F., 151, 154, *157*
Groebel, J. 146, *156*
Grouws, D.A., 132, *156*, 170, 172, *184*
Grove, E.L., 249, *276*

H

Halevi, T., 217, 218, *234*, 333, *335*
Hamley, H.R., 135, 137, 140, *157*, 166, *184*
Harel, G., 162, 163, 177, *185*, 193, 233, *234*, *235*, 269, *276*
Hart, K.M., 200, 202, 203, 204, *235*
Hart, W.W., 140, 141, *158*
Harvey, J.G., 227, 228, *235*
Harvey, W., 293, *312*
Hawkes, H.E., 139, *157*
Hawks, J., 72, *99*, 163, *183*
Hedrick, E.R., 137, 138, 145, *157*
Heid, M.K., 43, *67*, 211, 214, 228, *235*, 237, 268, *276*
Heidegger, M., 322, 327, *335*
Henderson, E., 170, *185*
Henderson, K.B., 143, *157*
Herscovics, N., 11, 17, 20, 26, *38*, 44, *67*, 164, *184*
Hershkowitz, R., 45, *68*, 167, *187*
Hiebert, J., 314, 328, 330, *335*
Hight, D.W., 193, *235*
Hooten, J.R., 178, *184*
Hotta, J., 289, *311*
Howe, A., 103, *130*
Hunter, B., 291, *310*

I

Izzo, J.A., 245, 248, *276*

J

Jablonower, J., 247, *276*
Jackiw, N., 292, 300, *311*
Jacobs, H.R., 143, *157*
Janvier, C., 19, *38*, 43, *67*, 168, *182*, *184*, 199, 202, *235*, 294, *311*, 330, *336*
Jensen, R.J., 168, *187*
Jones, D., 131, *156*, 132, 155, *156*, 172, *184*
Jones, P.S., 243, *276*

K

Kahn, H.F., 246, *276*
Kaput, J.J., 11, 24, 34, *38*, 169, *184*, 190, 193, *235*, 270, *276*, 284, 285, 286, 293, 295, 298, 300, 301, 302, 306, *311*
Karplus, R., 19, *38*, 103, *130*, 208, 217, *236*, 330, *336*
Keedy, M.L., 144, *158*
Keen, L., 308, *310*
Kennedy, J., 142, *157*
Kepner, H.S., 203, 204, *234*
Kerslake, D., 18, 19, *38*, 44, *67*, 168, *184*, 200, 204, *236*, 330, *336*
Kesler, R., 152, *157*
Kessel, C., 74, 86, *100*
Kibbey, D., 109, 110, 114, 120, 124, 129, *130*, 220, *234*
Kieran, C., 42, 59, *67*, 80, 87, *99*, *100*, 164, 172, *184*, *185*, 197, 231, *235*, *236*, 270, *277*
Klein, J., 285, *311*
Kleiner, I., 4, *9*, 132, 133, 134, 135, 136, *157*, 192, 193, *236*
Kliebard, H.M., 246, 247, *277*, 327, *336*
Kliman, M., 268, *276*
Kline, M., 133, 134, *157*, 285, *311*
Klinger, D.L., 143, *157*
Klotz, E., 300, *311*
Konshak, A.L., 168, *184*
Krabbendam, H., 200, 201, 202, *236*
Krickenberger, W.R., 141, *158*

L

Laborde, J.-M., 300, *311*
Lamon, S., 3, *8*, 160, 175, *183*
Lampert, M., 327, 328, *336*
Lave, J., 315, 316, 334, *336*
Lawson, A.E., 103, *130*

AUTHOR INDEX

Leinhardt, G., 1, 8, *9*, 11, 12, 17, 18, 24, 26, *38*, 196, *236*, 318, 331, *336*
Leitzel, J., 21, *38*, 299, *310*
Lennes, N.J., 246, 248, 249, *277*
Leonard, M., 74, 86, *100*
Lesh, R., 46, *67*, 162, 163, 177, *185*
Leveridge, M.E., 209, *237*
Lewin, P., 329, *335*
Lindquist, M.M., 203, 204, *234*
Lochhead, J., 196, *237*
Loef, M., 173, *183*
Lovell, K., 19, *38*, 138, *157*
Luby, W.A., 139, *157*
Luke, C., 300, *311*

M

Magidson, S., 76, 78, 89, *99*
Malik, M.A., 134, *157*, 167, *184*
Mallory, V.S., 140, *157*
Mandelbrot, B., 308, *311*
Markovits, Z., 19, 39, 167, *184*, 196, 216, 217, *236*, 330, *336*
Marnyanskii, I.A., 167, *184*
Marsh, W.R., 139, *157*
Mason, D.A., 170, *184*
Matras, M., 211, *235*
May, K.O., 135, 138, *157*
Mayberry, J.W., 162, *185*
McArthur, D., 289, 300, *311*
McDermott, L., 304, *311*
McGalliard, W.A., 152, *157*
McKenzie, D.L., 168, *186*
Menasian, J., 211, *235*
Merzbach, U., 286, *310*
Mokros, J.R., 109, 113, 114, 127, 128, *130*
Monk, G.S., 168, *184*
Moschkovich, J., 75, *100*, 233, *236*
Moser, J.M., 3, *8*, 325, 327, *335*
Mullikin, A.M., 249, *276*
Mullis, I.V., 70, *100*
Musser, G.L., 225, 226, *234*

N

Nachmias, R., 45, *67*
National Council of Teachers of Mathematics (NCTM), 20, 21, 35, 36, 37, *39*, 69, 70, *100*, 144, 145, 146, 154, *157*, 171, 182, *185*, 233, *236*, 266, 274, *277*, 315, 322, 325, 333, 334, *336*
Neill, H., 191, *237*

Nemirovsky, R., 301, 308, *311*
Newman, S.E., 333, *335*
Nichols, D., 72, *99*, 163, *183*
Norman, D.A. 333, *336*
Norman, F.A., 162, 163, 165, 170, 179, *185*, *186*

O

Oganowski, M., 308, *311*
Orbach, R., 74, 86, *100*
Orleans, J.S., 246, *277*
Orton, A., 60, 64, *67*, 166, *185*
Osborne, A.R., 140, *157*, 243, 244, 246, *277*, 299, *310*
Owen, E.H., 70, *100*
Owens, J.E., 152, *158*, 170, *185*

P

Padilla, M.J., 168, *186*
Pea, R.D., 323, *336*
Perry, W.G., 152, 154, *158*
Peters, P.C., 103, *130*
Peterson, P.L., 3, 7, *8*, 131, *156*, 132, 148, 149, 154, *156*, 161, 173, 174, *183*, 327, 328, *336*
Phillips, G.W., 70, *100*
Phillips, R.J., 104, 105, 106, 107, 108, 127, *130*, 207, 208, *236*, 318, *336*
Pietgen, H., 308, *311*
Pimm, D., 92, *100*
Pingry, R.E., 143, *157*
Poholsky, J., 300, *311*
Polanyi, M., 304, *311*
Porter, A., 12, *39*
Post, T.R., 162, 163, 177, *185*
Pratt, D.L., 170, *185*
Preece, J., 209, 210, *236*, 268, *277*
Prichard, M.K., 170, 179, *185*, *186*
Putnam, R.T., 327, 328, *336*

R

Rachlin, S.L., 168, *187*
Ragan, E., 142, *157*
Rath, J.N., 267, *276*
Rector, J., 170, *185*
Resnick, Z., 84, *100*
Reys, R.E., 203, 204, *234*
Rich, B.S., 25, 30, 34, *39*, 271, *277*
Richards, J., 300, 302, *311*

Richter, P., 308, *311*
Romberg, T.A., 3, 4, *8*, *9*, 160, 180, *185*, 248, 273, *277*
Rosenquist, M., 304, *311*
Rosskopf, M.F., 140, *158*
Royster, D.C., 170, *186*
Rubin, A., 304, 308, *311*
Rumelhart, D.E., 333, *336*
Rusch, J., 103, *130*
Rüthing, D., 192, *236*

S

Säljö, R., 1, *9*
Sandefur, J., 308, *312*
Saxe, G.B., 316, *336*
Sayer, A., 300, *311*
Scardamalia, M., 291, *312*
Schliemann, A.D., 316, *335*
Schmidt, W., 12, *39*
Schoenfeld, A.H., 19, *39*, 57, *67*, 73, 74, 75, 86, 88, *99*, *100*, 124, *130*, 164, 179, *185*, 233, *236*, 268, 270, *277*, 332, *336*
Schoeps, N.B., 170, *186*
School Mathematics Study Group (SMSG), 142, 143, *158*
Schorling, R., 138, 140, 141, *158*
Schwartz, J.L., 45, 48, 49, *68*, 71, 72, *100*, 169, *186*, 233, *236*, 293, 295, 300, *312*, 321, *336*
Schwarz, B., 215, 216, 217, *236*, 333, *336*
Schwille, J., 12, *39*
Senk, S.L., 249, *277*
Sfard, A., 71, *100*, 146, *158*, 193, 194, 195, 197, 198, 206, 231, 232, 233, *236*, *237*, 329, *336*
Sharron, S., 143, *157*
Shaughnessy, J.M., 226, *234*
Shavelson, R.J., 161, *186*
Shaw, E.L., 168, *186*
Sheets, C., 211, 228, *235*, *237*
Shuard, H., 191, *237*
Shulman, L.S., 160, 161, 162, 173, *186*
Sigurdson, S.E., 245, *277*
Skemp, R., 162, *186*
Slaught, H.E., 249, *277*
Smith, D.E., 246, *277*
Smith, J.P., 73, 74, 75, *100*, 124, *130*, 233, *236*, 332, *336*
Smith, R.R., 141, *158*
Smith, S.A., 144, *158*
Soloway, E., 196, *237*

Sowder, J.T., 3, *9*, 172, *186*, 314, *336*
Stanic, G.M.A., 246, 247, *277*
Stasz, C., 289, *311*
Steen, L.A., 325, *336*
Stein, E.I., 249, *277*
Stein, M.K., 1, 8, *9*, 11, 12, 17, 18, 24, 26, *38*, 196, *236*, 318, 331, *336*
Stephens, W.M., 271, *277*
Stern, P., 161, *186*
Sullivan, F., 103, *130*
Swan, M., 104, 106, 107, *130*, 199, 202, 204, 205, 232, *237*, 304, *312*, 318, 331, *336*
Sweller, J., 293, *312*
Symonds, P.M., 246, *277*

T

Tall, D., 42, 44, 60, *68*, 167, 169, *186*
Thaeler, J.S., 222, 223, 224, *237*
Thomas, G.B., 259, *277*
Thomas, H.L., 166, *186*, 196, *237*
Thompson, A.G., 131, *158*, 170, *186*
Thompson, D.R., 249, *277*
Thompson, P.W., 55, *68*, 169, *186*, 193, *237*, 288, 293, 300, *312*
Thorndike, E.L., 246, *277*
Thorpe, J.A., 5, *9*, 144, 146, *158*, 267, *277*, 317, 318, 325, *337*
Tierney, C., 308, *311*
Tinker, R.F., 109, 113, 114, 127, 128, *130*
Tirosh, D., 45, *67*, 170, *184*
Touton, P.B., 139, *157*
Townsend, E.J., 134, *158*
Tranter, J.A., 209, *237*
Travers, R.M.W., 161, *186*

U

Upton, C.B., 249, *278*
Usiskin, Z., 283, 288, *310*

V

Van Engen, H., 135, 138, *157*
Van Zee, E., 304, *311*
Vergnaud, G., *237*, 326, 330, *337*
Verstappen, P., 198, *237*
Viktora, S.S., 249, *277*
Vinner, S., 44, 45, *67*, *68*, 144, 146, *158*, 167, 171, *186*, *187*, 197, 216, 218, *234*, *237*, 296, *312*, 318, 330, *337*

W

Wagner, L.J., 120, 124, 129, *130*
Wagner, S., 80, *100*, 168, 172, *187*
Waits, B.K., 14, 24, 28, 35, *38*, 145, *156*, 227, 228, 229, *235*, *237*, 249, 261, 264, *276*, *278*
Wald, E., 246, *277*
Walen, S.B., 332, *337*
Ward, M., 293, *312*
Wearne, D., 314, 328, 330, *335*
Welchons, A.M., 141, *158*
Wells, W., 140, 141, *158*
Wenger, R.H., 58, *68*, 165, *187*, 330, *337*
Wertheimer, M., 164, *187*
West, M.M., 293, 298, 300, 306, *311*
Williams, S.R., 273, *277*, 332, *337*
Willoughby, S.S., 138, 139, *158*
Winograd, T., 306, *312*, 321, 323, 324, *337*
Winters, L.J., 267, *276*
Wollman, W.T., 103, *130*
Woodyard, E., 246, *277*
Wooten, W., 142, 143, *156*, *157*
Wren, F.L., 140, *156*

Y

Yerushalmy, M., 45, 47, 48, 49, *68*, 71, 72, *100*, 214, 215, 233, *236*, *237*, 292, 293, 295, 300, *312*, 321, *336*
Young, J.W.A., 140, *158*
Youschkevitch, A.P., 133, 134 *158*

Z

Zarinnia, E.A., 273, *277*
Zaslavsky, O., 1, 8, *9*, 11, 12, 17, 18, 24, 26, *38*, 196, *236*, 318, 331, *336*

Subject Index

A

Algebra
  as a corollary of functions, 287-288
  history of (in U.S. schools; see also
    Functions, Mathematics),
    243-250, 285
  prior to secondary school, 287
  role in schools, 69-75, 283, 287
    functions curriculum, 41-45
    research implications, 65-66
    student (postsecondary) learning of,
      159, 162, 163-166
    teacher knowledge of, 160, 161-166,
      177, 179-180
    visual, 46-52
Algebra standard, 69-71
Algebra with Computers project, 43,
  211
Algebraic Proposer, The, 169
American Mathematical Society, 245
Analyzer, The (software), 214
Apple Classroom of Tomorrow Program,
  287
Archimedes, 285
Assessment, 1, 3, 97-99, 247-248
  new methods, 181
Assessment of Performance Unit Study
  (U.K.), 204

B

Baire, 135
Bernoulli, 134
Bombelli, 192
Borel, 135
Bourbaki, 135-136, 144, 192-193, 194

C

CABRI, 200
Calculator and Computer Precalculus
  (C^2PC) Project, 227-228
Calculator/computer-based technology,
  189, 196, 198, 206, 239, 250-264
  advantages of, 6-7, 254-264, 272-275
  as basis for curriculum restructuring,
    273-275
  bit-mapped graphics systems, 284
  in classroom, 26-37, 75-79, 98-99,
    104-117, 117-123, 181, 206-230,
    271-273, 282, 284-287, 298-307
  curve fitting, 6, 296
  curve sketching, 204-205, 259, 264
  in curriculum, 4-8, 20-27, 27-37,
    52-65, 84-97, 239, 250-270,
    266-267

345

dangers/problems, 34-35, 320-324
impact of, 145-147, 169-170
integration into classroom, 286-287
new representational capacity of, 279
software, 46-52
in textbooks, 11-20
uses in teaching functions and graphs, 206-230
Cartesian Connection, 73-74, 75, 79, 83, 84, 329
Cartesian plane, 70, 72, 88, 90, 196
Chicago School Mathematics Project, 249
Cognitively Guided Instruction (CGI) project, 159, 173-182, 298
College Entrance Examination Board (CEEB), 139, 142, 145, 248
Commission on Mathematics (see College Entrance Examination Board [CEEB])
Computer environments, 42, 45-62
curriculum and, 52-62
Expressions, 55
ISETL, 45
uses of, 52-62
Visualizing Algebra Series, 47, 48, 49-52, 56, 58, 60, 64
Computer graphing tools, 189, 205, 214, 225-228
outcomes, 220
Computer Intensive Algebra, 214-215
Computers--teaching trigonometric identifies, 225
Concepts in Secondary Mathematics & Science (CSMS [U.K.]), 203-204
Constructivist concept of learning, 164-165, 173, 174
Continental philosophy, 327, 328
Continuous versus discrete misconception, 18
Curriculum and Evaluation Standards for School Mathematics (NCTM), 144-145, 171, 182, 274, 315-316, 322, 334
Curriculum development, 76, 86-97, 97-99, 173, 175-177
Curriculum of 21st century, 308
Curriculum (software-based), 207-230
role of teacher in, 217-220
Curriculum theory of
mental discipline, 246-247
social efficiency, 246-247

D

Descartes, René, 2, 134, 192
Dewey, 298
Diagram-constructing software (also see Chapter 9), 300-301
implications for teaching and learning, 300
Difference equations, 64-65
Dirichlet, 134-135, 144, 192, 194
Domain competence, 74-75, 86, 97
Domain knowledge strategy, 3

E

Education Development Center, 214
Equations and inequalities, 32-34, 48, 57-59
Euclid's *Elements*, 193
Euler, 134, 192, 194
Eureka Program, 104-110, 207-208, 318
Explorer's Toolkit, 302

F

Fermat, 192
Flexible competence (understanding), 86, 94, 97
Fourier, 134
Framework (for understanding functions), 74, 79-86, 94
Function Probe, 293
Functions, defined, 2, 4-6, 132-133, 133-147
category theory, 136
as central to mathematics instruction, 133-145, 190, 192-193
cognitive obstacles, 20, 42-45, 164, 206
complete graph of, 27-30
and computer technology, 282
connections, 69, 71, 73-74, 83-84, 94, 97-99
content domain, 325-328
in current curriculums, 1, 3, 137-145, 280, 282-288, 292-297
dependency perspective, 190-191, 199, 206
dependency relationships, 199, 294-295, 298
end-behavior models, 262
as entities, 45, 47, 49, 52-62, 62-65, 71, 83, 87

function-plotting tools, 101, 102, 114–123, 128–129
  common misconceptions in using, 101, 123–126
  impact of, 102, 117–123
function transformations, 116–117, 120–123, 128
historical development, 132–139, 137–145, 146, 192–206
learning, 1, 3
nature of zeros, 261–262, 272
notion of limit, 261
operational/structural conceptions, 195–198
process/object model, 71–75, 79–81, 83–97, 189–195, 206, 329
real-world contexts, 17, 19, 30–31, 36–37, 74, 137–138, 144–146, 152
  experiential linkages, 297
set-theoretic perspective, 190, 191, 196–197, 206
sets, 136, 138, 143–144
single/multivariable, 307
teaching of, 1, 3, 137–139
in technology-supported environments, 190
three-phase model of concept development (Sfard), 194
trigonometric identities, 117–119, 126, 128, 259
  Supplemented Traditional Treatment, 117–118
  Graphical Foundations Treatment, 117–119
as a unifying factor, 137–139, 140, 144, 145–147, 151
visual illusions, 124
Functions Group, 75–76
Function Probe, 293
Function Supposer, 47, 48, 50–52, 54, 56, 58, 64, 293
Fundamental Theorem of Calculus, 304

G

Galileo, 134, 192
Geometer's Sketchpad, The, 292
Geometric Supposer, 45, 62, 321
GRAPHER, 75, 88, 89, 98
Graphical reasoning, 101, 102, 118, 126, 128
  qualitative representation, 103, 127, 128

Graphing activities, categorized (Leinhardt), 12–14
  construction, 13, 19
  in curriculum, 20–27, 27–37, 264–275
  preparation for calculus, 28–30
Graphing calculators, introduction of, 6–7, 20, 25, 26–37
  impact on teaching, 2–4, 26–37, 102, 114–123
  pitfalls, 34–35
Graphs
  global meanings, 101, 104, 114, 127–128, 230
  interpretation of graphs, 12, 19, 102–114, 127, 317
    multiple graphs, 110–113
    student problems with, 103–104
  as pictures, 109–110, 114, 127, 128, 200, 202, 330
  qualitative understanding, 104, 118, 128
  real-time, 113–114
  real-world contexts/problems, 101, 102, 104–110, 127
  student interpretations, 101, 114, 127–128
    resolving contradictions, 101–102, 124–125, 129
  student strategies, 110–113, 127–128
    improving, 127–129
Graphs and graphing, 206–230
  in assessment, 267–268, 273
  calculators, 225–228
  in algebra, 225–228
  in calculus, 225–228, 228–229
  compared with graphing software, 225
  content analysis, 267–268
  in curriculum, 250–264, 264–267
  games, 220–221
  input-output, 221–224
  interpretation (by students), 200–203, 207–210
  new uses of, 230–231
  process/object perspective, 269–270
  process-object research model, 193–198, 206–207, 231–233
  qualitative feedback, 209–210, 214
  quantitative feedback, 211, 214
  real-world (applied) problems, 200–202, 230, 250, 256, 317–318, 326–328, 332–334
  for teaching trigonometric identities, 225
  in textbooks, 248–250, 272–273
Green Globs, 45, 76, 114–116, 169, 220–221

SUBJECT INDEX

Guess My Rule, 80, 89
Guided discovery, 76-78

H

Handbook of Research on Teaching, 160, 161, 171
Hawaii Algebra Conference, 172, 181

I

IEA Computers-In-Education survey, 284
Inactive learning, 44-45
Instruction
  and cognitive change, 314
Interpreting Graphs (software), 207
  Relating Graphs to Events, 207-208
  Escape, 207
Invisible college, 172, 181

K

Kepler, 134, 192
Klein, 137, 245
Knowledge profile assessments, 177

L

Learning as enculturation, 316
Lebesque, 135
Leibniz, 134, 286
Linearity, 69, 72-74, 75, 79-80, 98, 165, 168, 196, 215-217

M

MathCars, 302, 318
Mathematical Association of America (MAA), 137, 140, 145
Mathematical models, 241, 256, 265-267
Mathematics
  and authentic experience, 297-307
  as doing, 173
  dynamic relationships, 285
  history (see also Algebra; Functions), 284-286
  linkage with authentic experience, 297-307
  quantitative relationships, 284, 286, 292-293, 309
    general, 286, 293
    particular, 286, 293
  static relationships, 285
  uses of, 315
Microcomputer-Based Laboratory (MBL), 113-114, 295, 303-304
MINNEMAST Project, 294
Montessori, 298
muMath, 228

N

National Committee on Mathematical Requirements (see Mathematical Association of America)
National Council of Education, 245
National Council of Teachers of Mathematics (NCTM), 247, 266
National Education Association, 245
National Institute for Curriculum Development (The Netherlands), 200-202
Newton, 134, 285

O

Ohio State group (also see Chapter 2), 298-299
Oresme, 285

P

Parametrization, 71, 75, 98
Paris Problems, 314, 315
Pathfinder, 110
Point plotting, 26, 27, 37
Polynomials, 58-59, 119-125, 129, 221, 250, 260-261, 293, 296
Pond (software), 209-210
Probeware, 113-114
Procedural competence, 98
  knowledge, 74, 86, 94
Process/product research, 160-161
Professional Standards for Teaching Mathematics (NCTM), 154, 333
Proleptic, defined, 279-280, 309-310
  assumptions and alternatives, 290-297, 309
  research, 279-280, 282-285, 310, 322-324, 334
Proportional reasoning curriculum, 294

SUBJECT INDEX 349

Q

Quadfun, 217–220

R

Rational functions, 114–115
Rational Zeros Theorem, 35
Recorde, 285
Reform needed, 286
Relational and Functional Thinking in Mathematics, 166–167
Representation of functions (models of), 198–206
   algebraic (symbolic), 2–3, 6–8, 22–24, 27, 30, 42–43, 46–50, 52–53, 55–57, 60–62, 65, 69, 71–73, 79, 83–84, 97, 101, 102, 116–117, 124, 128–129, 146, 151, 153, 168–169, 189, 197–198, 203–204, 206, 211, 215–217, 220, 228, 254, 264, 270, 272, 286, 288, 289, 293, 299, 306, 318–319, 325
   graphical (visual), 2–3, 6–8, 20–26, 31, 42–43, 46–48, 52–53, 60–62, 64–65, 69, 71, 73, 79, 84, 86, 97, 101, 102, 115–117, 119–122, 124, 128–129, 137, 139–141, 143–145, 145–147, 148, 149, 168–169, 189, 194, 196–198, 206, 215–217, 220, 228, 254–255, 263–264, 270, 272, 292, 293, 299, 306, 307–308, 318–319, 325
     dilation, 42, 52, 55
     PAR (Parallel axis representations), 45
     reflection, 42, 52
     SAM (same axis representations), 46
   linking of, 41, 52–62, 169, 288, 292–293, 298
   multiple-linked, 318
   ordered pairs (tabular, numerical), 2–3, 6–8, 18, 20–26, 43, 46, 52–53, 64, 69, 71–73, 79, 97, 120–122, 123, 135–136, 137–139, 142–144, 146–147, 169, 189, 194, 196–197, 215–217, 228, 270, 286, 293, 299, 318–319
   set-theoretic approach, 136, 146, 148, 149
   as simulations, 301–307
Research
   on concept of function, 166–169
   current classrooms, 280–284, 285–290
   future/future-defining, 131, 147–155, 172, 179–182, 232–233, 266, 273–275, 280, 307–309
   on learning, 159–160, 160–180, 256, 333–334
   functions and graphs, 166–170
   practice-extending, 280, 285–286, 287–290, 309
   in quantitative relationships, 279
     curriculum, 279, 283–284, 287–288, 292–294
     representational, 279, 283–284, 288–290
     technological, 279, 283–284, 288, 290–292
   questions/issues, 36–37, 65–66, 153–155, 171, 178, 214, 232, 270–271, 275, 282, 292, 300, 304, 308–309, 313, 314–317, 320, 322, 327
   synthesis of teaching and learning, 180, 181–182
   on teacher beliefs re. functions and graphs, 159, 170–171
   cognition, 160–166, 169–170
   knowledge of algebra, 160, 161–166, 169–170, 179–180
   knowledge of functions and graphs, 159, 161–166, 169–170, 180, 182
   thinking, 148–153, 160, 172–180
   on teaching, 7–8, 131–132, 160–161, 171–182, 333–334
   process-product paradigm, 7
   student/teacher thinking paradigm, 7–8
   and technological change, 280–282
Research Agenda Project (NCTM), 171–182, 314
Research framework needed, 232–233
Resolver, The (software), 214–215
Russell, 135

S

Scaling, 24–26, 49, 253, 255, 264, 293–294, 295, 306–307
Scholastics, 304
School mathematics, 11
   algebraic manipulation, 35–36
   factoring, 35
   graphing content of, 11–20, 36–37
   problem solving as a focus of, 21
School Mathematics Study Group (SMSG), 142–143
Sfard's 3-phrase model of conceptual development, 194–196

Shell Centre for Mathematical Education
  (U.K.), 104-108, 200, 205, 207-208,
  304
Starburst problem, 89-94, 98
Stifel, 192
Student
  cognition, 328-333
  difficulties/misconceptions, 18, 42,
    44-45, 48, 59, 64-65, 70, 74, 94-97,
    101, 102-104, 109, 123-126, 129,
    164-166, 167-169, 200, 204, 208-209,
    216-220, 255, 268-271, 275, 307,
    324, 330-331
  and functions and graphs, 166-170
  interaction with classmates, 117-119
  learning, 159, 160-161, 166-170
  ownership of ideas and methods, 119
  qualitative interpretation of graphs, 101,
    102-114

T

Teachers
  as consultants, 219
  influence on algebra curriculum, 248
  as learners, 176, 178-179
  pedagogical and content knowledge,
    162-164, 174, 177
  reliance on textbooks, 248
  as researchers, 177-178
  as source of information/knowledge, 175,
    219, 321
  and student performance, 159-161, 163,
    173-176
  thinking and beliefs, 270-272
Teachers and functions and graphs
  beliefs, 131-132, 150-153, 170
  instructional practice, 137-139, 146-150,
    151-153
  knowledge, 131-132, 150-153
  research on, 131-132, 147-155, 170-172
  thinking, 131-132, 145-147, 148-149

Teaching
  with computers, 211-230
  current practice, 282-287
  process approach to graphing, 232
  process-product paradigm, 7
  student-teacher thinking paradigm, 7-8
Technical Education Research Centers
  (TERC), 113, 308
Technology, 189, 196, 198, 206, 313,
    316-322
  bit-mapped graphics systems, 284
  impact of, 145-147, 169-170
  integration into classroom practice,
    286-287
  new representational capacity of, 279
  uses in teaching functions and graphs,
    206-230
Technology-based instruction, 75, 98-99
Transition to Algebra, 299
Translation, 42, 52, 190, 196, 198-206,
    215-217, 232, 319
Triple Representation Model (TRM),
    215-217
Tutoring curriculum, 75, 86-97

U

University of Georgia Conference, 171-172

V

Van Hiele analysis, 30
Verstappen's categories of functional relations, 198-199
Viéte, 134, 192
Visual algebra, 46-52

W

Wingspread Conference, 3, 314
Wisconsin Center for Education Research
  (WCER), 172
Word Problem Assistant, 169